Undergraduate Topics in Computer Science

W0192971

'Undergraduate Topics in Computer Science' (UTiCS) delivers high-quality instructional content for undergraduates studying in all areas of computing and information science. From core foundational and theoretical material to final-year topics and applications, UTiCS books take a fresh, concise, and modern approach and are ideal for self-study or for a one- or two-semester course. The texts are authored by established experts in their fields, reviewed by an international advisory board, and contain numerous examples and problems, many of which include fully worked solutions.

The UTiCS concept centers on high-quality, ideally and generally quite concise books in softback format. For advanced undergraduate textbooks that are likely to be longer and more expository, Springer continues to offer the highly regarded *Texts in Computer Science* series, to which we refer potential authors.

Patricio Bulić

Understanding Computer Organization

A Guide to Principles Across RISC-V,
ARM Cortex, and Intel Architectures

 Springer

Patricio Bulić
Faculty of Computer Science
University of Ljubljana
Ljubljana, Slovenia

ISSN 1863-7310   ISSN 2197-1781 (electronic)
Undergraduate Topics in Computer Science
ISBN 978-3-031-58074-1   ISBN 978-3-031-58075-8 (eBook)
https://doi.org/10.1007/978-3-031-58075-8

This Springer imprint is published by the registered company Springer Nature Switzerland AG
The registered company address is: Gewerbestrasse 11, 6330 Cham, Switzerland

If disposing of this product, please recycle the paper.

To my Dearest Wife and Precious Sons,

In the world of bits and bytes, algorithms, and circuits, your love has been the unshakable foundation upon which I've built this journey into the realm of technology. Your unwavering support, patience, and understanding have fueled my passion and sustained my spirit.

To my Wife, your patience and understanding have been the guiding light through countless hours of writing and exploring the depths of technology. Your unwavering encouragement and belief in my endeavors have been the compass guiding me through the complexities of computer systems.

To my Sons, you are the future, the next generation of dreamers. Your boundless curiosity and the joy you bring have inspired the creativity embedded in every line of this book. May these pages serve as an invitation to explore the wonders of technology and encourage you to dream and create.

With love and appreciation, Patricio

Preface

In today's digital age, where computing technologies are advancing at an unprecedented pace, understanding the fundamental principles that govern computer organization and the intricate interplay between hardware and software is essential for students and professionals alike. Welcome to "Understanding Computer Organization: A Guide to Principles Across RISC-V, ARM Cortex, and Intel Architectures." This book aims to serve as a comprehensive guide to the intricate world of computer organization, offering practical insights and real-world examples from RISC-V, ARM, and Intel-based computer systems. As technology continues to advance rapidly, it becomes increasingly crucial for both students and professionals to grasp the fundamental principles that govern computer organization.

Our journey begins with exploring memory-mapped I/O, where we unravel the mechanisms by which the CPU communicates with peripherals through memory addresses.

From there, we venture into the realm of interrupts, examining how various architectures handle asynchronous events and prioritize tasks efficiently. Real-world examples from RISC-V, ARM, and Intel-based systems illuminate these concepts, providing practical insights into their implementation and significance.

Chapter 3 delves into the intricacies of Direct Memory Access (DMA), shedding light on the controllers responsible for managing data transfers between peripherals and main memory.

We then focus on the main memory, exploring the different types of memory technologies, including DRAMs, SDRAMs, and DDR SDRAMs, and their roles in storing and accessing data effectively.

Moving forward, we delve into the critical role of cache memory in improving system performance through data locality and access speed optimizations.

Finally, we explore the concept of virtual memory, unraveling how modern operating systems manage memory resources efficiently by leveraging secondary storage devices.

Throughout this journey, our goal is to provide readers with a comprehensive understanding of computer organization principles, grounded in real-world examples and practical applications. Whether you are a student embarking on your journey into the world of computer architecture or a seasoned professional seeking to deepen your knowledge, this book is designed to be your companion in unraveling the complexities of modern computing systems. So, without further ado, let us unravel the mysteries of hardware and software and pave the way into the field of computer organization.

Ljubljana, Slovenia   Patricio Bulić
February 2024

Contents

About the Author

Patricio Bulić is a computer engineering professor at the University of Ljubljana, Slovenia, with over 25 years of experience in computer science and engineering. As a university teacher, Patricio Bulić has inspired and mentored countless students, instilling in them a passion for computer systems and technology. His dynamic teaching style, pedagogical techniques, and profound knowledge of computer architecture and organization have garnered him recognition as a beacon of excellence in higher education. Patricio Bulić has been honored multiple times as Teacher of the Year, a testament to his unwavering commitment to teaching.

In the realm of research, Patricio Bulić has contributed to advancing the frontiers of computer science, particularly in digital design, embedded systems, parallel processing, computer arithmetics, and approximate computing. Drawing on his experience as an educator and researcher, this book offers insights into the principles, design methodologies, and emerging trends in computer system organization.

Outside academia, Patricio Bulić is a passionate cross-country mountain biker who finds solace and inspiration in the great outdoors. Whether navigating challenging trails or conquering rugged terrain, he approaches each ride with the same determination that defines his professional endeavors. He finds balance and perspective in the exhilarating rush of adrenaline and the breathtaking beauty of nature, fueling his creativity and drive to excel in all aspects of life.

As the author of a forthcoming book on computer systems, Patricio Bulić invites readers on a journey through the intricacies of computer organization and design, offering insights that are both enlightening and engaging.

Memory-Mapped Input/Output

1

CHAPTER GOALS

Have you ever wondered how your computer seamlessly interacts with external devices, such as storage drives, network interfaces, and peripherals? The answer lies in a clever memory-mapped I/O (Input/Output) technique. Unlike traditional I/O methods that involve separate address spaces for memory and I/O devices, memory-mapped I/O integrates device communication directly into the memory address space. A memory-mapped I/O device is a computer hardware component that uses a portion of the system's memory address space and is accessible by load and store instructions, while address decoding determines which device or peripheral in a computer system should respond to a particular load or store instruction. This chapter will explore the fascinating world of memory-mapped I/O, uncovering its principles, benefits, and applications.

Upon completion of this chapter, you will be able to:

- Understand and explain memory mapping.
- Understand and explain address decoding.
- Gain a clear understanding of the fundamental concept of memory-mapped I/O and its role in computer systems.
- Learn about memory-mapped registers and control structures used for configuring and interacting with I/O devices directly through memory addresses.
- Explore real-world case studies showcasing the use of memory-mapped I/O in diverse embedded devices.
- Able to program a memory-mapped general-purpose IO device.

© The Author(s), under exclusive license to Springer Nature Switzerland AG 2024 1
P. Bulić, *Understanding Computer Organization*, Undergraduate Topics in Computer Science, https://doi.org/10.1007/978-3-031-58075-8_1

1.1 Introduction

Recall that the only way for modern processors (e.g., RISC-V) to access data (read or write) is by using memory load and store instructions. These instructions are a fundamental part of an instruction set architecture (ISA) and allow the processor to interact with various types of memory. An important consequence of this principle is that if we want the CPU to read or write data from input/output (I/O) devices, all I/O devices should be visible to the CPU as a set of memory words. We also say that I/O devices should be memory-mapped. A memory-mapped I/O (MMIO) device is a computer hardware component that uses a portion of the system's memory address space for data transfer and control. This approach simplifies device interaction for software developers and is commonly used in modern computer systems.

In particular, these MMIO devices incorporate a small memory (actually, the memory size depends on the device type and its functionality, but in general, this memory has only a few memory words). Each memory word within this on-device memory is assigned a memory address within the system's memory map and, consequently, is accessible through load and store instructions. Besides, each word within this on-device memory has a dedicated meaning (for example, it can be used by the CPU to monitor the device status, to set some features of the device or to read/write data). Because these on-chip memory words have distinct meanings, each memory location is called **a register**. In other words, these registers control various aspects of the device's operation, such as configuration settings, data transfer, and status checks.

In order to assign a unique memory address to each I/O device and its registers, computer systems rely on **address decoding**. Address decoding is crucial in computer architecture and design, particularly in systems that use memory-mapped I/O, as it determines which device or peripheral in a computer system should respond to a particular load or store instruction provided by the CPU. To determine which device should respond to a specific address provided within a load or store instruction, address decoding logic involves a combination of simple digital logic gates, such as AND gates, OR gates, and NOT gates, or even the usage of decoders. When the CPU wants to read from or write to a specific I/O device register or memory location, it places the desired address on the address bus. For example, RISC-V would place this address on the address bus in the fourth pipeline stage (MEM stage) after the memory address has been calculated from the base address and offset within its third pipeline stage (EXE stage). Besides the address, the CPU would also activate the **read-write (R/W signal)**, which tells the addressed device whether the CPU would like to read from or write to. The address decoding logic continuously monitors the address bus. It compares the incoming address on the bus to predefined address ranges for each device. When it detects a match between the incoming address and one of the assigned address ranges, it generates a so-called **chip-select (CS) signal** for that device. When the chip-select signal for a specific device becomes active (typically pulled low), that device knows it should respond to the CPU's request. It **enables** its data bus interface so that data can be read from or written to the device according to the R/W signal. Once the correct device is selected, data can be transferred between the CPU and the selected device through the data bus. The CPU reads from or writes to the device's registers or memory locations based on the operation it wants to perform. When the CPU puts another address onto the address bus, the address decoding logic deactivates the select signal for the device, allowing other devices to respond to subsequent address requests. In the following subsection, we will explain this important concept in detail using a simple example.

1.2   A Memory-Mapped Register

Suppose we would like to connect a single 32-bit register to a 32-bit CPU (e.g., RISC-V). Also, suppose that the register has chip-enable (CE), output-enable (OE) and chip-select (CS) signals besides the standard data input, data output and clock signals. Such a register is presented in Fig. 1.1.

A register with chip-enable, output-enable and chip-select signals is a common component in digital systems, especially those that involve memory-mapped I/O. These signals control the register's behaviour, specifying when it should be enabled or disabled for data read and write operations. Here's how such a register typically works:

1. The output-enable signal (OE) connects or disconnects the output data signal to/from the data bus. When the OE signal is active (high), the register output becomes connected to the data bus. Data stored within the registers appears on the data bus and is accessible for reading. When the OE signal is inactive (low), the data output is disconnected from the data bus, and other components in the computer system can use the data bus.
2. The chip-enable signal (CE) enables the clock signal connected to the register. Hence, when the CE signal is active, data from the data bus will be stored in the register on the rising edge of the clock signal.
3. The chip select signal (CS) is used to activate or deactivate the register. When the CS signal is active, the register is selected and becomes available for read and write operations. When the CS signal is inactive, the register is deselected, and any access to it is disabled.

Fig. 1.1 A 32-bit register with the CS, OE and CE signals

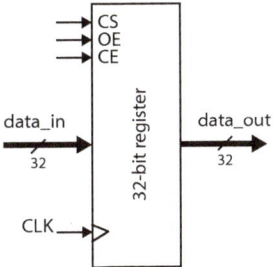

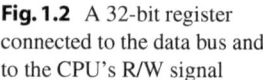

Fig. 1.2 A 32-bit register connected to the data bus and to the CPU's R/W signal

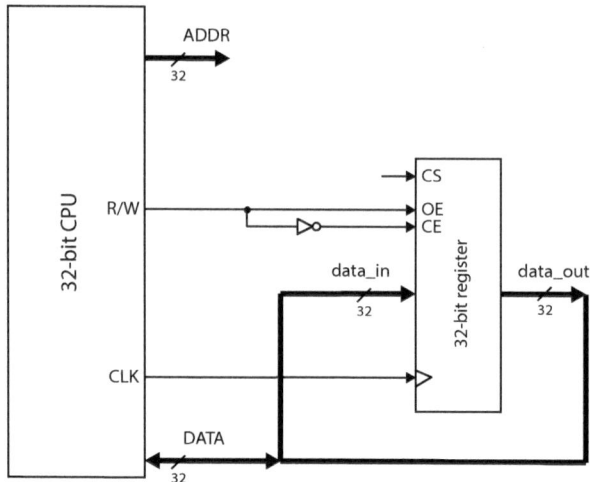

Using enable and chip select signals provides fine-grained control over register access. It allows us to isolate specific registers or components in a digital system, preventing unintended or erroneous data transfers. These signals are particularly useful in memory-mapped I/O scenarios, where multiple registers or devices share the same address bus.

This control mechanism is essential in digital systems to prevent unintended data transfers and efficiently manage communication with various components. A register with CE, OE and CS signals reads data from or outputs data to the data bus only when it is addressed (selected), and the proper combination of the CE and OE signals is present. But how do we know when to activate these signals? Well, the OE and CE signals depend solely on the R/W signal from the CPU. When the CPU executes a load instruction (reads from memory), the R/W signal is high, and we should activate the OE signal and deactivate the CE signal. On the contrary, when the CPU executes a store instruction, the R/W signal is deactivated, and we should activate the CE signal and deactivate the OE signal. This simple logic is implemented in Fig. 1.2.

But how about the CS signal? Well, this signal should be active only when the register is selected. But wait, what does this mean? Who selects the register? The register is selected when the CPU addresses it. Hence, the CS signal depends solely on the content on the address bus. Previously, we said that each memory-mapped register is assigned its own unique address from the CPU address space. The CPU address space is the set of all possible addresses that the CPU can generate. For a 32-bit CPU, the address is 32-bit long, and the CPU can issue any address from 0x00000000 to 0xFFFFFFFF.

Suppose that we would like to connect a register at address 0x80000000. Now, we can use a 32-input AND logical gate to compare the address lines with the desired address. For our example, we should create a logic expression that activates the chip select signal when the address matches 0x800000000. Figure 1.3 shows the solution. This AND gate activates the CS signal when all the specified address lines match their

Fig. 1.3 A 32-input AND gate used to decode address 0x80000000

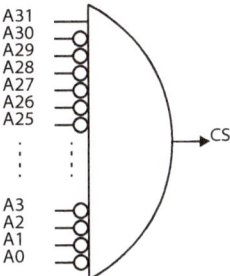

Fig. 1.4 A 32-input AND gate used to decode address 0x80000000

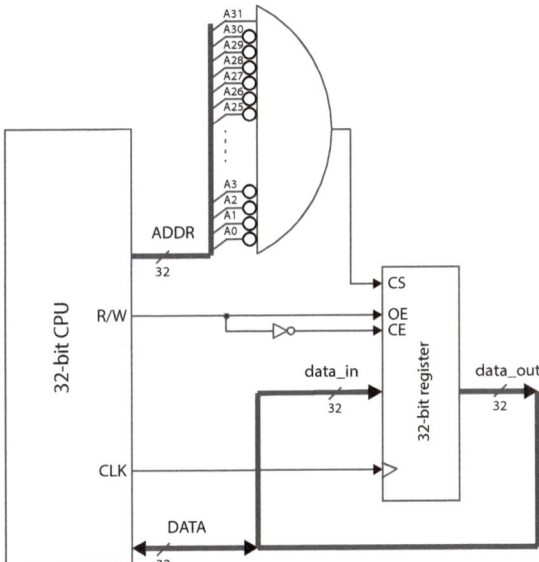

respective logic levels (high or low) as in the assigned address. This process is called **address decoding**. Figure 1.4 presents the final digital circuitry used to connect the register to the CPU. Now, the register is memory-mapped into the CPU address space, and it is accessible for reading and writing at the address 0x800000000.

We see that address decoding involves constantly comparing the addresses on the address bus and generating the CS signal when the address on the address bus matches the address assigned to an I/O device.

1.3 Two Memory Mapped Registers

Now that we understand how a register can be memory-mapped into the CPU address space and which signals are used in this process, we can try to memory-map and connect two registers to a CPU. Suppose we connect one register at address 0x80000000 and the other to address 0xC0000000. Actually, this task is straightforward and is

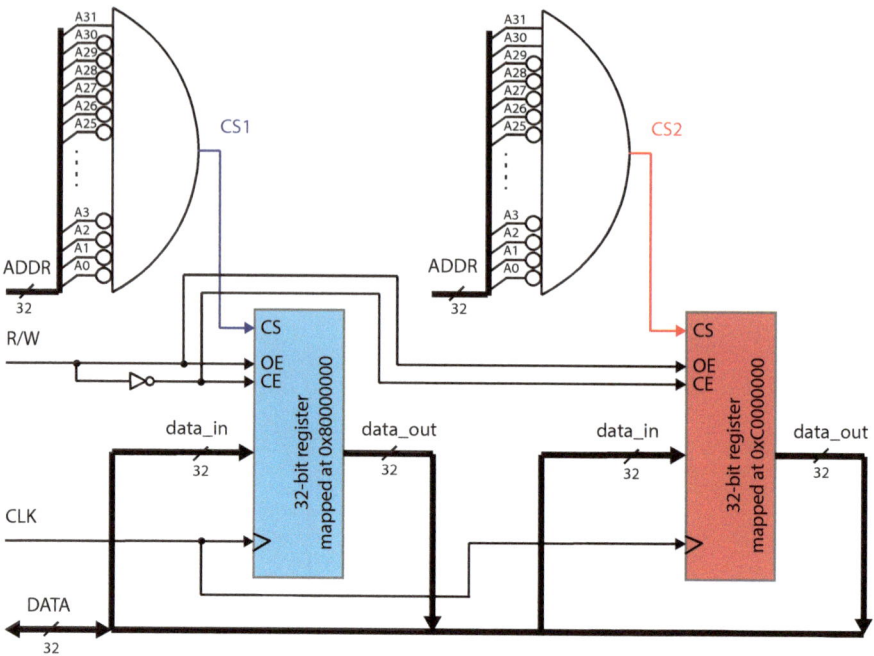

Fig. 1.5 Two memory-mapped registers at addresses 0x80000000 and 0xC0000000, respectively

depicted in Fig. 1.5. We should use two AND gates to decode two addresses. One AND gate decodes address 0x80000000 and selects the first register, while the second AND gate decodes address 0xC0000000 and selects the second register. Both registers can share the address, CE and OE signals because address decoding logic ensures that registers cannot be selected (active) simultaneously. Hence, we can see that address decoding isolates two or more registers or components in a computer system and is a crucial concept in memory-mapped I/O systems, where multiple registers or devices share the same address bus.

Although the presented address decoding with AND gates seems very simple, it has a serious drawback. In CMOS technology used to implement basic logic gates, we can usually implement only 2- or 3-input logic gates. In our example, we used 32-input AND gates that do not exist in the real world. Hence, in the real world, we would use tens of 2-input AND gates to implement the address decoding for only one address. In real-world computer systems with tens of I/O devices and hundreds of memory-mapped registers, this solution would be very inefficient in terms of the number of logic gates used. Hence, we should use a different solution to decode the addresses and select the I/O devices and their registers.

Let us return to our simple example with two memory-mapped registers. Recall that one register is accessible at address 0x80000000 and the other at address 0xC0000000. The two addresses differ only in address bit A30. Hence, we could select the registers based only on this bit and ignore all other address bits. We could

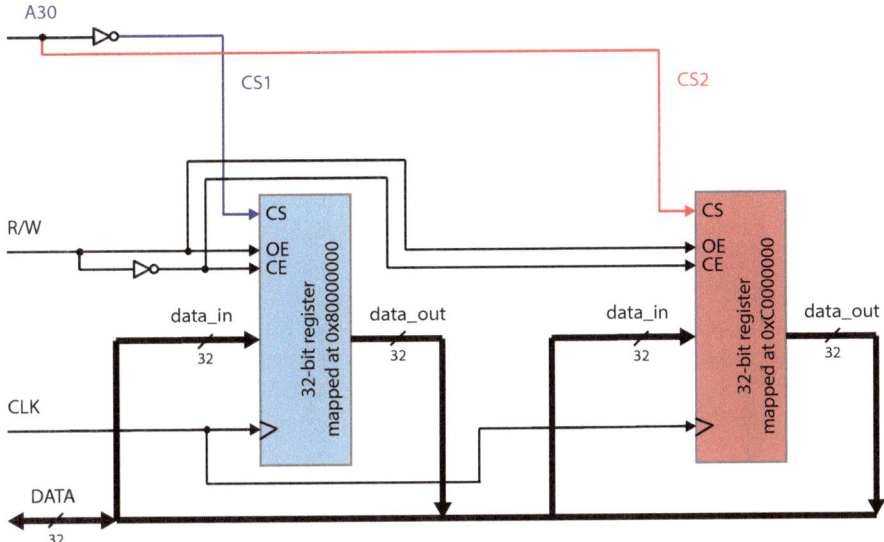

Fig. 1.6 Partial address decoding

select the first register when the address bit A30 is low and the second register when the address bit A30 is high. This solution is depicted in Fig. 1.6.

But wait! The CS signal for the first register will now be active when CPU issues addresses 0x00000000, 0x80000000 or 0xA03F0147. Actually, it is selected whenever the CPU issues an address with bit A30 set low. The second register will be selected when the CPU issues any address with the address bit A30 set high. In other words, each register is assigned exactly half of the CPU address space and not only one particular address! But this is not a problem at all if we have only these two registers in the system. Even now, they can be selected with their previously assigned addresses 0x80000000 and 0xC0000000. This method of address decoding is called **partial address decoding**. This is contrary to the previously presented method, called **full address decoding**, where each register is assigned only one address from the address space. Here, using partial address decoding, both 0x80000000 or 0xA03F0147 addresses point to the same memory location (the first register). In general, a set of memory addresses that point to the same memory location or an I/O device is called **aliases**. Modern computer systems use partial address decoding whenever possible to reduce the number of logic gates required to implement address decoding logic.

1.4   Several Memory Mapped Registers

This time, we want to connect eight registers to a CPU and map them into the CPU address space. Again, we will use partial address decoding to simplify the logic required to decode the addresses and select the registers. For this purpose, we will use an address decoder. Address decoders are fundamental logic components in digital systems, often used for selecting input/output devices. Recall that a 3-to-8 address decoder is a combinational logic circuit that takes a 3-bit binary input and activates one of its eight output lines based on the input value. Figure 1.7 depicts a 3-to-8 address decoder. The decoder has three input lines (A0, A1, and A2), representing a 3-bit binary number. These input lines can be either high (1) or low (0), creating eight possible binary combinations: 000 to 111. The decoder has eight output lines (Y0 to Y7), and each output corresponds to one of the possible input combinations. The operation of a 3-to-8 decoder can be described using the following truth table:

A2	A1	A0	Y7	Y6	Y5	Y4	Y3	Y2	Y1	Y0
0	0	0	0	0	0	0	0	0	0	1
0	0	1	0	0	0	0	0	0	1	0
0	1	0	0	0	0	0	0	1	0	0
0	1	1	0	0	0	0	1	0	0	0
1	0	0	0	0	0	1	0	0	0	0
1	0	1	0	0	1	0	0	0	0	0
1	1	0	0	1	0	0	0	0	0	0
1	1	1	1	0	0	0	0	0	0	0

When we provide a 3-bit binary number as input to the decoder, it decodes that binary value and activates the corresponding output line while setting all other output lines to 0. This operation allows us to select one of the eight output lines based on the input value. A 3-to-8 address decoder simplifies selecting devices based on a 3-bit binary input value and is used in address decoding to select one of eight memory-mapped I/O devices in a digital system.

Figure 1.8 shows the application of a 3-to-8 address decoder to select one of eight registers mapped into the CPU memory space. Each of the eight registers is assigned 1/8 of the CPU memory space in this case. For example, the first register will be accessible at addresses 0x00000000 to 0x1FFFFFFF. The second register is accessible at addresses 0x20000000 to 0x3FFFFFFF and so on, until the last one,

Fig. 1.7 A 3-to-8 address
decoder

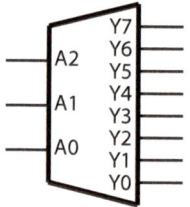

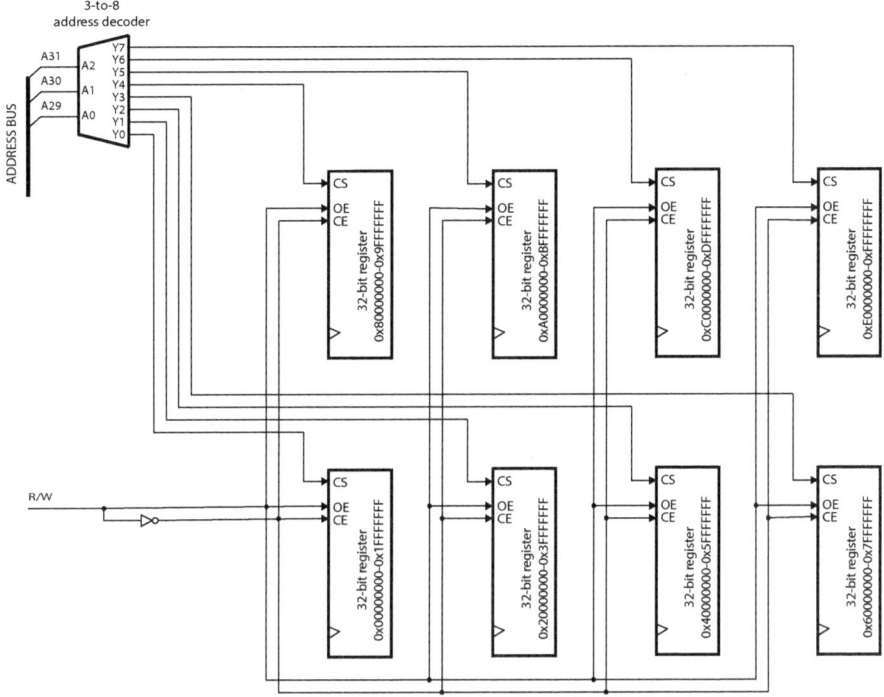

Fig. 1.8 Partial address decoding using a 3-to-8 address decoder to select eight registers

which is accessible at addresses 0xE0000000 to 0xFFFFFFFF. Because of partial address decoding, the registers do not have only one address; instead, each register is assigned 512 MB (one-eighth of a 4 GB) of address space.

1.5   Registers Mapped at Consecutive Addresses

In the previous section, we have learned how to memory map a set of registers over the whole memory space using partial address decoding. However, we often aim to map several registers belonging to the same IO device at consecutive memory addresses. Suppose we want to map eight 32-bit registers at the following addresses: 0x80000000, 0x80000004, 0x80000008, 0x8000000C, 0x80000010, 0x80000014, 0x80000018, and 0x8000001C. To decode the registers' addresses, we would use:

1. 2-input AND gates to decode whether the most significant bit A31 is set, and
2. a 3-to-8 address decoder to decode the address bits A4, A3 and A2 and select a particular register.

Figure 1.9 illustrates the solution.

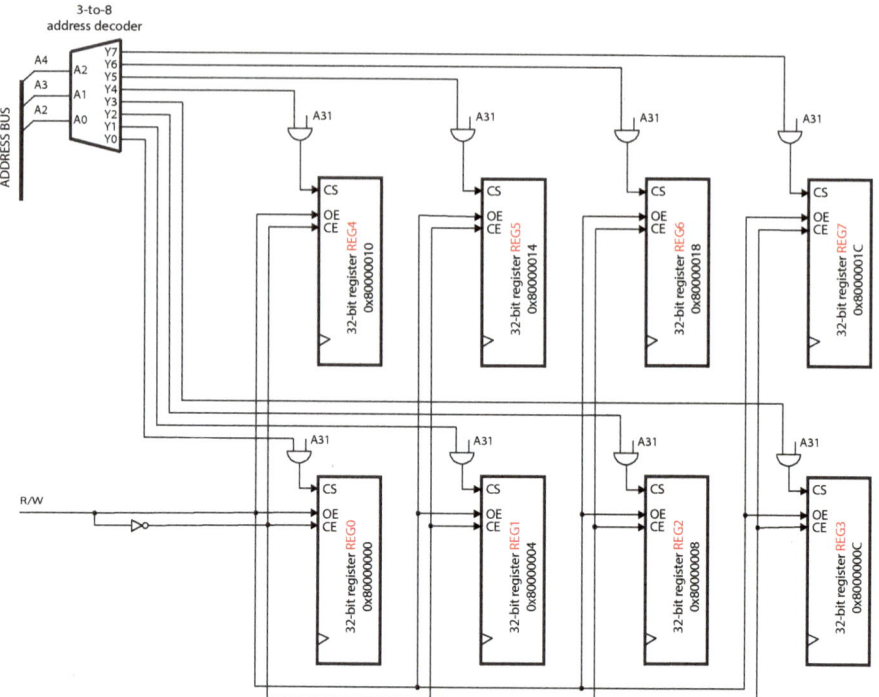

Fig. 1.9 Eight 32-bit registers mapped at consecutive addresses

There is a positive side-effect of consecutively memory-mapped registers that belong to the same IO device. Using a C structure and pointers, we can conveniently work with the consecutively memory-mapped registers as if they were C structure members, making the IO device's driver code well-organized and readable. This approach is commonly used when working with memory-mapped peripherals and hardware registers in embedded systems and microcontroller programming. To represent consecutively mapped registers using a C structure, we define a C structure where each member corresponds to a specific register at consecutive addresses. Here's an example for the registers from Fig. 1.9:

```
#define BASE_ADDRESS 0x80000000

// Define a structure to represent the memory-mapped registers
typedef struct {
    volatile uint32_t REG0;
    volatile uint32_t REG1;
    volatile uint32_t REG2;
    volatile uint32_t REG3;
    volatile uint32_t REG4;
    volatile uint32_t REG5;
    volatile uint32_t REG6;
    volatile uint32_t REG7;
} Registers_t;
```

```
15 // Define a pointer to the base address of the memory-mapped registers
   Registers_t *pMMIORegs = ((Registers_t *)BASE_ADDRESS);
17
   int main() {
19     // Access and manipulate the registers:
       pMMIORegs->REG0 = 0x12345678;        // Write to REG0
21     pMMIORegs->REG2 |= 0x01 << 13;       // Set bit 13 in REG2
       pMMIORegs->REG7 &= ~(0x01 << 27);    // Clear bit 27 in REG2
23     uint32_t value = pMMIORegs->REG6;    // Read from REG6
       ...
25     return 0;
   }
```

Listing 1.1 Representing and manipulating consecutively memory-mapped registers in C.

In the above code, we define a structure type named `Registers_t`, where each member represents a specific register at consecutive addresses. Then, we create the pointer `pMMIORegs` to the `Registers_t` type structure. We assume that the registers are memory-mapped to the base address 0x80000000, and we set this address to the pointer `pMMIORegs`. Finally, as shown in the above example, we can access and manipulate the registers using the `pMMIORegs` pointer and the structure members.

1.6 Partial Versus Full Address Decoding

Let us summarize what we have learned so far. Partial address decoding and full address decoding are two different methods used in computer memory and memory-mapped I/O systems to determine which memory locations or I/O devices are accessed at a particular address.

Full address decoding involves all address lines generated by the CPU or processing unit to select a specific memory location or I/O device. It is usually used when we need to uniquely identify and select individual memory locations or I/O devices, each with a distinct address. Full address decoding provides precise control over memory or I/O access but requires more complex hardware, especially when dealing with a large number of unique addresses.

On the contrary, partial address decoding involves examining only a portion of the address lines generated by the CPU to decode an address and select a memory location or an I/O device. For example, suppose you have a memory system with 16 memory locations or I/O devices. In that case, we may use partial address decoding and compare only four higher-order address lines (e.g., A31-A28) to determine which device is being accessed. The lower-order address lines (e.g., A27-A0) are ignored. This method is more efficient regarding hardware complexity than full address decoding because it reduces the number of logic gates required to decode an address.

In partial address decoding, **aliases** occur. Aliases are multiple addresses that map to the same memory location or I/O device. Aliases occur because only a portion of the address lines is used to select a specific memory location or device, allowing multiple addresses to access the same location due to address overlap. In general, aliases are not a problem. If address decoding is carefully designed and with appropriate software handling, aliases do not lead to conflicts in accessing memory or I/O resources. Despite aliases, partial address decoding offers several advantages over full address decoding. It reduces hardware complexity, lowers power consumption, simplifies PCB (printed circuit board) design and enables faster decoding.

The choice between partial address decoding and full address decoding depends on the specific requirements of the system design. Partial address decoding is often used when memory banks or I/O devices are organised in a structured way, with common prefixes, while full address decoding is necessary when each memory location or I/O device must have a unique address.

1.7 Case Study: Using the GPIO Interface in FE310-G002 RISC-V Based System-on-Chip

GPIO stands for General Purpose Input/Output, and it refers to a type of interface on a microcontroller that is used for simple digital input or output operations. GPIO interface controls GPIO pins that can be configured to serve various purposes, such as reading digital signals (input) or sending digital signals (output). GPIO pins are "general purpose" because they are not dedicated to a specific function. Instead, we can program them to perform various tasks based on the needs of your project.

GPIO pins can be configured as either input or output. Through input pins, the GPIO interface can detect whether the logical level on the pin is high (usually 3.3 V or 5V) or low (0V). Through output pins, the GPIO interface can set the logic level on the pin to high or low, which we often use for tasks such as controlling external devices like LEDs, controlling actuators (motors, relays), and interfacing with other digital devices.

A GPIO interface comprises a set of memory-mapped registers. These registers allow us to set the pin direction (input or output), read or write values to the pins, and handle events triggered by changes in the pin's state.

The SiFive Freedom FE310 is a microcontroller-based system-on-a-chip (SoC) developed by SiFive. The FE310 is built around the RISC-V E31 CPU core. The E31 CPU 32-bit core is based on the RISC-V RV32IMAC instruction-set architecture (ISA), which is an open-source and royalty-free ISA. RISC-V is gaining popularity in the embedded and processor design communities due to its flexibility, simplicity, and extensibility. The E31 RISC-V CPU comprises a single-issue, in-order pipeline. The pipeline comprises five stages: instruction fetch, instruction decode and register fetch, execute, data memory access, and register writeback. The pipeline has a peak execution rate of one instruction per clock cycle and is fully bypassed so that most instructions have a one-cycle result latency.

The FE310 includes on-chip memory components such as SRAM for program and data storage. Besides, it offers various peripherals and I/O options, including GPIO pins, UART (serial communication), SPI (Serial Peripheral Interface), I2C (Inter-

Table 1.1 SiFive FE310 GPIO peripheral register offsets. All registers are reset to 0

Offset	Name	Description
0x00	**GPIO_INPUT_VAL**	Pin input value
0x04	**GPIO_INPUT_EN**	Pin input enable
0x08	**GPIO_OUTPUT_EN**	Pin output enable
0x0C	**GPIO_OUTPUT_VAL**	Pin output value

Integrated Circuit), and timers. These peripherals enable the FE310 to interface with other hardware components and sensors.

The SiFive FE310 microcontroller has 32 GPIO pins. The GPIO interface in the SiFive FE310 comprises a set of special registers. Each bit in these registers manages the state and behaviour of a corresponding individual GPIO pin. These registers are part of the microcontroller's memory-mapped I/O (MMIO) address space. The GPIO interface is mapped at address `0x10012000` and comprises 19 data and control registers. To keep the description simple, we will focus only on four data and control registers. The memory map for the selected four GPIO control and data registers is shown in Table 1.1. Each register is 32 bits wide.

Figure 1.10 presents a simplified structure of the GPIO interface in the SiFive FE310. Several registers that are present in the GPIO interface are omitted for the sake of simplicity and clarity.

There are several key registers involved in configuring and controlling GPIO pins on the SiFive FE310:

1. **GPIO_INPUT_VAL**: This register stores the current input values of all GPIO pins. Each bit in this register corresponds to a specific pin, with '1' indicating a high voltage (logic level 1) and '0' indicating a low voltage (logic level 0).
2. **GPIO_OUTPUT_VAL**: This register stores the values to be output on the GPIO pins when they are configured as outputs.
3. **GPIO_OUTPUT_EN**: This register controls whether a GPIO pin is enabled as output by driving its tri-state buffer. When a bit in this register is '1', the corresponding bit in the **GPIO_OUTPUT_VAL** register is connected to the GPIO pin through the corresponding tri-state buffer. When the specific GPIO pin is output enabled, the content of the corresponding bit in the **GPIO_OUTPUT_VAL** register appears at the GPIO pin.
4. **GPIO_INPUT_EN**: This register controls whether a GPIO pin is enabled as input. When a bit in this register is '1', the GPIO pin is connected to the corresponding bit in the **GPIO_INPUT_VAL** register through the tri-state buffer. When the specific GPIO pin is input enabled, the content of the GPIO pin is stored in the corresponding bit in the **GPIO_INPPUT_VAL** register.

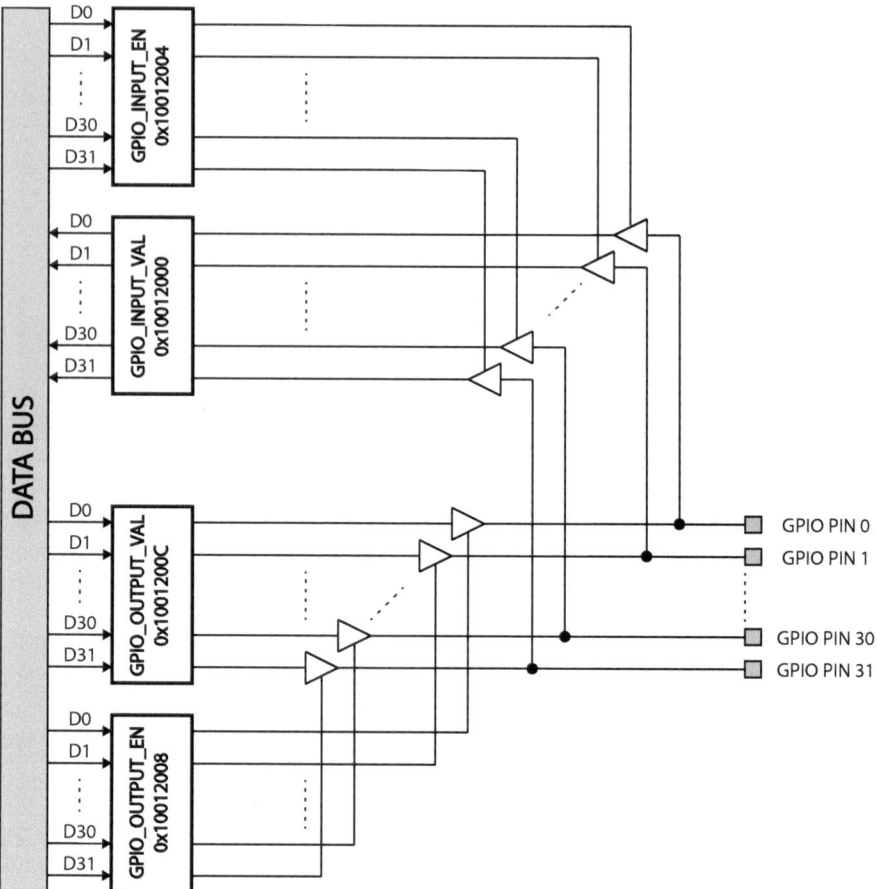

Fig. 1.10 A simplified structure of the GPIO interface in SiFive Freedom FE310

1.7.1 Program GPIO in Assembly

Using the GPIO interface to control the pins on the SiFive FE310 microcontroller in assembly language involves configuring the GPIO registers to control the behaviour of individual pins. To enable a GPIO pin as an output on the SiFive FE310 micro-controller using assembly language, we need to configure the **GPIO_OUTPUT_EN** register appropriately. Below is an example of enabling a GPIO pin as an output in assembly for the SiFive FE310. The pin number is given as the function parameter in the register a0:

```
; /*   GPIO output enable
;     Input: a0 - pin number
;     Output: None */
.align 2
.global gpio_output_en
.type gpio_output_en , @function
```

```
 7  gpio_output_en:
 8      # prologue:
 9      addi sp, sp, -16      # Allocate the routine
10                           #    stack frame
11      sw ra, 12(sp)        # Save the return address
12      sw fp, 8(sp)         # Save the frame pointer
13      sw s1, 4(sp)
14      sw s2, 0(sp)
15      addi fp, sp, 16      # Set the framepointer
16
17      # function body :
18      li t0, 0x10012000    # load GPIO base address
19      lw t1, 0x08(t0)      # read GPIO_OUTPUT_EN
20      li t2, 0x01
21      sll t2, t2, a0       # shift 1 to pin position
22      or t1, t1, t2        # set the bit @ pin position
23      sw t1, 0x08(t0)      # Store back
24
25      # epilogue:
26      lw s2, 0(sp)
27      lw s1, 4(sp)
28      lw fp, 8(sp)         # restore the frame pointer
29      lw ra, 12(sp)        # restore the return address
30      addi sp, sp, 16      # de-allocate the routine
31                           #    stack frame
32      ret
```

Listing 1.2 Assembly code used to implement the function that enables output on a GPIO pin.

Similarly, to enable a GPIO pin as an input on the SiFive FE310 microcontroller using assembly language, we need to configure the **GPIO_INPUT_EN** register appropriately. Below is an example of enabling a GPIO pin as an input in assembly for the SiFive FE310. The pin number is given as the function parameter in the register a0:

```
 1  ; /*   GPIO input enable
 2  ;     Input: a0 - pin number
 3  ;     Output: None */
 4  .align 2
 5  .global gpio_input_en
 6  .type gpio_input_en , @function
 7  gpio_input_en:
 8      # prologue:
 9      addi sp, sp, -16      # Allocate the routine
10                           #    stack frame
11      sw ra, 12(sp)        # Save the return address
12      sw fp, 8(sp)         # Save the frame pointer
13      sw s1, 4(sp)
14      sw s2, 0(sp)
15      addi fp, sp, 16      # Set the framepointer
16
17      # function body :
18      li t0, 0x10012000    # load GPIO base address
19      lw t1, 0x04(t0)      # read GPIO_INPUT_EN
20      li t2, 0x01
21      sll t2, t2, a0       # shift 1 to the pin position
22      or t1, t1, t2        # set the bit @ pin position
23      sw t1, 0x04(t0)      # Store back
24
25      # epilogue:
```

```
26    lw  s2,  0(sp)
27    lw  s1,  4(sp)
28    lw  fp,  8(sp)          # restore the frame pointer
29    lw  ra,  12(sp)         # restore the return address
30    addi sp, sp, 16         # de-allocate the routine
31                            #    stack frame
32    ret
```

Listing 1.3 Assembly code used to implement the function that enables input on a GPIO pin.

To set a GPIO pin, we need to set the corresponding bit in the **GPIO_OUTPUT_VAL** register:

```
1  ; /*  GPIO set pin
2  ;     Input: a0 - pin number
3  ;     Output: None */
4  .align 2
5  .global gpio_set_pin
6  .type gpio_set_pin, @function
7  gpio_set_pin:
8      # prologue:
9      addi sp, sp, -16       # Allocate the routine
10                            #    stack frame
11     sw  ra, 12(sp)         # Save the return address
12     sw  fp, 8(sp)          # Save the frame pointer
13     sw  s1, 4(sp)
14     sw  s2, 0(sp)
15     addi fp, sp, 16        # Set the framepointer
16
17     # function body :
18     li  t0, 0x10012000     # load GPIO base address
19     lw  t1, 0x0C(t0)       # read GPIO_OUTPUT_VAL
20     li  t2, 0x01
21     sll t2, t2, a0         # shift 1 to pin position
22     or  t1, t1, t2         # set the bit @ pin position
23     sw  t1, 0x0C(t0)       # Store back
24
25     # epilogue:
26     lw  s2, 0(sp)
27     lw  s1, 4(sp)
28     lw  fp, 8(sp)          # restore the frame pointer
29     lw  ra, 12(sp)         # restore the return address
30     addi sp, sp, 16        # de-allocate the routine
31                            #    stack frame
32     ret
```

Listing 1.4 Assembly code used to implement the function for setting a GPIO pin.

To reset a GPIO pin, we need to reset the corresponding bit in the **GPIO_OUTPUT_VAL** register:

```
1  ; /*  GPIO clear pin
2  ;     Input: a0 - pin number
3  ;     Output: None */
4  .align 2
5  .global gpio_clear_pin
6  .type gpio_clear_pin, @function
7  gpio_clear_pin:
8      # prologue:
```

```
 9    addi sp, sp, -16      # Allocate the routine
10                          #   stack frame
11    sw ra, 12(sp)         # Save the return address
12    sw fp, 8(sp)          # Save the frame pointer
13    sw s1, 4(sp)
14    sw s2, 0(sp)
15    addi fp, sp, 16       # Set the framepointer
16
17    # function body :
18    li t0, 0x10012000     # load GPIO base address
19    lw t1, 0x0C(t0)       # read GPIO_OUTPUT_VAL
20    li t2, 0x01
21    sll t2, t2, a0        # shift 1 to pin position
22    not t2, t2            # 1' complement
23    and  t1, t1, t2       # clear pin
24    sw t1, 0x0C(t0)       # Store back
25
26    # epilogue:
27    lw s2, 0(sp)
28    lw s1, 4(sp)
29    lw fp, 8(sp)          # restore the frame pointer
30    lw ra, 12(sp)         # restore the return address
31    addi sp, sp, 16       # de-allocate the routine
32                          #   stack frame
33    ret
```

Listing 1.5 Assembly code used to implement the function for resetting a GPIO pin.

A handy function is to toggle a GPIO pin. To toggle a GPIO pin, we need to EXOR the corresponding bit in the **GPIO_OUTPUT_VAL** register with '1':

```
 1  ; /*   GPIO clear pin
 2  ;     Input: a0 - pin number
 3  ;     Output: None */
 4  .align 2
 5  .global gpio_toggle_pin
 6  .type gpio_toggle_pin, @function
 7  gpio_toggle_pin:
 8      # prologue:
 9      addi sp, sp, -16      # Allocate the routine
10                           #   stack frame
11      sw ra, 12(sp)        # Save the return address
12      sw fp, 8(sp)         # Save the frame pointer
13      sw s1, 4(sp)
14      sw s2, 0(sp)
15      addi fp, sp, 16      # Set the framepointer
16
17      # function body :
18      # function body :
19      li t0, 0x10012000    # load GPIO base address
20      lw t1, 0x0C(t0)      # read GPIO_OUTPUT_VAL
21      li t2, 0x01
22      sll t2, t2, a0       # shift 1 to pin position
23      xor t1, t1, t2       # toggle the bit @ pin position
24      sw t1, 0x0C(t0)      # Store back
25
26      # epilogue:
27      lw s2, 0(sp)
28      lw s1, 4(sp)
29      lw fp, 8(sp)         # restore the frame pointer
30      lw ra, 12(sp)        # restore the return address
```

```
31    addi sp, sp, 16      # de-allocate the routine
32                         #   stack frame
33    ret
```

Listing 1.6 Assembly code used to implement the function for toggling a GPIO pin.

1.7.2 Program GPIO in C

We can also program a memory-mapped I/O device in C. We abstract an MMIO device with a C structure that represents and mirrors the layout of the registers in the MMIO device. We will present this concept using the GPIO Interface in FE310-G002 RISC-V based System-On-chip. To abstract GPIO registers with a C structure, we create a structure that mirrors the layout of the GPIO registers:

```
typedef struct
{
    volatile int GPIO_INPUT_VAL;
    volatile int GPIO_INPUT_EN;
    volatile int GPIO_OUTPUT_EN;
    volatile int GPIO_OUTPUT_VAL;
} GPIO_Registers_t;
```

Listing 1.7 A C structure that mirrors the GPIO registers layout.

This abstraction makes it easier to access and manipulate GPIO registers and pins and control their behaviour. Each member of the structure corresponds to a specific register in the GPIO interface, such as the input value register, output value register, etc. The layout of the members of the structure exactly mirrors the layout of the registers in memory, i.e., the members are in the same order as the registers in memory space. Recall that in C, the `volatile` keyword is used to indicate to the compiler that a variable can change its value at any time, even if it doesn't appear to be modified by the program. It informs the compiler that the variable should always be fetched from memory when needed rather than relying on cached values or optimizations that could result in unexpected behaviour. When working with hardware peripherals, we often access memory-mapped registers that control or represent hardware components. These registers can be modified by the hardware (e.g., GPIO pins) at any time outside our program, and the compiler might not be aware of these changes. By declaring such registers as volatile, you ensure the compiler generates code that correctly reflects the behaviour of hardware registers, making it suitable for hardware interaction.

Next, we define a pointer (in our example, the pointer is named `GPIO`, but you are free to use any name you wish) that holds the base address of the GPIO interface:

```
#define GPIO_BASEADDR      0x10012000

GPIO_Registers_t *GPIO = (GPIO_Registers_t*) GPIO_BASEADDR;
```

Listing 1.8 A pointer that holds the base address of the GPIO interface.

This pointer is used to access the GPIO registers as if they were part of a C structure. For example, we set pin 19 as output in the output enable register and toggle the state of pin 19 in the output value register:

```
GPIO->GPIO_OUTPUT_EN  |= (0x01 << 19);
GPIO->GPIO_OUTPUT_VAL ^= (0x01 << 19);
```

Listing 1.9 Enabling and setting a GPIO in C.

Instead of using the above `GPIO` pointer to access the GPIO registers directly, we usually define an initialization structure and implement several access functions. This is especially true when we implement the hardware abstraction layer (HAL) of a MMIO device that other users will use. In the hardware abstraction layer, we try to provide more user-friendly way to configure the peripheral without forcing the programmers to know how to configure its registers in detail. For example, to configure and use GPIO, we define several other constants and the `GPIO_InitTypeDef` structure in C:

```
#define GPIO_MODE_INPUT 0x00U
#define GPIO_MODE_OUTPUT 0x01U

/* GPIO pins define
 *
 */
#define GPIO_PIN_0                 ((uint32_t)0x00000001)
#define GPIO_PIN_1                 ((uint32_t)0x00000002)
#define GPIO_PIN_2                 ((uint32_t)0x00000004)
#define GPIO_PIN_3                 ((uint32_t)0x00000008)

...

#define GPIO_PIN_30                ((uint32_t)0x40000000)
#define GPIO_PIN_31                ((uint32_t)0x80000000)

typedef struct
{
  uint32_t Pin;      /* GPIO pins to be configured. */
  uint32_t Mode;     /* Operating mode for the selected pins */
} GPIO_InitTypeDef;
```

Listing 1.10 Pins definition and a C structure used to configure the GPIO.

The meaning of each field of the struct is:

1. `Pin`: it is the position of the pin in a 32-bit word, starting from 0, of the pins we will configure. For example, for pin 22 it assumes the value `GPIO_PIN_22`. Take note that the `GPIO_PIN_x` is a bit mask, where the i-th pin corresponds to the i-th bit of a `uint32_t` datatype. For example, the `GPIO_PIN_9` has a value of `0x00000200`. We can use the same `GPIO_InitTypeDef` instance to configure several pins at once, doing a bitwise OR (e.g., `GPIO_PIN_1|GPIO_PIN_21|GPIO_PIN_22`).

2. `Mode`: it is the operating mode of the pin, and it can be `GPIO_MODE_INPUT` or `GPIO_MODE_OUTPUT`.

We can now write HAL functions in C that will provide manipulation routines to initialize and change the state of GPIO pins. For example, to initialize and toggle GPIO pins, we implement the following C functions:

```
void HAL_GPIO_Init(GPIO_Registers_t *GPIO, GPIO_InitTypeDef *GPIO_Init){

  if (GPIO_Init->Mode == GPIO_MODE_INPUT) {
    GPIO->GPIO_INPUT_EN  |= GPIO_Init->Pin;
    GPIO->GPIO_OUTPUT_EN &= ~(GPIO_Init->Pin);
  }

  else if (GPIO_Init->Mode == GPIO_MODE_OUTPUT) {
    GPIO->GPIO_OUTPUT_EN |= GPIO_Init->Pin;
    GPIO->GPIO_INPUT_EN &= ~(GPIO_Init->Pin);
  }
}

void HAL_GPIO_TogglePin(GPIO_Registers_t *GPIO, uint32_t GPIO_Pin){
  GPIO->GPIO_OUTPUT_VAL ^= GPIO_Pin;
}
```

Listing 1.11 Hardware abstraction layer functions for the GPIO.

The `HAL_GPIO_Init` function accepts the GPIO register and the initialization structures. For example, to initialize pins 19, 21 and 22 as outputs, we use the following C code:

```
GPIO_Registers_t *GPIO = (GPIO_Registers_t*) GPIO_BASEADDR;
GPIO_InitTypeDef GPIO_InitStruct;

GPIO_InitStruct.Mode = GPIO_MODE_OUTPUT;
GPIO_InitStruct.Pin = GPIO_PIN_19 | GPIO_PIN_21 | GPIO_PIN_22;
HAL_GPIO_Init(GPIO, &GPIO_InitStruct);
```

Listing 1.12 GPIO pins initialization.

1.8   Case Study: Using the GPIO Interface in ARM Cortex-M Based STM32H7 System-on-Chip

1.8.1   Cortex-M Fixed Memory Address Space

ARM defines a fixed memory address space common to all Cortex-M cores. This fixed memory space ensures code portability among different silicon manufacturers, which incorporate ARM Cortex-M cores into their systems-on-chip. The ARM Cortex-M processors have a fixed default memory map that provides up to 4 GB of addressable memory. The memory map of ARM Cortex-M processors is split into regions with different aims. Figure 1.11 shows the memory map of a Cortex-M processor. All on-chip peripherals should be mapped to a specific region, starting from `0x4000 0000` and lasting up to `0x5FFF FFFF`. This region is further divided into several

Fig. 1.11 Cortex-M fixed memory address space

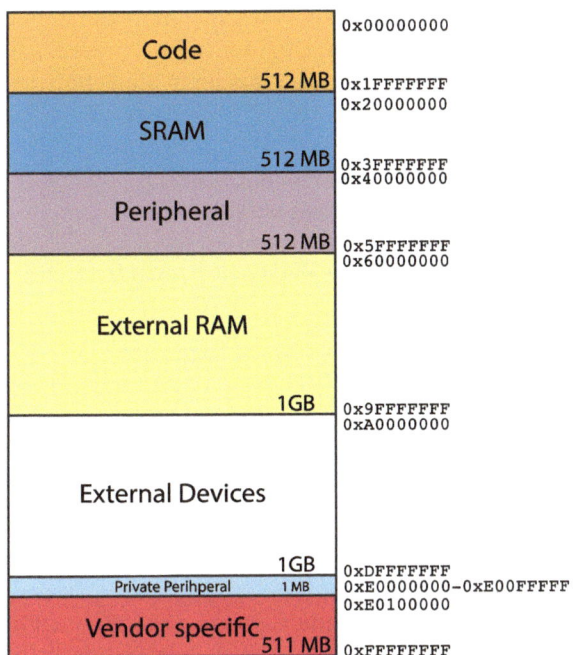

sub-regions, each one mapped to a specific peripheral. For example, STM32H750 system-on-chip comprises 11 16-bit GPIO interfaces, GPIOA to GPIOK. Each GPIO interface is mapped to a 1 kB memory region using partial address decoding. For example, the GPIOI interface is mapped to `0x58022000-0x580223FF`. The reader should look for more details in a reference manual for a specific SoC.

1.8.2 GPIO Interface in STM32H7

The STM32H7 microcontrollers come with 176 GPIO pins organized into 11 GPIO interfaces (or ports), named GPIOA to GPIOK. Each GPIO port contains 16 pins that can be configured individually. We can configure GPIO pins for various **modes**, including input, output, alternate function, and analog. Additionally, depending on our requirements, we can configure GPIO output speed to low, medium, or high. Similar to FE310, many GPIO pins in STM32H7 microcontrollers have alternate functions, allowing them to be used for tasks other than general-purpose I/O. These alternate functions may include UART, SPI, I2C, PWM, or other peripherals.

In STM32H7 microcontrollers, GPIO functionality is managed through a set of registers that control the configuration, state, and behaviour of individual GPIO pins and their associated ports. Each GPIO interface in an STM32H7 microcontroller has:

1. four 32-bit configuration registers (GPIOx_MODER, GPIOx_OTYPER, GPIOx_OSPEEDR and GPIOx_PUPDR),

2. two 32-bit data registers (GPIOx_IDR and GPIOx_ODR),
3. a 32-bit set/reset data register (GPIOx_BSRR), and
4. two 32-bit alternate function selection registers (GPIOx_AFRH and GPIOx_AFRL).

The reader should reference the STM32H7 reference manual and the HAL (Hardware Abstraction Layer) library provided by STMicroelectronics for detailed information on using these registers. Here, we provide only a brief description of GPIO registers. Table 1.2 provides the memory map for a GPIO interface data and control and registers in STM32H7.

GPIOx_MODER (Port Mode Register):
This register (Fig. 1.12) configures the mode of each pin within a GPIO port (x). Each pin can be set to one of the following modes: input, output, alternate function, or analog mode. For example, setting two bits for a pin to '01' configures it as a general-purpose output pin.

GPIOx_OTYPER (Output Type Register):
This register (Fig. 1.13) configures the output type of each pin within a GPIO port. We can select either push-pull (default) or open-drain output modes for each pin. In open-drain mode, the pin can only pull the signal low; it needs an external pull-up resistor for logic high.

Let us simply and briefly explain the difference between output configurations without diving into the electrical details of both. **Push-pull and Open-Drain** are two different output configurations for digital pins in electronics, particularly in microcontrollers and digital logic circuits. These configurations determine how a pin drives or controls an external circuit and have different characteristics and use cases.

In the **push-pull output configuration**, a **digital pin can actively pull the output to either a high voltage level or a low voltage level**. A push-pull pin can output both high and low logic levels, making it suitable for driving both high and low states.

Table 1.2 STM32H7 GPIO peripheral register offsets

Offset	Name	Description
0x00	**GPIOx_MODER**	GPIO port mode register
0x04	**GPIOx_OTYPER**	GPIO port output type register
0x08	**GPIOx_OSPEEDR**	GPIO port output speed register
0x0C	**GPIOx_PUPDR**	GPIO port pull-up/pull-down register
0x10	**GPIOx_IDR**	GPIO port input data register
0x14	**GPIOx_ODR**	GPIO port output data register
0x18	**GPIOx_BSRR**	GPIO port bit set/reset register
0x20	**GPIOx_AFRL**	GPIO alternate function low register
0x24	**GPIOx_AFRH**	GPIO alternate function high register

31	30	29	28	27	26	25	24	23	22	21	20	19	18	17	16
MODER15[1:0]		MODER14[1:0]		MODER13[1:0]		MODER12[1:0]		MODER11[1:0]		MODER10[1:0]		MODER9[1:0]		MODER8[1:0]	
rw	rw	rw	rw	rw	rw	rw	rw	rw	rw	rw	rw	rw	rw	rw	rw
15	14	13	12	11	10	9	8	7	6	5	4	3	2	1	0
MODER7[1:0]		MODER6[1:0]		MODER5[1:0]		MODER4[1:0]		MODER3[1:0]		MODER2[1:0]		MODER1[1:0]		MODER0[1:0]	
rw	rw	rw	rw	rw	rw	rw	rw	rw	rw	rw	rw	rw	rw	rw	rw

Bits 31:0 **MODER[15:0][1:0]**: Port x configuration I/O pin y (y = 15 to 0)
These bits are written by software to configure the I/O mode.
00: Input mode
01: General purpose output mode
10: Alternate function mode
11: Analog mode (reset state)

Fig. 1.12 GPIOx_MODER (Port Mode Register)

31	30	29	28	27	26	25	24	23	22	21	20	19	18	17	16
15	14	13	12	11	10	9	8	7	6	5	4	3	2	1	0
OT15	OT14	OT13	OT12	OT11	OT10	OT9	OT8	OT7	OT6	OT5	OT4	OT3	OT2	OT1	OT0
rw	rw	rw	rw	rw	rw	rw	rw	rw	rw	rw	rw	rw	rw	rw	rw

Bits 31:16 Reserved, must be kept at reset value.

Bits 15:0 **OTy**: Port x configuration bits (y = 0..15)
These bits are written by software to configure the I/O output type.
0: Output push-pull (reset state)
1: Output open-drain

Fig. 1.13 GPIOx_OTYPER (Output Type Register)

Push-pull configuration is commonly used for general-purpose digital signal output. It can drive both high and low signals directly, providing faster signal transitions. The main disadvantage of the push-pull configuration is that it can consume considerably high power when actively driving a high state.

In an **open-drain output configuration**, a **digital pin can only actively pull the output to a low voltage level**. It relies on an external pull-up resistor to pull the output to a high voltage level. Open-drain output configuration is often used when multiple devices need to share a single bus (e.g., I2C or SMBus). It allows multiple devices to communicate without conflicts. The main disadvantage of the open-drain output configuration is that it is slower and consumes more power when actively pulling the output low because it depends on the external pull-up resistor. Let us summarize the key differences between the two output configurations:

31	30	29	28	27	26	25	24	23	22	21	20	19	18	17	16
OSPEEDR15 [1:0]		OSPEEDR14 [1:0]		OSPEEDR13 [1:0]		OSPEEDR12 [1:0]		OSPEEDR11 [1:0]		OSPEEDR10 [1:0]		OSPEEDR9 [1:0]		OSPEEDR8 [1:0]	
rw	rw	rw	rw	rw	rw	rw	rw	rw	rw	rw	rw	rw	rw	rw	rw

15	14	13	12	11	10	9	8	7	6	5	4	3	2	1	0
OSPEEDR7 [1:0]		OSPEEDR6 [1:0]		OSPEEDR5 [1:0]		OSPEEDR4 [1:0]		OSPEEDR3 [1:0]		OSPEEDR2 [1:0]		OSPEEDR1 [1:0]		OSPEEDR0 [1:0]	
rw	rw	rw	rw	rw	rw	rw	rw	rw	rw	rw	rw	rw	rw	rw	rw

Bits 31:0 **OSPEEDR[15:0][1:0]**: Port x configuration I/O pin y (y = 15 to 0)
These bits are written by software to configure the I/O output speed.
00: Low speed
01: Medium speed
10: High speed
11: Very high speed

Fig. 1.14 GPIOx_OSPEEDR (Output Speed Register)

Output Range: Push-pull can drive both high and low states, while open-drain can only drive low states.

Speed: Push-pull typically provides faster signal transitions.

Applications: Open-drain is often used in bidirectional communication buses (e.g., I2C), while push-pull is commonly used for general-purpose digital signal output.

The choice between push-pull and open-drain configurations depends on the application's specific requirements.

GPIOx_OSPEEDR (Output Speed Register):
This register (Fig. 1.14) configures the output speed of each pin within a GPIO port. We can choose from various output speed options, such as low, medium, or high. The speed setting affects the rise and fall times of the output signal. A higher GPIO speed increases the EMI noise from STM32 and the power consumption. It is good to adapt the GPIO speed to the peripheral speed. For example, low speed is optimal for toggling GPIO at 1 Hz, while fast serial communication at 45 MHz requires a very high-speed setting.

GPIOx_PUPDR (Pull-Up/Pull-Down Register):
This register (Fig. 1.15) configures the pull-up and pull-down resistors for each pin within a GPIO port. We can select pull-up, pull-down, or neither (floating) for each pin. Pull-up and pull-down resistors are useful for input pins to provide defined logic levels when no external signal is applied.

Pull-up and pull-down resistors are commonly used in digital electronics to set the default logic state of a digital input pin when it is not actively being driven to a logic high or low level. They are often used to ensure that an input pin has a defined logic state in situations where no active signal is present.

A pull-up resistor is connected between a digital input pin and the positive voltage supply. It *"pulls up"* the voltage level of the input pin to a high logic state when there

31	30	29	28	27	26	25	24	23	22	21	20	19	18	17	16
PUPDR15[1:0]		PUPDR14[1:0]		PUPDR13[1:0]		PUPDR12[1:0]		PUPDR11[1:0]		PUPDR10[1:0]		PUPDR9[1:0]		PUPDR8[1:0]	
rw	rw	rw	rw	rw	rw	rw	rw	rw	rw	rw	rw	rw	rw	rw	rw
15	14	13	12	11	10	9	8	7	6	5	4	3	2	1	0
PUPDR7[1:0]		PUPDR6[1:0]		PUPDR5[1:0]		PUPDR4[1:0]		PUPDR3[1:0]		PUPDR2[1:0]		PUPDR1[1:0]		PUPDR0[1:0]	
rw	rw	rw	rw	rw	rw	rw	rw	rw	rw	rw	rw	rw	rw	rw	rw

Bits 31:0   **PUPDR[15:0][1:0]**: Port x configuration I/O pin y (y = 15 to 0)
These bits are written by software to configure the I/O pull-up or pull-down
00: No pull-up, pull-down
01: Pull-up
10: Pull-down
11: Reserved

Fig. 1.15 GPIOx_PUPDR (Pull-Up/Pull-Down Register)

is no other signal driving the pin. It prevents floating or undefined states on input pins, which can lead to unreliable circuit behaviour. Besides, it is also often used in conjunction with open-drain outputs (common in communication interfaces like I2C and UART) to provide a defined high state of the communication line when there is no active communication.

A pull-down resistor is connected between a digital input pin and the ground. It *"pulls down"* the voltage level of the pin to a low logic state when there is no other signal driving the pin. It ensures that digital input is in a known low state when not actively being driven high. It is often used in scenarios where a digital input should default to a low state in the absence of an external signal or when a pin is unconnected.

The choice between pull-up and pull-down resistors depends on the desired default state for the input and the specific requirements of the circuit. Without pull-up or pull-down resistors, digital inputs can be left in a floating or undefined state when not actively driven high or low, which can lead to erratic behaviour or excessive power consumption. Both pull-up and pull-down resistors are essential tools for maintaining stable and predictable logic states in digital circuits and ensuring reliable operation, especially in microcontroller-based systems, digital communication interfaces, and any situation where digital signals need to be defined in the absence of active signals.

GPIOx_IDR (Input Data Register):
This register (Fig. 1.16) reads the input state of each pin within a GPIO port. Each bit corresponds to a pin, reflecting the pin's logic level (0 or 1). Reading this register allows us to determine the state of input pins.

GPIOx_ODR (Output Data Register):
This register (Fig. 1.17) controls the output state of each pin within a GPIO port. Each bit corresponds to a pin, and writing a 0 or 1 to a bit sets the pin's output state accordingly.

31	30	29	28	27	26	25	24	23	22	21	20	19	18	17	16
15	14	13	12	11	10	9	8	7	6	5	4	3	2	1	0
IDR15	IDR14	IDR13	IDR12	IDR11	IDR10	IDR9	IDR8	IDR7	IDR6	IDR5	IDR4	IDR3	IDR2	IDR1	IDR0
r	r	r	r	r	r	r	r	r	r	r	r	r	r	r	r

Bits 31:16   Reserved, must be kept at reset value.

Bits 15:0   **IDRy:** Port input data bit (y = 0..15)
These bits are read-only. They contain the input value of the corresponding I/O port.

Fig. 1.16   GPIOx_IDR (Input Data Register)

31	30	29	28	27	26	25	24	23	22	21	20	19	18	17	16
15	14	13	12	11	10	9	8	7	6	5	4	3	2	1	0
ODR15	ODR14	ODR13	ODR12	ODR11	ODR10	ODR9	ODR8	ODR7	ODR6	ODR5	ODR4	ODR3	ODR2	ODR1	ODR0
rw	rw	rw	rw	rw	rw	rw	rw	rw	rw	rw	rw	rw	rw	rw	rw

Bits 31:16   Reserved, must be kept at reset value.

Bits 15:0   **ODRy:** Port output data bit (y = 0..15)

Fig. 1.17   GPIOx_ODR (Output Data Register)

31	30	29	28	27	26	25	24	23	22	21	20	19	18	17	16
BR15	BR14	BR13	BR12	BR11	BR10	BR9	BR8	BR7	BR6	BR5	BR4	BR3	BR2	BR1	BR0
w	w	w	w	w	w	w	w	w	w	w	w	w	w	w	w
15	14	13	12	11	10	9	8	7	6	5	4	3	2	1	0
BS15	BS14	BS13	BS12	BS11	BS10	BS9	BS8	BS7	BS6	BS5	BS4	BS3	BS2	BS1	BS0
w	w	w	w	w	w	w	w	w	w	w	w	w	w	w	w

Bits 31:16   **BR[15:0]:** Port x reset I/O pin y (y = 15 to 0)
These bits are write-only. A read to these bits returns the value 0x0000.
   0: No action on the corresponding ODRx bit
   1: Resets the corresponding ODRx bit
Note: If both BSx and BRx are set, BSx has priority.

Bits 15:0   **BS[15:0]:** Port x set I/O pin y (y = 15 to 0)
These bits are write-only. A read to these bits returns the value 0x0000.
   0: No action on the corresponding ODRx bit
   1: Sets the corresponding ODRx bit

Fig. 1.18   GPIOx_BSRR (Bit Set/Reset Register)

GPIOx_BSRR (Bit Set/Reset Register):

This register (Fig. 1.18) allows us to atomically set or reset individual bits (pins) in the ODR register. Writing a '1' to a bit in this register sets the corresponding bit in ODR, and writing a '1' to the upper 16 bits resets the corresponding bits in ODR.

GPIOx_AFRL and GPIOx_AFRH (Alternate Function Registers Low/High):

These registers configure the alternate function of pins when they are used for alternate functions (e.g., UART, SPI). Each pin's alternate function is selected by writing a specific value to the corresponding bits in these registers.

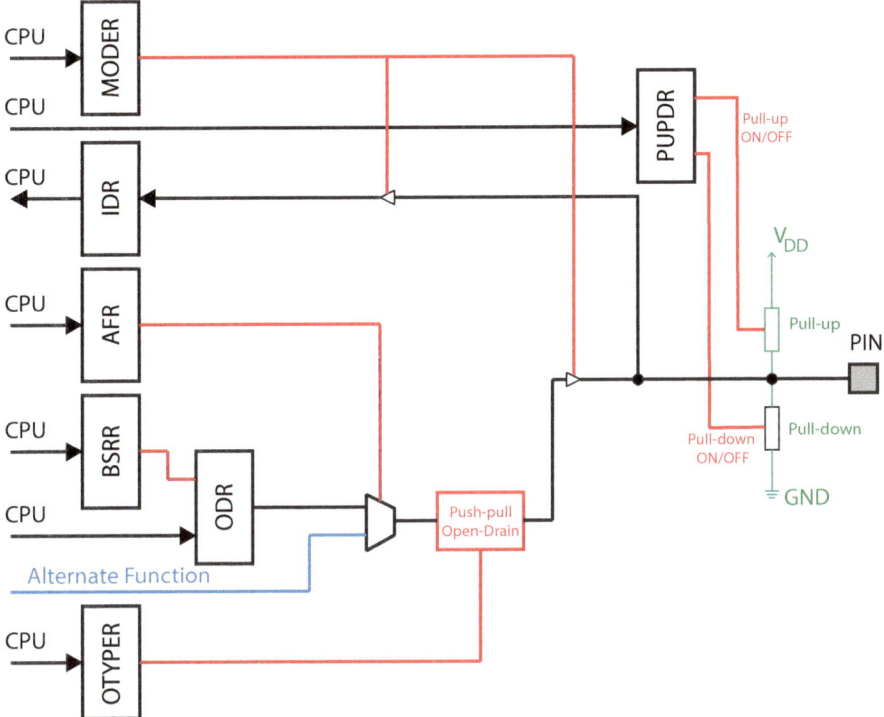

Fig. 1.19 Basic structure of an I/O port bit

1.8.3 Functional Description of the GPIO Interface in STM32H7

STM32 MCUs provide flexible GPIO management. Figure 1.19 shows the simplified hardware structure of a single I/O port of an STM32 microcontroller. Each port bit of the GPIO ports can be individually configured by software in several modes:

1. Input floating.
2. Input pull-up.
3. Input pull-down.
4. Output open-drain with pull-up or pull-down capability.
5. Output push-pull with pull-up or pull-down capability.
6. Alternate function output push-pull with pull-up or pull-down capability.
7. Alternate function output open-drain with pull-up or pull-down capability.

Each I/O port bit is freely programmable; however, the I/O port registers have to be accessed as 32-bit words. Changing the memory-mapped GPIO registers, the MCU changes the way the hardware of an I/O bit (pin) works. Let us have a look at the main modes.

To configure an I/O pin as input, we should write '00' to the corresponding bits in the MODER register. When the I/O is configured as input the output is disabled. The pull-up and pull-down resistors are activated depending on the value of the PUPDR register. The data present on the I/O pin are sampled into the input data register (ADR) every clock cycle. A read access to the IDR register provides the I/O state.

To configure an I/O pin as output, we should write '01' to the corresponding bits in the MODER register. The output buffer is enabled when the I/O is configured as output according to the OTYPER. If the corresponding bit in OTYPER is 0, the output is push-pull; otherwise, it is open-drain. Besides, The pull-up and pull-down resistors are activated depending on the value of the PUPDR register. The CPU can set the state of a pin by writing to the ODR register. The CPU can also control the state of an output pin through the BSRR register. The purpose of the BSRR register is to allow atomic read/modify accesses to the ODR register, avoiding the risk of an interrupt request occurring between the read and the modify access.

1.8.4　Program GPIO in C Using HAL

To configure and use peripherals in STM32H7 systems on a chip, we often use the hardware abstraction library. The HAL (Hardware Abstraction Layer) library for STM32 microcontrollers is a high-level, platform-independent library provided by STMicroelectronics. It serves as an abstraction layer between the low-level hardware of STM32 microcontrollers and the application code, making it easier to develop embedded applications for STM32 devices. The HAL library simplifies hardware initialization, peripheral configuration, and low-level I/O operations, enabling developers to focus on application-level code. The HAL library provides a consistent and abstracted API for STM32 peripheral control. This allows developers to write portable code that can be used across different STM32 microcontroller families. Besides, HAL provides drivers for various STM32 peripherals, including GPIO, UART, SPI, I2C, timers, and more. These drivers are used to configure and control the peripherals. STMicroelectronics provides extensive documentation, including reference manuals, datasheets, application notes, and examples, to help developers use the HAL library effectively. The description of the HAL library is beyond the scope of this textbook; however, we will try to explain its usage in a simple example of using the GPIO interface.

The STM32H7 GPIO interface is abstracted in HAL in the same way as we abstracted the GPIO interface for FE310 SoC in Sect. 1.7.2. HAL uses two structures: GPIO_TypeDef for GPIO registers abstraction (Listing 1.13) and GPIO_InitTypeDef to facilitate the initialization of GPIO (Listing 1.14).

```
typedef volatile __IO;
typedef struct
{
  __IO uint32_t MODER;     // Address offset: 0x00
  __IO uint32_t OTYPER;    // Address offset: 0x04
  __IO uint32_t OSPEEDR;   // Address offset: 0x08
  __IO uint32_t PUPDR;     // Address offset: 0x0C
  __IO uint32_t IDR;       // Address offset: 0x10
```

```
     __IO uint32_t ODR;         // Address offset: 0x14
     __IO uint32_t BSRR;        // Address offset: 0x18
     __IO uint32_t LCKR;        // Address offset: 0x1C
     __IO uint32_t AFR[2];      // Address offset: 0x20-0x24
   } GPIO_TypeDef;
```

Listing 1.13 A C structure that mirrors the GPIO registers layout.

```
#define GPIO_PIN_0                 ((uint16_t)0x0001)
#define GPIO_PIN_1                 ((uint16_t)0x0002)
#define GPIO_PIN_2                 ((uint16_t)0x0004)
#define GPIO_PIN_3                 ((uint16_t)0x0008)
...
#define GPIO_PIN_15                ((uint16_t)0x8000)
#define GPIO_PIN_All               ((uint16_t)0xFFFF)

typedef struct
{
  uint32_t Pin;         // GPIO pins to be configured.
  uint32_t Mode;        // Pin mode
  uint32_t Pull;        // Pull-up or Pull-Down
  uint32_t Speed;       // Speed
  uint32_t Alternate;   // Alternate function
} GPIO_InitTypeDef;
```

Listing 1.14 Pins definition and a C structure used to configure the GPIO.

Using the HAL library in STM32 microcontrollers to work with GPIO pins is a common and convenient approach for configuring and controlling GPIO pins. The HAL library provides a set of functions and macros that abstract the hardware-specific details, making it easier to work with GPIO pins. Suppose we want to configure pin 2 in the GPIOJ interface in the STM32H7 microcontroller as output. Here are the general steps:

1. Include the appropriate HAL header for the STM32H7 microcontroller, initialize the HAL Library and enable Clock for GPIOJ Peripheral in the main() function:

```
#include "stm32h7xx_hal.h"

int main(void)
{
  /* Init HAL: reset of all peripherals,
     initialize the Flash interface and the Systick. */
  HAL_Init();

  /* GPIO Ports Clock Enable */
  __HAL_RCC_GPIOJ_CLK_ENABLE();

  ...
}
```

Listing 1.15 Include HAL header, initialize HAL and enable GPIO clock.

2. Create a `GPIO_InitTypeDef` structure, configure the desired parameters for the GPIO pin and initialize the GPIO pin:

```
/* Configure GPIO pin : Pin 2 */
GPIO_InitStruct.Pin   = GPIO_PIN_2;
GPIO_InitStruct.Mode  = GPIO_MODE_OUTPUT_PP; // push-pull
GPIO_InitStruct.Pull  = GPIO_PULLUP; // pull up
GPIO_InitStruct.Speed = GPIO_SPEED_FREQ_VERY_HIGH;

/* Initialize GPIO pin : Pin 2 */
HAL_GPIO_Init(GPIOJ, &GPIO_InitStruct);
```

Listing 1.16 Configure and initialize GPIO pin.

3. Read, Write or Toggle the GPIO Pin:

```
GPIO_PinState bitstatus;
bitstatus = HAL_GPIO_ReadPin(GPIOJ, GPIO_PIN_2);

HAL_GPIO_WritePin(GPIOJ, GPIO_PIN_2, GPIO_PIN_SET);
HAL_GPIO_WritePin(GPIOJ, GPIO_PIN_2, GPIO_PIN_RESET);

HAL_GPIO_TogglePin(GPIOJ, GPIO_PIN_2);
```

Listing 1.17 Read, Write or Toggle pin.

1.9 Case Study: Using the UART Interface in FE310-G002 RISC-V Based System-on-Chip

Here, we will show how to program another handy memory-mapped IO device, Universal Asynchronous Receiver Transmitter (UART), but we will use only C this time. This is indeed possible for all memory-mapped IO devices, and there is no need to use an assembler. UART is a commonly used serial communication interface that allows asynchronous data transfer between a microcontroller, such as the SiFive FE310, and external devices like sensors, displays, other microcontrollers or even desktop computers. The SiFive FE310 microcontroller features two memory-mapped UART interfaces that provide serial communication capabilities.

1.9.1 Universal Asynchronous Receiver Transmitter

Before we start explaining the Universal Asynchronous Receiver Transmitter (UART) provided in SiFive FE310, let us briefly describe the UART interface and its communication protocol. When we want to exchange data between two devices, we generally have two alternatives. Firstly, we can simultaneously transmit all bits **in parallel** using a number of GPIO lines. The number of GPIO lines would be equal to the size of the data word (e.g., eight GPIO lines for a word made of eight bits). Secondly, we can transmit each bit, constituting a data word, one by one **serially**, i.e., in a continuous stream of bits flowing on a single wire. A UART is a device that translates parallel bits in a data word (usually grouped in a byte) into a continuous stream of bits and puts them one by one on a single wire. When the data flows between two devices serially (here, we refer to them as the **sender** and the **receiver**), they have to agree on the timing. Timing defines how long it takes to transmit each individual bit

Fig. 1.20 Two UARTs
directly communicate with
each other

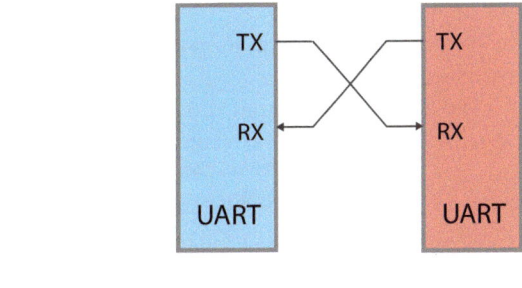

Fig. 1.21 UART frame
format

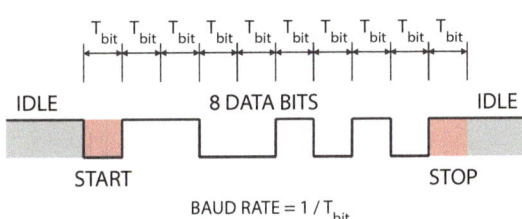

of the data. In **synchronous serial transmission**, the sender and the receiver share a common clock generated by the sender. The clock's frequency determines how fast we can transmit a single bit. But if both devices involved in data transmission agree on how long it takes to transmit a single bit and how to distinguish the start and finish of transmission, then we can avoid using a dedicated clock line. In this case, we have **asynchronous serial transmission**.

A Universal Asynchronous Receiver/Transmitter interface is a device able to transmit data word serially using two I/O lines, one acting as a transmitter (TX) and one as a receiver (RX) (Fig. 1.20). One of the big advantages of UART is that it is asynchronous—the transmitter and receiver do not share a common clock signal. Although this greatly simplifies the protocol, it does place certain requirements on the transmitter and receiver. Since they do not share a clock, both ends must transmit at the same agreed speed for the same bit timing. Communication in UART can be simplex (data is sent in one direction only), or full-duplex (both sides can transmit simultaneously).

Data in UART is transmitted in the form of frames. Figure 1.21 shows a UARTS's typical frame and the timing diagram. The high signal on the transmission line represents the idle state (that is, no transmission occurring). Because UART is asynchronous, the transmitter must signal that data bits are coming. This is accomplished by using the start bit. The start bit is a transition from the idle high state to a low state and is immediately followed by eight data bits. The data bits are the user data that come immediately after the start bit. There can be 5–9 user data bits, although 8 bits is most common. The least significant bit (LSB) is typically transmitted first. An optional parity bit is then transmitted (for error checking of the data bits). Often, this bit is omitted.

After the data bits are transmitted, the stop bit indicates the end of user data. The stop bit is either a transition back to the high or idle state or remaining at the high state for an additional bit-time. A second (optional) stop bit can be configured, usually to give the receiver time to get ready for the next frame, but this is uncommon in practice.

The time it takes to transmit a single bit determines the **baud rate**. The baud rate specifies how fast data is sent over a serial line. It's usually expressed in units of bits-per-second (bps). If we invert the baud rate, we can find out just how long it takes to transmit a single bit. This value determines how long the transmitter holds a serial line high/low or at what period the receiver samples its line. Baud rates can be just about any value within reason. The only requirement is that both devices agree upon the same rate. The standard baud rates are 1200, 2400, 4800, 19200, 38400, 57600, and 115200 bits per second.

1.9.2   The UART Interface in the SiFive FE310

The UART interface in the SiFive FE310 supports the following features:

1. frames formats: 8 data bits, no parity bit, 1 start bit, 1 or 2 stop bits,
2. 8-entry transmit and receive FIFO buffers with programmable watermark interrupts.

FE310 SoC contains two memory-mapped UART interfaces. The interface UART0 is mapped at address `0x10013000`, while the interface UART1 is mapped at address `0x10023000`. We will focus on UART0 only.

The UART0 interface in the SiFive FE310 comprises several memory-mapped data and control registers. Table 1.3 presents the memory map for the UART data and control and registers. The UART registers are 32-bit wide, requiring only naturally aligned 32-bit memory accesses.

Here, we will describe only a few registers required to transmit and receive data without using interrupts:

Table 1.3 Register offsets within UART memory map

Offset	Name	Description
0x00	**txdata**	Transmit data register
0x04	**rxdata**	Receive data register
0x08	**txctrl**	Transmit control register
0x0C	**rxctrl**	Receive control register
0x10	**ie**	UART interrupt enable
0x14	**ip**	UART interrupt pending
0x18	**div**	Baud rate divisor

Fig. 1.22 The **txdata** register

Fig. 1.23 The **rxdata** register

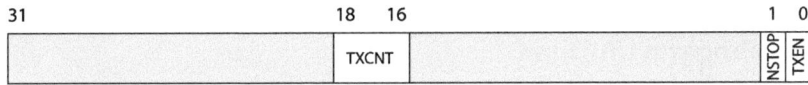

Fig. 1.24 The **txctrl** register

1. **Transmit Data Register (txdata)** (Fig. 1.22). Writing to the **txdata** register enqueues the character contained in the data field to the transmit FIFO if the FIFO is able to accept new entries. Reading from **txdata** returns the current value of the FULL flag and zero in the data field. The FULL flag indicates whether the transmit FIFO is able to accept new entries; when set, writes to data are ignored.
2. **Receive Data Register (rxdata)** (Fig. 1.23). Reading the **rxdata** register dequeues a character from the receive FIFO and returns the value in the data field. The EMPTY flag indicates if the receive FIFO was empty; when set, the data field does not contain a valid character. Writes to **rxdata** are ignored.
3. **Transmit Control Register (txctrl)** (Fig. 1.24). The read-write **txctrl** register controls the operation of the transmitter. The TXEN bit controls whether the transmitter is enabled. When cleared, the transmission is suppressed, and the TX pin is driven high. The NSTOP field specifies the number of stop bits: 0 for one stop bit and 1 for two stop bits.
4. **Receive Control Register (rxctrl)** (Fig. 1.25). The read-write **rxctrl** register controls the receiver's operation. The RXEN bit controls whether the receive is enabled. When cleared, the state of the RX pin is ignored.
5. **Baud Rate Divisor Register (div)** (Fig. 1.26). The read-write, **div** register specifies the divisor used by the baud rate generator to divide the CPU's clock frequency to generate a desired baud rate. For example, to set the baud rate of 115200 bits per second, the **div** register should be set to 139. We should refer to the SiFive FE310 documentation and reference manual for precise details on configuring the **div** register.

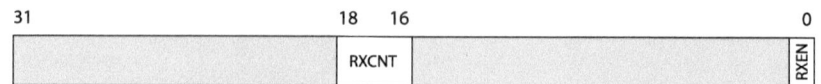

Fig. 1.25 The **rxctrl** register

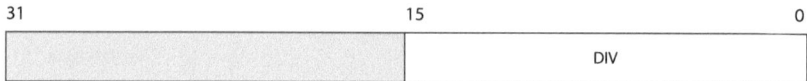

Fig. 1.26 The **div** register

1.9.3   Program UART in C

To abstract UART registers with a C structure, we create a structure that mirrors the layout of the UART registers:

```
typedef struct
{
  volatile int UART_TXDATA;
  volatile int UART_RXDATA;
  volatile int UART_TXCTRL;
  volatile int UART_RXCTRL;
  volatile int UART_IE;
  volatile int UART_IP;
  volatile int UART_DIV;
} UART_Registers_t;
```

Listing 1.18 A C structure that mirrors the UART registers layout.

Next, we define a pointer (in our example, the pointer is named UART0, but you are free to use any name you wish) that holds the base address of the UART0 interface:

```
#define UART0_BASEADDR      0x10013000

UART_Registers_t *UART0 = (UART_Registers_t*) UART0_BASEADDR;
```

Listing 1.19 A pointer that holds the base address of the UART interface.

This pointer is used to access the UART registers as if they were part of a C structure. Here, we present a few useful UART functions:

```
/*
 * Set Baud Rate to 115200
 * With tlclk at 16Mhz, to achieve 115200 baud,
 *  divisor should be 139. SiFive FE310-G002 Manual:, page 85
 * @arguments:
 *  uart: UART0 or UART1
 */
void uart_set_baud(UART_Registers_t *uart){
  uart->UART_DIV = 139;
}
```

```
11
12  /*
13   * Enable TX
14   * @arguments:
15   *   uart: UART0 or UART1
16   */
17  void uart_enable_tx(UART_Registers_t *uart){
18    uart->UART_TXCTRL |= 0x00000001;
19  }
20
21  /*
22   * Set No. stop bits
23   * @arguments:
24   *   uart: UART0 or UART1
25   *   nstop: UART_1_STOP_BIT or UART_2_STOP_BIT
26   */
27  void uart_set_nstop(UART_Registers_t *uart, unsigned int nstop){
28
29    if (nstop == UART_1_STOP_BIT) {
30      uart->UART_RXCTRL &= 0xfffffffd;
31    }
32    else if (nstop == UART_2_STOP_BIT) {
33      uart->UART_RXCTRL |= 0x00000002;
34    }
35  }
```

Listing 1.20 Enabling transmission and setting baud rate and the number of stop bits for UART in C.

1.9.4 UART Pins

Many GPIO pins on the FE310 can serve dual purposes. In addition to their basic input and output capabilities that we presented in Sect. 1.7, these pins can be controlled by other IO devices in the FE310 SoC. Each GPIO pin can implement up to two so-called IO functions (IOF) enabled with the GPIO_IOF_EN register. Which IOF is used is selected with the GPIO_IOF_SEL register. These alternative functions are often related to various peripherals or communication interfaces available on the microcontroller. IOF allows us to assign alternative functions to GPIO pins, such as enabling them as inputs or outputs for specific peripherals or features like UART. We should refer to the SiFive FE310 datasheet and reference manual for precise information on configuring IOF and alternative functions for GPIO pins on your particular hardware setup. For example, GPIO pin 17 can be used by UART0 transmitter (UART0_TX). Figure 1.27 shows all registers that control the behaviour of the GPIO pin 17. In Sect. 1.7, we have already explained the purpose of GPIO input, output and enable registers. These registres are depicted in light grey in Fig. 1.27. Besides these registers, there are two more registers, GPIO_IOF_SEL and GPIO_IOF_EN. These two registers enable and select an IO function for a particular pin. For example, for the GPIO pin 17, bit 17 in GPIO_IOF_SEL selects an IO function. If this bit is 0, the UART0 transmitter can drive GPIO pin 17. Bit 17 in GPIO_IOF_SEL enables the IO function on pin 17. If bit 17 is set, IOF is enabled for pin 17.

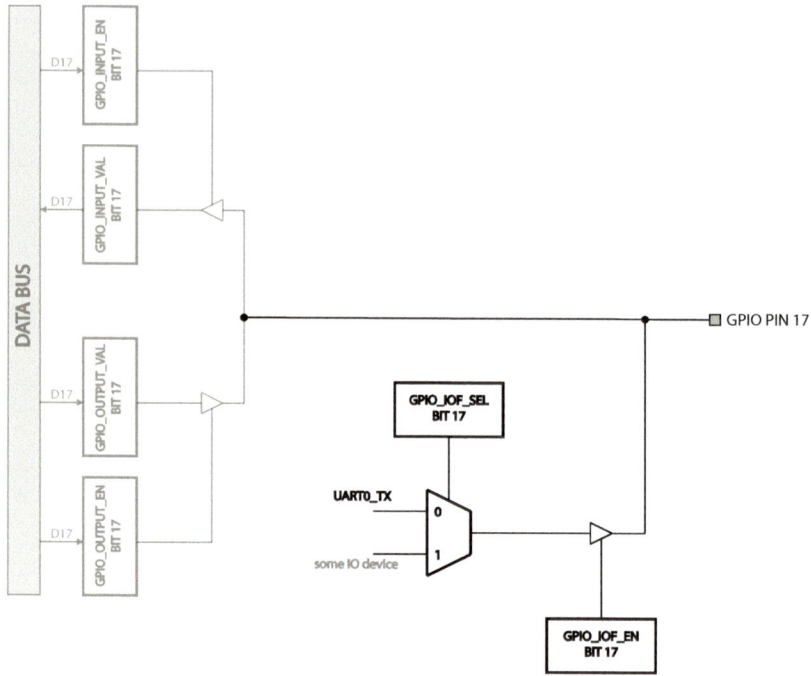

Fig. 1.27 The IO function for GPIO pin 17

In order to set the UART IO function for GPIO pin 17, we should implement a complete C data structure that mirrors all GPIO registers (refer to SiFive FE310 Manual):

```
typedef struct
{
  volatile int GPIO_INPUT_VAL;
  volatile int GPIO_INPUT_EN;
  volatile int GPIO_OUTPUT_EN;
  volatile int GPIO_OUTPUT_VAL;
  volatile int GPIO_PUE;
  volatile int GPIO_DS;
  volatile int GPIO_RISE_IE;
  volatile int GPIO_RISE_IP;
  volatile int GPIO_FALL_IE;
  volatile int GPIO_FALL_IP;
  volatile int GPIO_HIGH_IE;
  volatile int GPIO_HIGH_IP;
  volatile int GPIO_LOW_IE;
  volatile int GPIO_LOW_IP;
  volatile int GPIO_IOF_EN;
  volatile int GPIO_IOF_SEL;
  volatile int GPIO_OUT_XOR;
} GPIO_Registers_t;
```

Listing 1.21 A complete C structure for GPIO.

To set up UART0 IOF, we need to configure the GPIO_IOF_EN and GPIO_IOF_SEL registers appropriately. These registers control which alternative functions are enabled for specific GPIO pins. Below is an example of how to configure UART0 IOF for UART TX on GPIO pin 17:

```
GPIO->GPIO_IOF_SEL &= (1 << 17);
GPIO->GPIO_IOF_EN  |= (1 << 17);
```

Listing 1.22 A code for setting up UART0 IO function.

Interrupts and Interrupt Handling

2

CHAPTER GOALS

Have you ever wondered how computer components demand and get attention from the CPU? How do they tell the CPU or operating system that something important has just happened in the computer system, which requires an immediate response from the CPU, e.g., new data has just arrived at an I/O interface and should be processed immediately? This is done using so-called interrupts. This chapter will cover the theory and practice of interrupts and their handling. An interrupt is a hardware-initiated procedure that interrupts whatever program the CPU is currently executing and requests that the CPU immediately start running another program that is written to service the particular interrupt request.

In this chapter, you will:

- Explore the fundamental concepts of interrupts in computer systems.
- Distinguish between interrupts and traps.
- Examine the sequence of steps involved in processing interrupts.
- Learn about interrupt vector tables and their role in routing interrupts to the appropriate interrupt service routines (ISRs).
- Understand the interactions between hardware components, the operating system, and application software during interrupt handling.
- Explain the operation of the interrupt signals.
- Understand the mechanisms for handling nested interrupts and resolving interrupt conflicts.
- Understand the role of interrupts in operating system design and kernel development.
- Learn about interrupt-driven I/O and its impact on system architecture and performance.

© The Author(s), under exclusive license to Springer Nature Switzerland AG 2024
P. Bulić, *Understanding Computer Organization*, Undergraduate Topics in Computer, Science, https://doi.org/10.1007/978-3-031-58075-8_2

- Explore real-world examples and case studies illustrating the importance of interrupts in diverse computing scenarios.
- Explain the function of interrupt vectors and vector tabels.
- Explain the function of an interrupt controller.
- Learn from practical examples of interrupt-driven programming and system design in various domains, including networking, storage, and multimedia processing.
- Explain the interrupts and interrupt handling in the Intel, RISC-V, and ARM family of processors.

2.1 Introduction

In the dynamic landscape of computer systems, where tasks are executed concurrently and hardware resources are shared among multiple processes, interrupts and traps emerge as fundamental mechanisms for managing and responding to asynchronous events. As crucial elements of system operation, interrupts play a pivotal role in maintaining system stability, responsiveness, and reliability in the face of unforeseen circumstances and external stimuli.

At their core, interrupts and traps serve as mechanisms for diverting the processor's attention from its current task to handle urgent or exceptional events that require immediate attention. Whether triggered by external hardware devices, software instructions, or exceptional conditions within the processor itself, interrupts and traps provide a means for prioritizing and responding to critical events in real-time.

Interrupts are signals generated by hardware peripherals or devices to signal the processor of an event that requires attention. These events can range from user input (e.g., keyboard strokes, mouse movements) to hardware errors (e.g., disk errors, network interrupts) and are typically handled asynchronously, allowing the processor to respond promptly while continuing normal execution.

Traps, on the other hand, are internally generated signals that indicate exceptional conditions or errors encountered during the execution of program instructions. These conditions may include arithmetic errors (e.g., dividing by zero), memory access violations, or illegal instructions. Traps disrupt the normal flow of program execution and typically require special handling by the processor or operating system to resolve and recover from the error.

Throughout this exploration of interrupts in computer systems, we will delve into the fundamental principles, mechanisms, and implications of their operation. From their role in ensuring system responsiveness and reliability to their impact on software design and system architecture, we will unravel the intricacies of interrupts and their indispensable contribution to the robustness of modern computing environments.

2.2   Why Having Interrupts?

During my childhood, there were two powerful military blocs in Europe and the world: the Eastern (Soviet) Bloc and the Western (USA) Bloc. That was a period of geopolitical tension between the Soviet Union and the United States and their respective allies, the Eastern Bloc and the Western Bloc. The country where I grew up, former Yugoslavia, was not part of any of these military blocs, though politically, it was closer to the eastern bloc. In the 1970s, the former Yugoslav Air Force purchased a number of Soviet MIG-21 fighter aircraft from the USSR. The MIG-21 aircraft sold to the Yugoslav Air Force had virtually no modern electronic devices, and the Yugoslavian military wanted to install missile sensors in the planes. However, the USA and its allies have imposed an embargo on the purchase of electronic and computer components against Yugoslavia. Among all the universities in Yugoslavia, only the University of Ljubljana was allowed to purchase a few pieces (up to 20) of each chip that would be used only in the educational process. That's why the Yugoslav Army approached the University of Ljubljana to buy all the necessary electronic and computer components and develop a system that would be installed on the aircraft and would detect missiles. The system at the time had to be based on the modern Motorola 6800 microprocessors from the US. At its core, the system had a microcomputer built on the Motorola 6800 processor and a missile sensor. In addition to detecting missiles, the microcomputer also had to do other things. If the missile sensor detected a rocket, the computer system had to immediately stop whatever it was currently doing and alert the pilot to the approaching missile. But how would a missile sensor communicate with the CPU if the CPU could do nothing but fetch and execute instructions from memory? Remember that the CPU fetches and executes instructions every clock cycle. That's all it can do. So, there must be some mechanism by which the CPU can be immediately interrupted and required to start another program. In our case, the CPU would run another program (e.g., display the current altitude and speed of the aircraft). If the sensor detects a missile, it must, in some way, immediately suspend the currently running program and require the CPU to execute a program to flash the warning lights and alert the pilot. So, the CPU must have some mechanism in place to immediately stop the execution of one program and start another program. This mechanism is called **interrupts**, and the program that the CPU starts running in response to is called **interrupt service program (ISP)** or **interrupt handler**.

Interrupts and interrupt handling must be **transparent**. This means that the stopped (interrupted) program must not know that it has been stopped and must continue after the termination of the interrupt handler as if it had not been interrupted at all.

2.3 Interrupts

As we said in the previous section, we want to be able to service external interrupts. This is useful if a device external to the processor needs attention. Figure 2.1 illustrates a simplified system with a CPU and a peripheral device. To respond to interrupt requests from a peripheral device, a CPU usually has at least one interrupt request (IRQ) pin and one interrupt acknowledge (INTA) pin. The IRQ pin is the input a peripheral device uses to interrupt the processor (i.e., to interrupt the normal program flow in the CPU). Since the CPU should finish executing the current instruction(s) before servicing any external interrupts, the peripheral device may have to wait for several clock cycles before the CPU responds to the interrupt request. The INTA pin is the output used to signal the peripheral device, which has requested an interrupt via the IRQ signal, that the CPU has started servicing the interrupt request, and that the IRQ signal can be deactivated. Both pins in Fig. 2.1, IRQ and INTA, are active low. Two resistors are used to establish a logic one on IRQ and INTA signals (i.e., both signals are deactivated) when no one drives them.

In general, CPUs can respond to interrupts in two different ways: in either an **edge-sensitive** or **level-sensitive** manner. In an edge-sensitive manner, the interrupt signal input is designed to be triggered by a particular signal edge (level transition): either a falling edge (high to low) or a rising edge (low to high). In a level-sensitive manner, the interrupt signal input is designed to be triggered by a logic signal level. A peripheral device invokes a level-triggered interrupt by driving the signal to and holding it at the active level. We refer to this operation as **asserting the signal**. It de-asserts the signal when the processor signals it to do so. One advantage of level-triggered interrupt inputs is that they allow multiple devices to share a common interrupt signal. Most often, the active level of an interrupt input signal is LOW. In such a case, the interrupt signal is tied to the HIGH voltage level using a pull-up resistor. When multiple peripheral devices share one level-triggered interrupt input signal, the device that wants to assert the interrupt request simply connects the signal to the ground (pulls the signal LOW). The system in Fig. 2.1 uses level-sensitive interrupt signals.

Fig. 2.1 A simplified block diagram of a computer system with interrupt controlling signals

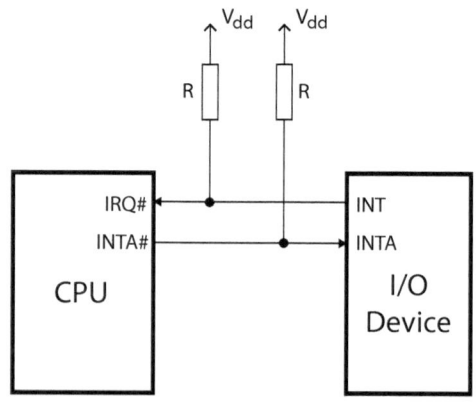

> **Summary: Assering and de-asserting a signal**
>
> Some signals are active high, and some signals are active low. To avoid the problem of high versus low and the fact that for some signals, active means high, and for some signals, active means low, we just say asserted (activated) versus de-asserted (deactivated).

When the device needs attention from the CPU, it activates (asserts) the IRQ pin on the CPU. During the normal execution flow through a program, the program counter increases sequentially through the address space, with branches to nearby labels or branches and links to subroutines. The CPU checks the status of the IRQ pin every time before a new instruction pointed to by the program counter is fetched from memory. When a peripheral device requests the interrupt, it is necessary to preserve the previous processor status while handling the interrupt, so that execution of the program that was running when the interrupt request occurred can resume when the appropriate interrupt handler has completed. We say that the interrupts must be 100% transparent. So, when an interrupt request occurs, the CPU completes the current instruction and asserts the INTA signal. When a peripheral device sees the INTA signal, it de-asserts the IRQ signal. Figure 2.2 shows the timing diagram for an external interrupt request for the simple system from Fig. 2.1.

Then, the CPU saves the part of the context of the interrupted program in the stack. A context is the state of the program counter, status register, stack pointer, and all other program-visible CPU registers. Some CPUs save the whole context in the stack, while others save only a part of the context in the stack. Since interrupts

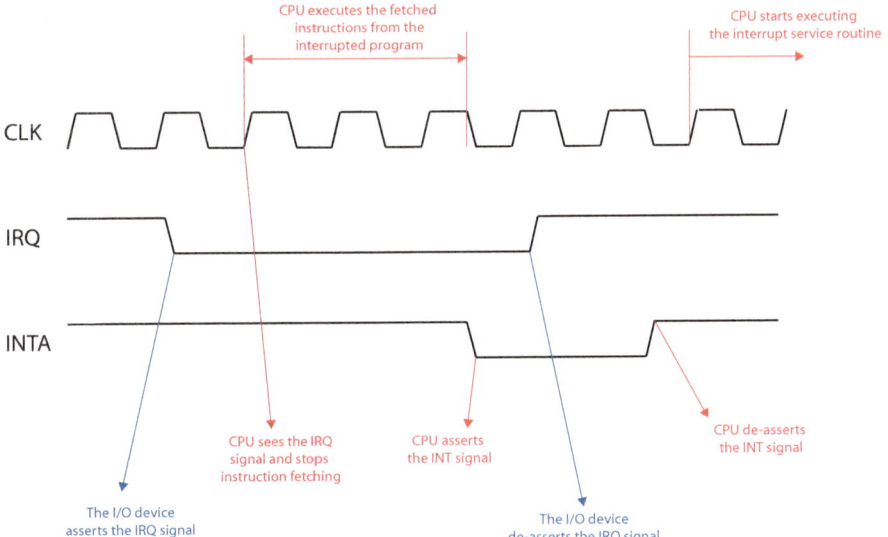

Fig. 2.2 A timing diagram for an external interrupt request

can happen at any time, there is no way for the active programs to prepare for the interrupt (e.g., by saving registers that the interrupt handler might write to). It is important to note that calling conventions do not apply when handling interrupts: the interrupt is not being "called" by the active program; it is interrupting the active program. Thus, the interrupt handler code must preserve the content and ensure it does not overwrite any registers the program may use before their content is saved. After the CPU has saved the context, the CPU automatically loads the address of the interrupt handler into the program counter. The interrupt handler is a program written by the user and depends on the peripheral device's functionality. Depending on how much of the context is automatically saved by the CPU, the interrupt handler must first save every register it intends to use in the stack or somewhere in memory.

Figure 2.3 shows the procedure involved in interrupts: the CPU executes the sequence of instructions from a user program until an interrupt request occurs at the time t_1. When the IRQ signal is asserted, the CPU stops executing the user code and starts executing the interrupt handler. But before executing the interrupt handler at time t_2, the CPU must finish the execution of already fetched instructions, save the (part of) context, and obtain the address of the interrupt handler. The time $t_2 - t_1$ required for this procedure is called **interrupt latency**. In general, interrupt latency is the time that elapses from when the IRQ signal is asserted to when the CPU starts to execute the interrupt handler. Interrupt latency duration is usually not predetermined and depends on how many instructions are already in the CPU's pipeline, how the CPU saves the context, and whether any new interrupt requests are temporarily disabled. Once the CPU completes the execution of the interrupt handler at time t_3, it returns to the execution of the user code at time t_4. Before returning to the user code, the CPU must automatically restore the previously saved context.

2.3.1 Types of Interrupts

There are typically three types of interrupts (also called exceptions) regarding the source of the interrupt (exception): **external** interrupts (or simply interrupts), **traps**, and **software interrupts**. An external device triggers external interrupts by activating the interrupt request pin on the CPU. Traps are activated internally in the CPU, usually as a result of some exceptional condition caused by instruction. For example, traps are caused when an illegal or undefined instruction is fetched or when the CPU attempts to execute an instruction that was not fetched because of the illegal address. A special instruction triggers software interrupts. Such instructions function similarly to subroutine calls, but the subroutine, in this case, the interrupt handler, is not being "called", but an interrupt-like sequence occurs. These software-interrupt instructions are useful when the user program does not know or is not allowed to know the address of the routine that it would like to "call," e.g., they are usually used for requesting operating system services and routines.

External interrupts are divided into two types: maskable and non-maskable interrupts. Maskable interrupts can be enabled or disabled by setting a bit in the CPU's control register or executing a special instruction. For example, Intel has the CLI instruction to mask the interrupts, and ARM has CPSID instruction for this purpose.

Fig. 2.3 The procedure
involved in interrupts

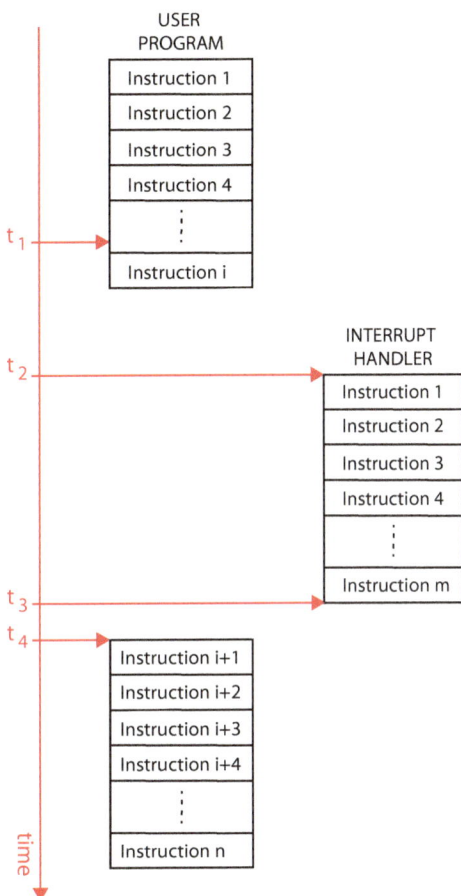

Non-maskable interrupts have a higher priority than maskable interrupts. That means that if both maskable and non-maskable interrupts are activated simultaneously, the CPU will service the non-maskable interrupt first.

2.3.2   Handling Interrupts

In a situation where multiple types of interrupts can occur, there must be a mechanism where different handler codes can be executed for different types of events. In general, there are two methods for handling this problem: polled interrupts and vectored interrupts.

In polled interrupts, the processor branches to a specific address that begins a sequence of instructions that check the cause of the interrupt or exception and branch to handler code for the type of interrupt/exception encountered. This is also called polled interrupt/exception handling.

In vectored interrupts, the processor branches to a different address for each type of interrupt or exception. Each exception address is separated by only one word, and these addresses form a table called **interrupt vector table**. Each entry of the interrupt vector table is called **interrupt vector**, and it is the address of an interrupt handler. Hence, the vector table contains the start addresses, called interrupt vectors, for all exception handlers. This method is called **vectored interrupt handling**. This concept is common across many processor architectures, although interrupt vector tables may be implemented in other architecture-specific fashions. For example, another common concept is to place a jump instruction (instead of vectors) at each entry in the table. Each jump instruction forces the processor to jump to the handler code for each type of interrupt/exception. In this case, the address of each table entry is considered an interrupt vector.

2.4 ARM Cortex-M7 Interrupts

In the terminology ARM uses, all events or conditions that can interrupt the normal program flow and transfer control to a specific handler (service) routine are referred to as exceptions. ARM Cortex-M7 processors support a variety of exceptions, and they are essential for handling events like interrupts, faults, and system calls. In general, exceptions can originate both by the hardware and the software.

2.4.1 ARM Cortex-M7 Programmer's Model

In this subsection, we will briefly describe the ARM Cortex-M7 programmer's model. The ARM Cortex-M7 processor core features a set of registers used for various purposes in program execution and system control. These registers can be categorized into two groups: register bank and special registers (see Fig. 2.4).

2.4.1.1 Register Bank
The register bank contains 16 32-bit registers. Thirteen of them are general-purpose registers, and the other three have special uses:

1. Registers R0 to R12 are general-purpose registers for data storage and data operations.
2. R13 is Stack Pointer (SP) for maintaining the stack, typically used for local variables and function call frames. The Cortex-M7 contains two physically different stack pointers for different privilege levels:
   a. The Main Stack Pointer (MSP) is the default Stack Pointer after reset and is mainly used when the processor runs in privileged or system mode.
   b. The Process Stack Pointer (PSP) can only be used in unprivileged or user mode.

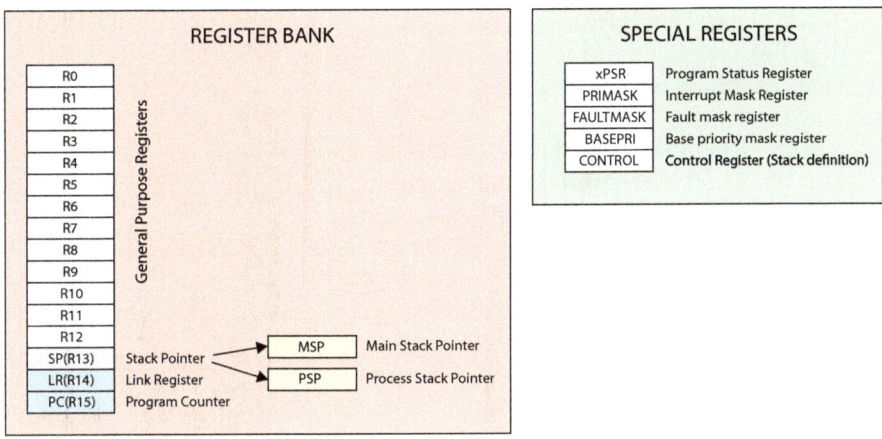

Fig. 2.4 ARM Cortex-M7 core registers

3. R14 is Link Register (LR), which stores the return address when calling subroutines or functions. On reset, the processor sets the LR value to 0xFFFFFFFF.
4. R15 is Program Counter (PC), which holds the memory address of the currently executing instruction.

Because the stack pointer register in ARM Cortex-M7 has two physical copies, we say it is **banked**. In the context of ARM Cortex processors, the term 'banked register' refers to a type of register that has multiple copies or 'banks', each associated with a specific execution mode or privilege level. These banks allow the processor to maintain separate register sets for different execution contexts, such as user mode, privileged mode, and exception modes. The selection of the stack pointer is determined by a special register called the CONTROL register, which is a part of the special register set.

2.4.1.2 Special Registers

Besides the registers in the register bank, there are several special registers. These registers contain the processor status and define the operation states and interrupt/exception masking. The special registers are:

1. xPSR is a 32-bit Program Status Register. Some of the bit fields in the xPSR register are N (negative flag), Z (zero flag), V (overflow flag), C (carry flag), T (Thumb state) and EXCEPTION NUMBER representing the number of the current exception (interrupt) (Fig. 2.5).
2. CONTROL register is a 32-bit register that allows the processor to manage privileged and unprivileged execution modes and select the active stack pointer. It includes the following fields: nPRIV (Privilege Level Bit) determines the privilege level of the processor (0 for privileged, 1 for unprivileged), and SPSEL

(Stack Pointer Select Bit) selects the active stack pointer (0 for MSP, 1 for PSP) (Fig. 2.6).

3. Three exception masking registers:

  a. The PRIMASK register is a 1-bit wide interrupt mask register. When set, it blocks all exceptions (including interrupts) apart from the Non-Maskable Interrupt (NMI) and the HardFault exception.
  b. The FAULTMASK register is very similar to PRIMASK, but it also blocks the HardFault exception.
  c. The BASEPRI register masks (blocks) exceptions or interrupts based on their priority level.

Special registers are not memory mapped and can be accessed using special register access instructions MSR and MRS:

MRS reg, special_reg

reads special register into general-purpose register, and

MSR special_reg, reg

writes to special register from general-purpose register.

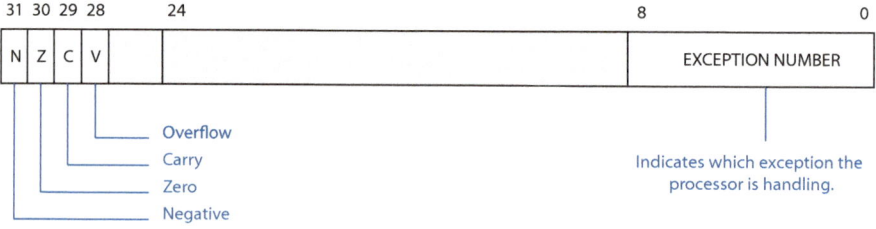

Fig. 2.5 xPSR register

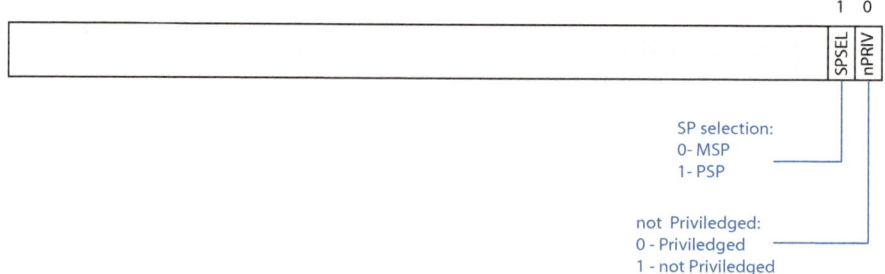

Fig. 2.6 CONTROL register

2.4.2   System Control Block

In addition to the registers we have just covered, ARM Cortex-M7 processors maintain another important register bank called System Control Block (SCB). The System Control Block is a crucial part of the processor's control and configuration. The SCB is a memory-mapped register bank that includes several registers and control bits that influence the processor's behaviour, manage exceptions, and provide system-level control. For example, the SCB registers for controlling processor configurations (e.g., low power modes), providing fault status information (fault status registers), relocating the vector table and controlling/obtaining the status of some interrupts. Here, we provide a brief description of only one CSB register related to interruptions and exceptions. This is the Interrupt Control and State Register (ICSR). This register provides bits for setting and clearing two software interrupts, PendSV and SysTick. The ICSR register is memory-mapped at address 0xE000ED04. For example, writing 1 to bit 28 in ICSR will set the PendSV exception to pending.

2.4.3   Exceptions

ARM architecture distinguishes between the two types of exceptions: **interrupts** originate from the external hardware, and **exceptions** originate from the CPU core or software (e.g., access to an invalid memory location or an SVC assembly instruction, which is commonly used as a convenient way to enter the operating system kernel). The following information identifies each ARM Cortex-M7 exception:

1. **Exception Number**—A unique number referencing a particular exception (starting at 1). This number is also used as the offset within the **vector table**, where the address of the handling routine for the exception is stored. This routine is usually referred to as the **exception handler** or **interrupt service routine (ISR)** and is the procedure which runs when an exception is triggered. The ARM hardware will automatically look up this function pointer (address of the exception handler) in the vector table when an exception is triggered and start executing the code. When the CPU is servicing an exception, its exception number is in the lower nine bits of the xPSR register.
2. **Priority Level/Priority Number**—Each exception has a priority associated with it. For most exceptions, this number is configurable. Counter-intuitively, the lower the priority number, the higher the precedence the exception has. So, for example, if two exceptions of priority level 2 and priority level 1 occur simultaneously, the exception with priority level 1 exception will be serviced first. When we say an exception has the "highest priority", it will have the lowest priority number. If two exceptions have the same priority number, the exception with the lowest exception number will run first.
3. **Synchronous or Asynchronous**—As the name implies, some exceptions will fire immediately after an instruction is executed (e.g., SVCall). These exceptions are referred to as synchronous. Exceptions that do not fire immediately after a

particular code path is executed are referred to as asynchronous (e.g., external interrupts).

ARM Cortex-M7 exceptions can be broadly categorised into four main types:

1. **Interrupts** are asynchronous events that can occur anytime and interrupt the normal program execution. They are typically generated by external peripherals (e.g., timers, UARTs, GPIO), and the processor responds to them by temporarily halting the current execution and transferring control to an interrupt service routine (ISR). For instance, a UART may use an interrupt request to indicate that new data have been received. A corresponding exception handler (ISR) is then executed that reads the received data. Interrupts can be divided into two main categories:

   a. **External Interrupts**: These are generated by external peripherals or devices to request the processor's attention. The Cortex-M7 processor supports a set of external interrupts (IRQs) that can be individually configured and prioritized.
   b. **NMI (Non-Maskable Interrupt)**: This is a special type of interrupt that has higher priority than regular interrupts and cannot be disabled or masked. NMIs are typically used for critical system functions. Like ordinary interrupt requests, Non-Maskable Interrupt (NMI) requests can be issued by either hardware or software (e.g., if errors happen in other exception handlers, an NMI will be triggered). The main difference is that their priority is extremely high, namely, the highest in the system below the reset exception.

   Two more exceptions also belong to this category and are generated within the processor rather than from external peripheral devices. They are:

   a. **SysTick** exception, generated periodically by the 24-bit count-down system timer and often used by operating systems to drive time slicing. If needed, the same exception can also be issued by software.
   b. **PendSV** exception can only be triggered by software. Operating systems often use it to indicate that a context switch is due and perform it in the future when no other exceptions are waiting to be handled. The PendSV exception can be triggered by writing 1 to bit 28 in the ICSR (a part of the System Control block), which is memory-mapped at address 0xE000ED04.

2. **Faults** are synchronous events generated due to an abnormal event detected by the processor, either internally or while communicating with memory and other devices. These exceptions are of great interest and concern because they indicate serious hardware or software issues that likely prevent the software itself from continuing with normal activities. The following faults are present in Cortex-M7 processors:

   a. **UsageFault** occurs when the processor detects an issue with the program's execution or when an instruction cannot be executed for various reasons. For instance, the instruction may be undefined or may contain a misaligned address that prevents it from accessing memory correctly. Another reason for raising a UsageFault exception is an attempt to divide by zero. Some of the faults mentioned above (like dividing by zero) can be masked in software, i.e., the processor can be instructed to just ignore them without generating any exception, whereas others (such as undefined instruction) cannot, for obvious reasons.

   b. **BusFault** triggers when an error occurs on the data or instruction bus while accessing memory. In other words, it can be generated as a consequence of an explicit memory access performed by an instruction during its execution and also by fetching an instruction from memory. BusFaults result from issues in memory access, most often as attempting to access a location with no valid memory. As Cortex-M7 is a memory-mapped input-output (I/O) architecture, whenever we refer to a memory address, we actually mean an address within the processor's address space that may refer to either a memory location or an I/O register.

   c. **MemManage** (Memory Management Fault) faults occur when there is a memory access violation, such as accessing restricted memory regions. In other words, this fault occurs when the memory protection mechanism blocks memory access. An optional Memory Protection Unit (MPU) provides a programmable way of protecting memory regions against data read and write operations, as well as instruction fetches. For instance, the processor's MPU can be programmed to forbid instruction fetch from address areas containing I/O registers.

   d. **HardFault** is a severe fault that can be generated when an error occurs during exception processing, thus disrupting the normal exception handling flow. HardFaults have a higher priority than any exception with configurable priority. HardFaults are typically unrecoverable, meaning the processor cannot continue the normal program execution from the point of the fault. Usually, the application or CPU must be reset. To prevent HardFaults, developers should follow best practices for writing robust and well-tested code. This includes avoiding undefined instructions, ensuring valid memory accesses, and monitoring stack usage to prevent stack overflows. Additionally, proper fault handling and diagnostics can help identify and address issues before they lead to a HardFault. Hard faults in Cortex-M7 processors are a critical part of system reliability and safety, as they help detect and report severe issues that could otherwise result in unpredictable or incorrect system behaviour.

3. **Supervisor call (SVC)** is a software-initiated exception. It is used to transition from the user or application mode to a more privileged mode, typically for making requests to the operating system or kernel. The execution of an SVC assembly instruction raises this exception. It is commonly used as a convenient way to enter the operating system kernel and request it to perform a function on behalf of the application.

4. **Reset Exception (Reset)** is invoked on power up or a warm reset. The exception model treats reset as a special form of exception. When reset is asserted, the operation of the processor stops, potentially at any point in an instruction. When reset is de-asserted, execution restarts from the address provided by the reset entry

in the vector table. It is handled as other exceptions for the most part, except that instruction execution can stop at an arbitrary point.

2.4.4 Exception Numbers and Priorities

Table 2.1 lists different types of exceptions with their priorities, exception numbers and vector addresses. All exceptions have an associated priority with a lower number value indicating a higher priority. The programmer (software) configures the priorities for most exceptions, except for Reset, NMI and HardFault. If the software does not configure any priorities, then all exceptions with a configurable priority have a priority of 0. Configurable priority values are in the range 0–15. Here is the rule of **order of execution** of exceptions:

1. If two or more exceptions are pending, the exception with the highest priority runs first.
2. If two or more exceptions with the same priority are pending, the exception with the lowest exception number runs first.
3. When the processor executes an exception handler, the exception handler is preempted if a higher-priority exception occurs. If an exception occurs with the same priority as the exception being handled, the handler is not preempted, irrespective of the exception number. However, the status of the new interrupt remains pending.

Table 2.1 Exception types in Cortex-M7

Exception number	Exception type	Priority	Vector address	Activation
1	Reset	−3 (Highest)	0x00000004	Asynchronous
2	NMI	−2	0x00000008	Asynchronous
3	HardFault	−1	0x0000000C	Synchronous
4	MemManage	Configurable	0x00000010	Synchronous
5	BusFault	Configurable	0x00000014	Synchronous
6	UsageFault	Configurable	0x00000018	Synchronous
7–10	*unused*	–	–	–
11	SVCall	Configurable	0x0000002C	Synchronous
12–13	*unused*	–	–	–
14	PendSV	Configurable	0x00000038	Asynchronous
15	SysTick	Configurable	0x0000003C	Asynchronous
16 and above	Interrupt (IRQ)	Configurable	0x00000040 and above	Asynchronous

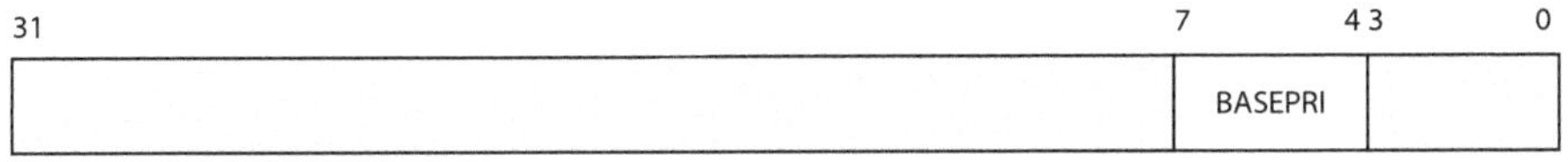

31		7	4 3	0
		BASEPRI		

Fig. 2.7 BASEPRI register

The exceptions with exception numbers 1–15 are so-called **built-in exceptions**. The built-in exceptions are a mandatory part of every ARM Cortex-M core. The ARM Cortex-M specifications reserve exception numbers 1–15, inclusive, for built-in exceptions.

ARM Cortex-M7 processors support a fixed-priority scheme where each interrupt source (or exception) can have a unique priority level assigned to it. **Each priority is associated with its priority value, where a lower priority value indicates a higher exception priority**. Cortex-M7 processors support up to 16 priority levels, where the value 0 represents the highest priority, and the value 15 represents the lowest priority. If the software does not configure any priorities, then all the exceptions with a configurable priority have a priority value of 0. A higher-priority (smaller priority level) exception can preempt a lower-priority (larger priority level) exception. Some exceptions (reset, NMI, and HardFault) have fixed priority levels. Their priority levels are represented with negative values to indicate that they are of higher priority than other exceptions. The BASEPRI (Base Priority) register (Fig. 2.7), which is a part of special registers in the ARM Cortex-M7 core registers block, provides a mechanism to set a threshold for exception priorities, allowing the processor to temporarily restrict the servicing of specific exceptions to prevent lower-priority interrupts from preempting critical tasks. The 4-bit BASEPRI field in the BASEPRI register defines a priority mask. The processor does not process any exception with a priority value greater than or equal to the value in the BASEPRI field.

2.4.5   Vector Table and Exception Handlers

The vector table contains the reset value of the stack pointer and the start addresses, also called **exception vectors**, for all exception handlers. On system reset, the vector table is at address 0x00000000. This is the default start address of the vector table, where Cortex-M7 expects to find it. This is usually a linker job that places the vector table at the beginning of the binary file we upload to the flash memory. Figure 2.8 shows how the vector table is organized in memory and the order of the exception vectors in the vector table. The first entry of this array is the value of the stack pointer. Note that the programmer is responsible for setting the first value into the stack pointer (which is the address of the beginning of the stack). Usually, this address corresponds to the end of the SRAM, as we often use the stack that expands in the direction of descending addresses. Starting from the second entry of this table, we can find the starting addresses for all exception handlers. This means that the vector table has a length of up to 256 for Cortex-7 and depends on the number of interrupts implemented. The silicon vendor that uses an ARM Cortex-M7 core can implement

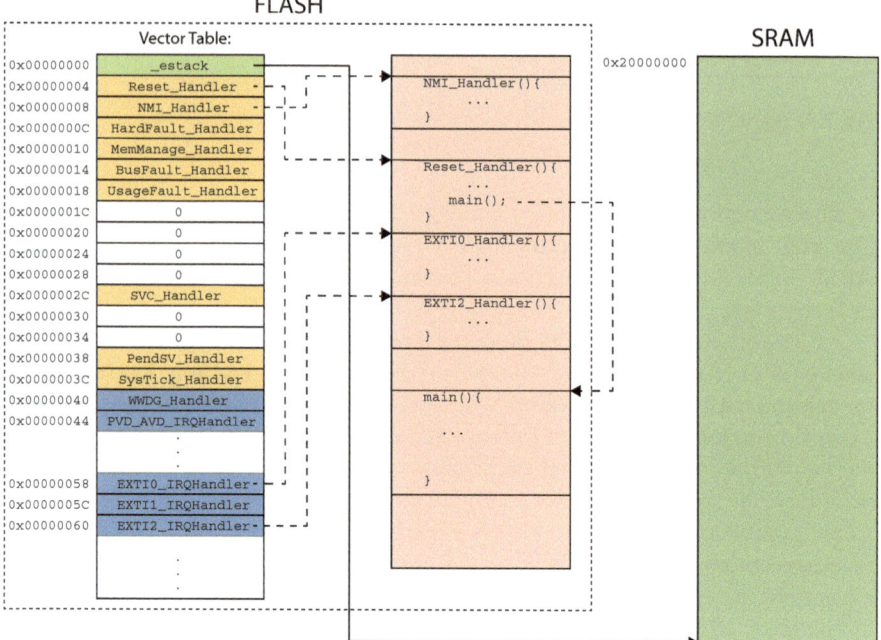

Fig. 2.8 The memory layout of the vector table and exception handlers in ARM Cortex-M7 cores

up to 240 interrupts. The silicon vendor must configure the top range value, which is dependent on the number of interrupts implemented. ARM requires that we always adjust the vector table's size by rounding up to the next power of two. For example, if there are 16 interrupts, the minimum size of the vector table is 32 words, enough for 16 built-in exceptions and up to 16 interrupts. If the user (silicon vendor) requires 21 interrupts, the size of the vector table must be 64 words because the required table size is 37 words, and the next power of two is 64. The name of the exception handlers in Fig. 2.8 is just a convention, and we are totally free to rename them if we like a different one. They are just symbols.

Defining a vector table for a Cortex-M7 processor involves setting up a table of exception handler addresses that the processor will jump to when specific exceptions occur. As said before, the vector table must be placed at the beginning of the flash memory, where the processor expects to find it. In ARM Cortex-M microcontroller development, the .isr_vector is a special section in the microcontroller's memory where the vector table for exceptions and interrupts is defined. The vector table contains addresses of exception and interrupt service routines (ISRs). The .isr_vector section is a label used in the linker script to specify the location of the vector table in memory. Commonly, the vector table is implemented in assembly code in the startup file (e.g., for the Cortex-M7-based STM32H753 microcontroller, the startup file would be startup_stm32h753xx.s) as:

```
 1  .section   .isr_vector
 2
 3  g_pfnVectors:
 4    .word   _estack
 5    /* Built-in Exceptions */
 6    .word   Reset_Handler
 7    .word   NMI_Handler
 8    .word   HardFault_Handler
 9    .word   MemManage_Handler
10    .word   BusFault_Handler
11    .word   UsageFault_Handler
12    .word   0
13    .word   0
14    .word   0
15    .word   0
16    .word   SVC_Handler
17    .word   DebugMon_Handler
18    .word   0
19    .word   PendSV_Handler
20    .word   SysTick_Handler
21    /* External Interrupts */
22    .word   WWDG_IRQHandler
23    .word   PVD_AVD_IRQHandler
24    ...
25    .word   EXTI0_IRQHandler
26    .word   EXTI1_IRQHandler
27    .word   EXTI2_IRQHandler
28    ...
29    .word   WAKEUP_PIN_IRQHandler
```

Listing 2.1 The vector table for Cortex-M7.

Then, the exception and interrupt handler functions should be implemented in the code. These functions are called when their corresponding exceptions or interrupts occur. The handler function names should match the names of the entries in the vector table for a very obvious reason:

```
 1  void Reset_Handler(void) {
 2      // Reset handler code
 3  }
 4
 5  void NMI_Handler(void) {
 6      // NMI handler code
 7  }
 8
 9  void HardFault_Handler(void) {
10      // HardFault handler code
11  }
12
13  void EXTI0_IRQHandler (void) {
14      // HardFault handler code
15  }
```

Listing 2.2 Exception handlers in C.

2.4.6   Exception Entry and Exit

Exception entry and exit in an ARM Cortex-M7 processor is a well-defined process that enables the CPU to handle various exceptions, including interrupts and faults while preserving the state of the currently executing program. This mechanism ensures that the system can respond to events without compromising the integrity of the application code. Here, we provide a detailed description of the exception entry and exit process in a Cortex-M7.

2.4.6.1 Exception Entry

The **exception entry** occurs when there is a pending exception with sufficient priority and either:

1. The processor is executing a normal program and the new exception terminates the currently executing program.
2. The processor executes the exception handler, and the new exception is of higher priority than the exception being handled, in which case the new exception preempts the original exception. When one exception preempts another, we say the exceptions are **nested**.

When the processor takes an exception, the processor pushes the **current execution context** onto the current stack. The execution context consists of eight 32-bit words: registers R0 through R3, R12, the link register LR (also accessible as R14), the program counter PC (R15), and the program status register xPSR, for a total of 32 bytes. This operation is referred to as **stacking**, and the structure of eight 32-bit data words is referred to as the **stack frame**. The reason behind automatically saving the execution context is that accepting and handling an exception should not necessarily prevent the processor from returning to its current activity later. This is particularly true for interrupts and other exception requests that occur asynchronously to current processor activities and are most often totally unrelated to them. Thus, the exceptions and interrupts should be transparent with respect to any code executing when they arrive. Figure 2.9 shows the exception stack frame after stacking. Immediately after stacking, the stack pointer indicates the lowest address in the stack frame. The reader will notice that Cortex-M processors use the **full-descending stack** (the stack grows downward in memory, and the stack pointer points to the lowest memory address in use). The stack frame includes the return address, as the PC is also saved dur-

Fig. 2.9 The layout of the stack frame after stacking in ARM Cortex-M7

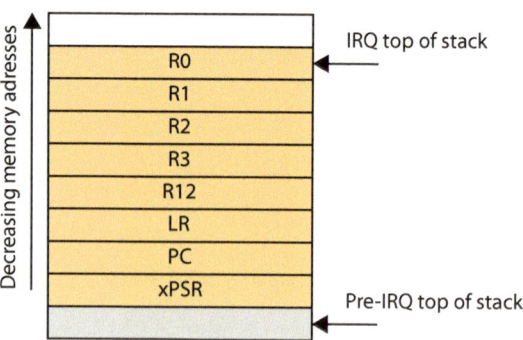

ing stacking. This is the address of the next instruction in the interrupted program. This value is restored to the PC at exception return so that the interrupted program resumes.

Here, we have to describe stack pointers and processing modes in ARM Cortex-M processors in more detail. In ARM Cortex-M processors, there are two registers used to access and manipulate stack: the **Main Stack Pointer (MSP)** and the **Process Stack Pointer (PSP)**. These stack pointers are critical in managing the execution context and handling exceptions in the processor. Additionally, the Cortex-M architecture defines two processing modes: **Thread mode** and **Handler mode**, each with distinct purposes and behaviours. The Main Stack Pointer (MSP) and Process Stack Pointer (PSP) can be accessed and manipulated through the stack pointer (SP), also known as register R13. Commonly, operating mode defines which of the two (MSP or PSP) is accessible through SP (i.e., visible as SP).

Thread mode is the typical execution mode for user/application code. The processor often uses the PSP (although it is possible to use MSP in this mode also) as the current stack pointer in this mode. The processor enters Thread mode after a reset or when returning from an exception or interrupt. User-level code runs in Thread mode, and the PSP is often used for function calls and managing thread-specific context. Handler mode is a privileged execution mode used for handling exceptions and interrupts. The processor switches from Thread mode to Handler mode when an exception or interrupt occurs. The processor automatically saves the current context onto the PSP or MSP stack (depending on the operation mode of the interrupted program) before executing the exception handler. The MSP is then used in Handler mode as the stack pointer. Handler mode is reserved for system-level tasks and ensures that critical operations can be carried out even when the application stack is compromised.

In parallel to the stacking operation, the processor writes an **exception return value** (called EXC_RETURN value in the ARM documentation) to the link register (LR). This indicates which stack pointer corresponds to the stack frame and what operation mode the processor was in before the entry occurred. The information provided by the EXC_RETURN value allows the processor to locate the stack frame to be restored upon returning from an exception, interpret it in the right way, and bring back the processor to the execution mode of the interrupted program. Table 2.2 shows the EXC_RETURN values and their meaning upon returning from an exception.

Table 2.2 Exception return values and their behaviour upon returning from an exception

EXC_RETURN[31:0]	Description
0xFFFFFFF1	Return to Handler mode, exception return uses the exception stack frame from the MSP and execution uses MSP after return
0xFFFFFFF9	Return to Thread mode, exception return uses the exception stack frame from the MSP and execution uses MSP after return
0xFFFFFFFD	Return to Thread mode, exception return uses the exception stack frame from the PSP and execution uses PSP after return

In parallel to the stacking operation, the processor also performs a vector fetch that reads the exception handler start address from the vector table. The processors determines the exception vector to be fetched into the PC by the exception number:

$$PC \leftarrow M[\texttt{0x0000 0000} + 4 \times (\text{exception number})].$$

When stacking is complete, the processor starts executing the exception handler, switching to Handler Mode. Associated with the execution mode switch, the processor may also use a new stack. As mentioned previously, handler mode execution always uses MSP, whereas thread mode execution may use either MSP or PSP, depending on processor configuration. The Reset exception is a deviation from this general rule. The Reset exception is handled in Thread mode instead. Upon reset, execution starts in Thread mode, and the processor is automatically configured to use MSP.

2.4.6.2   Exception Return

The **exception return** occurs when the processor is in Handler mode and executes an instruction which loads the EXC_RETURN value into the PC (for example `bx lr`). Recall that EXC_RETURN is the value loaded into the LR on exception entry. The exception mechanism relies on this value to detect when the processor has completed an exception handler. The lowest bits of this value provide information on the return stack and processor mode. When this value is loaded into the PC, it indicates to the processor that the exception is complete, and the processor should initiate the appropriate exception return sequence instead of fetching an instruction.

When an exception return value is loaded into the program counter PC as part of an exception handler epilogue, it directs the processor to initiate an exception handler return sequence instead of simply returning to the caller. In fact, the ARM Architecture Procedure Calling Standard (AAPCS) states that a function call must save into the link register LR the return address before setting the program counter PC to the function entry point. This is typically accomplished by executing a branch and link instruction `bl` with a PC-relative target address. In the epilogue of the called function, it is then possible to return to the caller by storing back into PC the value stored into LR at the time of the call. This can be done, for instance, by means of a branch and exchange instruction `bx`, using LR as argument.

This aspect of the exception return has been architected to permit any AAPCS-compliant function to be used directly as an exception handler. In this way, any AAPCS-compliant function can be used as an exception handler. This is especially important when exception handlers are written in a high-level language like C because compilers are able to generate AAPCS-compliant code by default, and hence, they can also generate exception-handling code without treating it as a special case. The exception handlers for ARM Cortex-M processors are thus implemented as regular C functions and do not require a special function declaration keyword. As a result, an exception handler return performed by hardware is indistinguishable from a regular software-managed function return.

The following code presents the exception handler for an exception triggered by GPIO Pin 13 through EXTI15_10 lines. The exception handler is implemented just as a regular C function without any special function declaration:

```
void EXTI15_10_IRQHandler(void)
{
  // Check if GPIO_PIN_13 triggered the interrupt:
  if (__HAL_GPIO_EXTI_GET_IT(GPIO_PIN_13) != 0x00U)
  {
    // Your code to handle the GPIO_PIN_13 interrupt goes here

    // Clear the GPIO_PIN_13 interrupt flag
    __HAL_GPIO_EXTI_CLEAR_IT(GPIO_PIN_13);
  }
}
```

Listing 2.3 The exception handler for EXTI15_10 interrupt implemented as a regular C function.

2.4.7 Case Study: A Simple Task Scheduler on ARM Cortex-M7

In the realm of computer systems and real-time operating systems (RTOS), the concept of context switching is the linchpin of multitasking and responsiveness. It's a finely tuned mechanism that orchestrates the efficient execution of multiple tasks, allowing a processor to handle numerous concurrent operations with precision and determinism. The ability to seamlessly transition between multiple tasks, known as **context switching**, lies at the heart of efficient and responsive systems. Context switching refers to the process where the state of one task is saved, allowing another task to take precedence and execute. These tasks may be threads of a single application or various concurrent applications. At its core, context switching is a process by which the processor transitions from executing one task to another. This transition involves the preservation of the current task's context, the loading of the new task's context, and the seamless continuation of the latter's execution. The context of each task includes the task's state of the processor–registers, program counter, stack pointer, and system variables.

Context switching begins with a trigger—typically a timer interrupt signalling the need to switch contexts. The processor diligently saves the current context onto a task's stack and retrieves the context of the next task to be executed from its stack. A successful context switch involves the preservation of the current execution context and restoration of a new execution context, enabling the next task to resume from precisely where it left off. This process demands meticulous stack management and the precise handling of interrupts and exceptions.

An RTOS relies on a task scheduler, interrupt handling mechanisms, and precise memory management to orchestrate this performance. The scheduler keeps a record of tasks and manages their execution, while the interrupt system plays a pivotal role in triggering context switches when a timer interrupt occurs.

Understanding the intricacies of context switching is paramount for engineers working with computer systems to create efficient, deterministic, and robust applications. So, let's raise the curtain and delve into the intricacies of context switching, where the processor seamlessly switches tasks, and the computer system transforms into a multitasking maestro.

2.4.7.1   Background

A simple round-robin task scheduler (Fig. 2.10) on Cortex-M7 processors effectively manages multiple tasks or threads in a cooperative multitasking environment. In this scheduler, each task is given a fixed time slice (quantum) during which it can execute. When its time slice expires, the scheduler switches to the next task in the queue. **The task scheduler relies on the interrupts and stacks to achieve context switching**. The SysTick and PendSV interrupts can both be used for context switching. The SysTick peripheral is a 24-bit timer that interrupts the processor each time it counts down to zero. This makes it well-suited to round-robin style context switching, and we are going to use the SysTick to perform a context switch.

When switching contexts, the scheduler needs a way to keep track of which tasks are doing what using a task table. Recall from the previous sections that the ARM Cortex-M7 processor has two separate stack pointers which can be accessed through a banked SP register: Main Stack Pointer (MSP), which is the default one after startup and is used in exception handlers running in the Handler mode, and Process Stack Pointer (PSP), which is often used in regular user procedures running in the Thread mode. In our application, tasks run in the Thread Mode with PSP, and the context-switcher (kernel) runs in the Handler Mode with MSP. This allows stack separation between the kernel and tasks (which simplifies the context switch procedure) and prevents tasks from accessing important registers and affecting the kernel.

Figure 2.11 shows the scheduler operations during a context switch in more detail. The scheduler relies on exception entry and exit mechanisms, which automatically save and restore the critical CPU context (registers R0-R3, R12, LR, PC and xPSR) using the exception frame on the stack. When a SysTick exception occurs, the Task1 critical registers are automatically saved into the Task1 exception stack frame. Once in the SysTick handler, the scheduler is responsible for pushing the interrupted task Task1 registers R4-R11 onto the task's stack and saving its PSP in the task's Task Control Block (TCB). Then, the scheduler selects the next task (Task2) in a round-robin fashion. Before returning from the SysTick handler, the scheduler is responsible for loading the Task2 SP into the PSP register and restoring the Task2 registers R4-R11 from the Task2 stack. Then, upon exception exit, the Task2 critical registers are restored from its exception stack frame, and the execution returns to the new task.

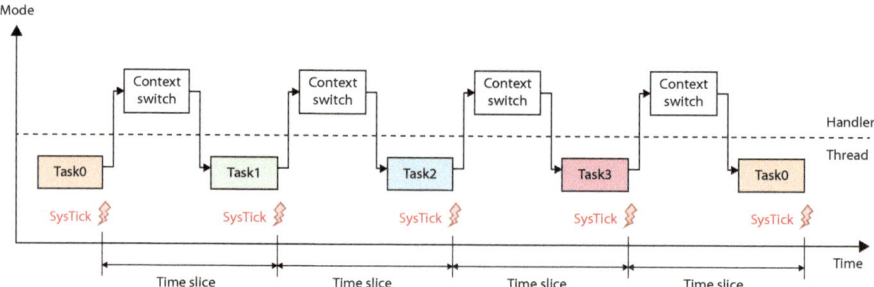

Fig. 2.10 A simple task scheduler

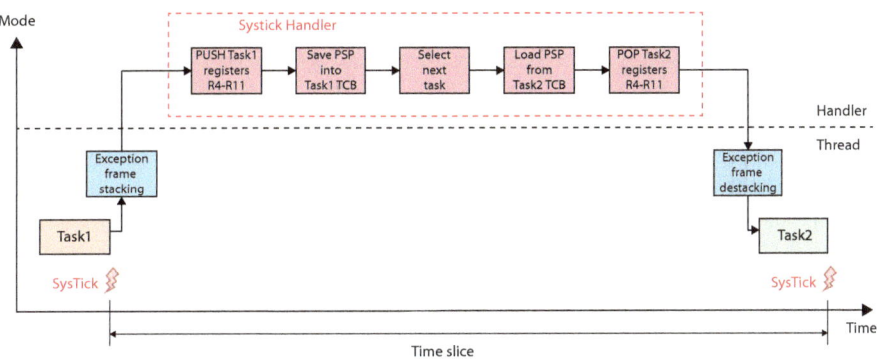

Fig. 2.11 Scheduler operations during a context switch

Usually, three routines are required to implement and run the scheduler: create new tasks, initialize tasks, and perform the context switch. Besides, several data structures are required to implement and manage the stack for each task and represent each task's state. In the following subsections, we provide a step-by-step description of implementing a very simple round-robin scheduler on a Cortex-M7 processor.

2.4.7.2 Tasks

A task is a piece of code or a function that does a specific job when it is allowed to run. Usually, a task is an infinite loop which can repeatedly do multiple steps. In our simple scheduler application, the tasks cannot be finished (they never return) and do not take any arguments. Here is a C implementation of a task:

```
void task() {
    // init task:
    ...
    // main loop
    while(1) {
        // do things over and over
    }
}
```

Listing 2.4 A task in C. In our application, a task never returns and does not take any arguments.

2.4.7.3 Stacks

In a multitasking environment, where multiple tasks are executed in a time-sharing manner, each task needs to have its own stack. Each task executes within its own context with no coincidental dependency on other tasks within the system or the scheduler itself. Each task's stack provides isolation between tasks. It ensures that local variables and function call frames of one task do not interfere with those of another task. This isolation is crucial for maintaining data integrity and preventing unintended side effects between tasks. Only one task within the application can execute at any point in time, and the scheduler is responsible for deciding which task this should be. As a task does not know of the scheduler activity, it is the scheduler's responsibility to ensure that the processor context (register values, stack contents, etc.) when a task is swapped in is exactly the same as when the same task was swapped out. In other words, each task's stack allows tasks to be reentrant. Reentrancy means that a task can be interrupted while executing and later resume from where it left off without corrupting its state. The stack stores the task's execution context, enabling reentrant behaviour. Besides, each task should be able to make function calls and put arguments on the stack without worrying about function call frames interfering with those of other tasks. Furthermore, allocating a fixed amount of stack space for each task makes it easier to predict memory usage and stack requirements for each task, simplifying system design and analysis.

To achieve this, **each task is provided with its own stack** in our simple task scheduler. The size of each task's stack is 1 kB (256 32-bit words). So, for four tasks we create a memory block that holds all four stacks as follows:

```
unsigned int stackRegion[NTASKS * TASK_STACK_SIZE];
```

Listing 2.5 Memory block for tasks' stacks. NTASKS equals 4 and TASK_STACK_SIZE equals 256.

2.4.7.4 Task Control Block

A Task Control Block (TCB), also known as a Task Control Structure (TCS), is a data structure used in real-time operating systems (RTOS) and multitasking environments to manage and control individual tasks or threads. The TCB holds essential information about a task's state, allowing the operating system or scheduler to manage and switch between tasks efficiently. The exact contents and structure of a TCB may vary depending on the operating system or RTOS, but it typically includes the following information: task identifier, task state (e.g., ready to run, blocked, suspended, etc.), task priority, stack pointer, task name, and additional task's parameters.

In our implementation, each task will always be ready to run, so we will omit the task state from TCB. Besides, all tasks in our scheduler will have the same priority and will be selected on a round-robin basis, so we will omit the task priority from TCB. Because each task should have its own stack to save its local variable and exception frame, our TCB must include the SP value, which points to the current stack pointer of the task. The scheduler will select the next task in a round-robin fashion and write its SP value into the PSP register. The scheduler will also copy the PSP register of the interrupted task into its SP value. Also, in our implementation, the Task Control Block will contain the start address of the task. Here is a minimal TCB implementation using struct in C:

```
typedef struct{
  unsigned int *sp;
  void (*pTaskFunction)();
} TCB_Type;
```

Listing 2.6 TCB structure.

In our simple implementation, our scheduler will contain only four tasks. It would be easy to add additional tasks later, but for now, we will keep the code as simple as possible. Each of the four tasks should have its TCB. Hence, we create a TCB table as:

```
TCB_Type TCB[NTASKS];
```

Listing 2.7 TCB table. NTASKS is a constant equal to 4.

2.4.7.5 Task Creation

The TaskCreate() function saves the address of the task's stack and the address of the task's function into the task's TCB.

The following code presents the function used to create a new task:

```
void TaskCreate(TCB_Type* pTCB,
               unsigned int* pTaskStackBase,
               void (*TaskFunction)()){

  pTCB->sp        = (unsigned int*) pTaskStackBase;
  pTCB->pTaskFunction = TaskFunction;
}
```

Listing 2.8 The function TaskCreate() that creates a new task.

The parameters of the above TaskCreate() function are:

- pTCB—a pointer to a task's TCB,
- pStackBase—pointer to a task's stack block,
- TaskFunction—address of a task's function.

Figure 2.12 illustrates the memory layout and the contents of the task's stack and TCB after creating Task1 using the TaskCreate() function.

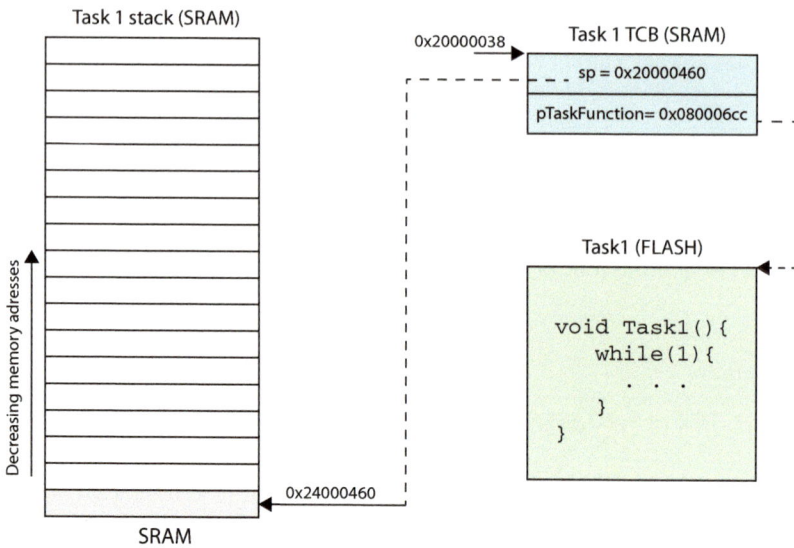

Fig. 2.12 Memory layout and content after calling the `TaskCreate()` function

2.4.7.6 Task Initialisation
The following code presents the function used to initialize a new task:

```
void TaskInit(TCB_Type* pTCB){
  HWSF_Type* pHWStackFrame;
  SWSF_Type* pSWStackFrame;


  // Set pointers to HWSF and SWSF:
  pHWStackFrame = (HWSF_Type*)((void*)pTCB->sp - sizeof(HWSF_Type));
  pSWStackFrame = (SWSF_Type*)((void*)pHWStackFrame
                                      - sizeof(SWSF_Type));


  // populate HW Stack Frame
  pHWStackFrame->r0   = 0;
  pHWStackFrame->r1   = 0;
  pHWStackFrame->r2   = 0;
  pHWStackFrame->r3   = 0;
  pHWStackFrame->r12  = 0;
  pHWStackFrame->lr   = 0xFFFFFFFF; // (reset val - task never exits)
  pHWStackFrame->pc   = (unsigned int) (pTCB->pTaskFunction);
  pHWStackFrame->psr  = 0x01000000;  // Set T bit (bit 24) in EPSR.
                                     // The Cortex-M4 processor only
                                     // supports execution of
                                     // instructions in Thumb state.
                                     // Attempting to execute
                                     // instructions when the T bit
                                     // is 0 (Debug state)
                                     //    results in a fault.
  // populate SW Stack Frame
  pSWStackFrame->r4 = 0x04040404;
  pSWStackFrame->r5 = 0x05050505;
```

```
pSWStackFrame->r6  = 0x06060606;
pSWStackFrame->r7  = 0x07070707;
pSWStackFrame->r8  = 0x08080808;
pSWStackFrame->r9  = 0x09090909;
pSWStackFrame->r10   = 0x0a0a0a0a;
pSWStackFrame->r11   = 0x0b0b0b0b;

// Set task's stack pointer in the TCB to point at the top
    // of the task's SW stack frame
pTCB->sp       = (unsigned int*) pSWStackFrame;

}
```

Listing 2.9 The function `TaskInit()` that initializes a new task.

The only parameter of the above `TaskInit()` function is a pointer to a task's TCB. The `TaskInit()` function performs the following steps:

1. Initializes two pointers to two stack frames that hold the exception stack frames: so-called hardware stack frame and the so-called software stack frame. The hardware stack frame will hold eight registers saved by the CPU during exception entry. Besides these eight registers, we need to save the remaining eight registers from the task's context (R4-R11) onto the software stack frame. We need to prepare these stack frames for each new task so that when the task switch occurs, both frames will be ready for de-stacking and, hence, entering a new task. To make this task easier, we will abstract the frames with two structures:

```
typedef struct{
   unsigned int r0;
   unsigned int r1;
   unsigned int r2;
   unsigned int r3;
   unsigned int r12;
   unsigned int lr;
   unsigned int pc;
   unsigned int psr;
} HWSF_Type;

typedef struct{
   unsigned int r4;
   unsigned int r5;
   unsigned int r6;
   unsigned int r7;
   unsigned int r8;
   unsigned int r9;
   unsigned int r10;
   unsigned int r11;
} SWSF_Type;
```

Listing 2.10 Structures used to abstract the hardware and software stack frames.

The hardware stack frame resides at the bottom of the task's stack, and the software stack frame resides above the hardware stack frame.

2. Now, as two pointers to stack frames, `pHWStackFrame` and `pHWStack Frame`, are set, we can populate both frames with initial values. The hardware stack frame is populated as follows:

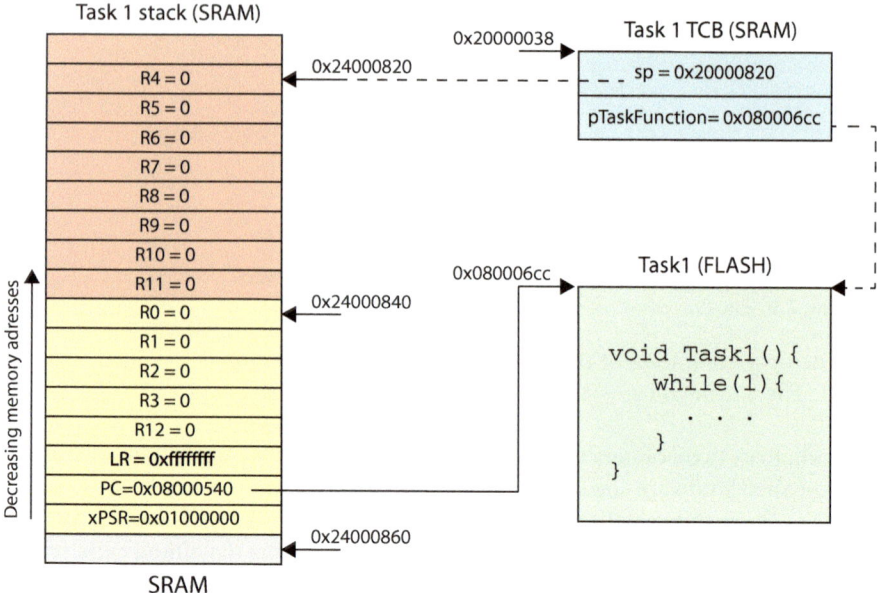

Fig. 2.13 Memory layout and content after calling the `TaskInit()` function

- PSR = 0x01000000—this is the default reset value in the program status register,
- PC = the address of the task,
- LR = 0xFFFFFFFF—in our case, tasks never finish, so LR = 0xFFFFFFFF (reset value),
- r12, r3 – r0 = 0x00000000—we may also pass the arguments into the task via r0-r3, but this is not the case in our simple scheduler.

3. Finally, it saves the address of the top of the software stack frame into the task's SP entry in the task's TCB.

After these steps, a new task is ready to be executed for the first time when the task switch occurs, and the task is selected for execution. Figure 2.13 illustrates the memory layout and the contents of the task's stack and TCB after creating Task1 using the `TaskInit()` function.

2.4.7.7 Scheduler Initialisation

The following code presents the function used to initialize all four tasks:

```
void InitScheduler(unsigned int* pStackRegion,
                   TCB_Type pTCB[],
                   void (*TaskFunctions[])()){
  unsigned int* pTaskStackBase;

  // 1. create all tasks:
  for(int i=0; i<NTASKS; i++){
    pTaskStackBase = pStackRegion + (i+1)*TASK_STACK_SIZE;
    TaskCreate(&pTCB[i], pTaskStackBase, TaskFunctions[i]);
  }
  // 2. initialize all tasks except the Task0.
  //    Task0 will be called by main()
  //    and will be the first task interrupted.
  //    Its HWSF and SWSF will be created upon
  //    interrupt/contecxt switch
  for(int i=1; i<NTASKS; i++){
    TaskInit(&pTCB[i]);
  }
  // set PSP to Task0.SP:
  __set_PSP((unsigned int)pTCB[0].sp);
}
```

Listing 2.11 The function `InitScheduler()` creates all tasks and initializes all tasks except the first one (Task0). At the end, it sets the top of the stack of the first task (Task0) into the PSP register.

The function `InitScheduler()` performs the following steps:

1. Creates all tasks.
2. Initializes all tasks except the first one (Task0). Task0 will be called from the main function and will be the first task interrupted by the SysTick timer. Hence, its stack frames will be populated during the context switch.
3. Saves the top of the stack of the first task (Task0) into the PSP register.

To read or write the PSP register, which is not memory-mapped, requires the usage of special CPU instructions MSR and MRS. Hence, in order to access the PSP register, we are forced to use assembly. To make programming easier, the above code relies on the `__set_PSP` function defined in the Cortex Microcontroller Software Interface Standard (CMSIS) library to write into the PSP register. CMSIS is a vendor-independent hardware abstraction layer (HAL) for ARM Cortex-M processors. It simplifies software development for a wide range of microcontroller devices, promoting code portability and reusability across various microcontroller families and manufacturers. CMSIS defines two inline assembly functions to read or write the PSP register:

```
/**
  \brief    Set Process Stack Pointer
  \details Assigns the given value to the Process Stack Pointer (PSP)
  \param [in]    topOfProcStack   Process Stack Pointer value to set
 */
__attribute__((always_inline))
static inline void __set_PSP(uint32_t topOfProcStack)
{
  __asm volatile ("MSR psp, %0" : : "r" (topOfProcStack) : );
}

/**
  \brief    Get Process Stack Pointer
  \details Returns the current value of the Process Stack Pointer (PSP)
  \return   PSP Register value
 */
__attribute__((always_inline))
static inline void uint32_t __get_PSP(void)
{
  uint32_t result;

  __asm volatile ("MRS %0, psp"  : "=r" (result) );
  return(result);
}
```

Listing 2.12 The CMSIS definition of inline assembly functions for accessing the PSP register.

After these steps, everything is set up for the first context switch. Figure 2.14 illustrates the memory layout and the task's stack after initializing the scheduler using the `InitScheduler()` function.

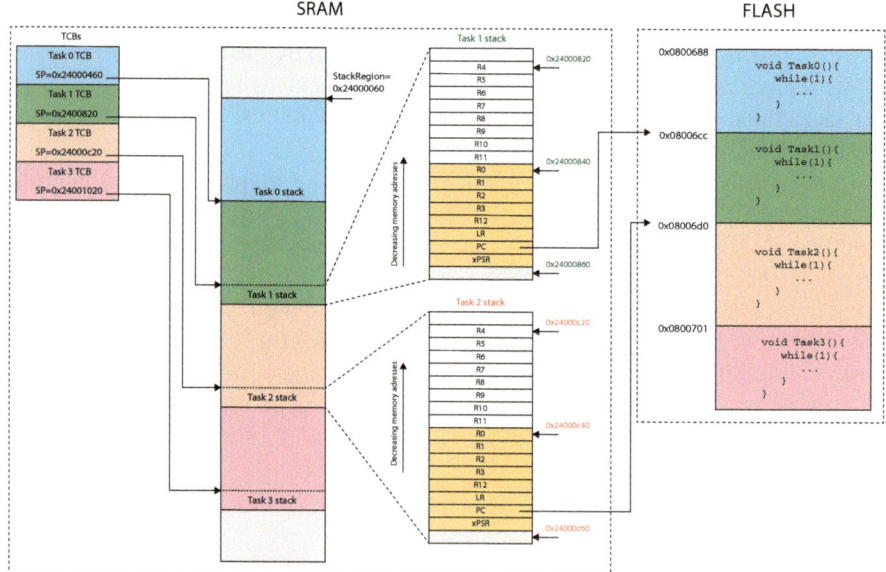

Fig. 2.14 Memory layout and content after initializing four tasks during the scheduler initialization

2.4.7.8 Context Switch

Context switching in multitasking environments can be performed using stack pointer (SP) swapping. The process involves saving the current task's context onto its stack and then loading the context of the next task to be executed by swapping the SP. Figure 2.15 shows the process of context switching using stack pointer swapping. Here's a step-by-step description of how context switching is accomplished using this method:

1. When a trigger for context switching occurs (the trigger is a timer interrupt), the CPU saves the exception stack frame onto the Task1 stack using the PSP stack pointer and enters the timer's interrupt handler.
2. The remaining eight registers (R4-R11) are saved the onto the Task1 stack. The context switcher saves the current PSP into the Task1 TCB.
3. The context switcher determines which task should run next. The scheduler considers the round-robin scheduling policy to make this decision.
4. The context switcher retrieves the SP of Task2 from the Task2 TCB and saves it into the PSP register. The PSP now points to the stack where the context of Task2 is saved.
5. The eight registers (R4-R11) of Task2 are popped from stack.
6. The timer handler exits; hence, the de-stacking operation performed by the CPU retrieves the exception frame from the Task2 stack. As the PC of Task2 is part of its exception frame, the CPU returns to Task2.

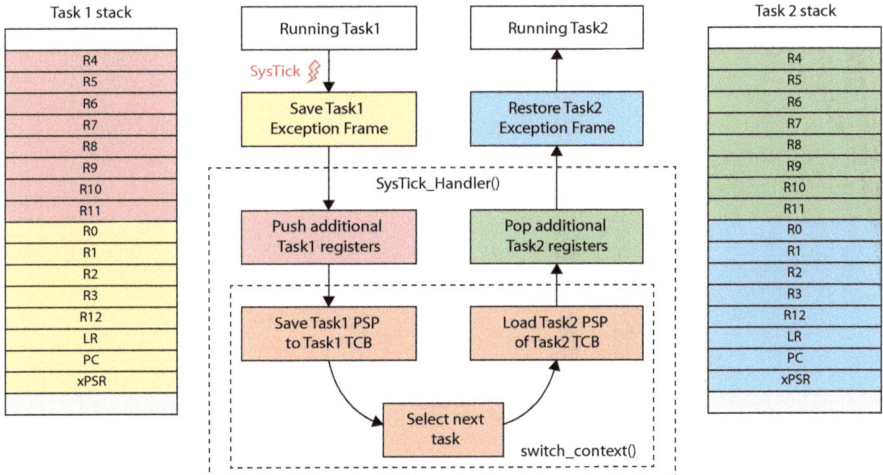

Fig. 2.15 Context switching using stack pointer swapping

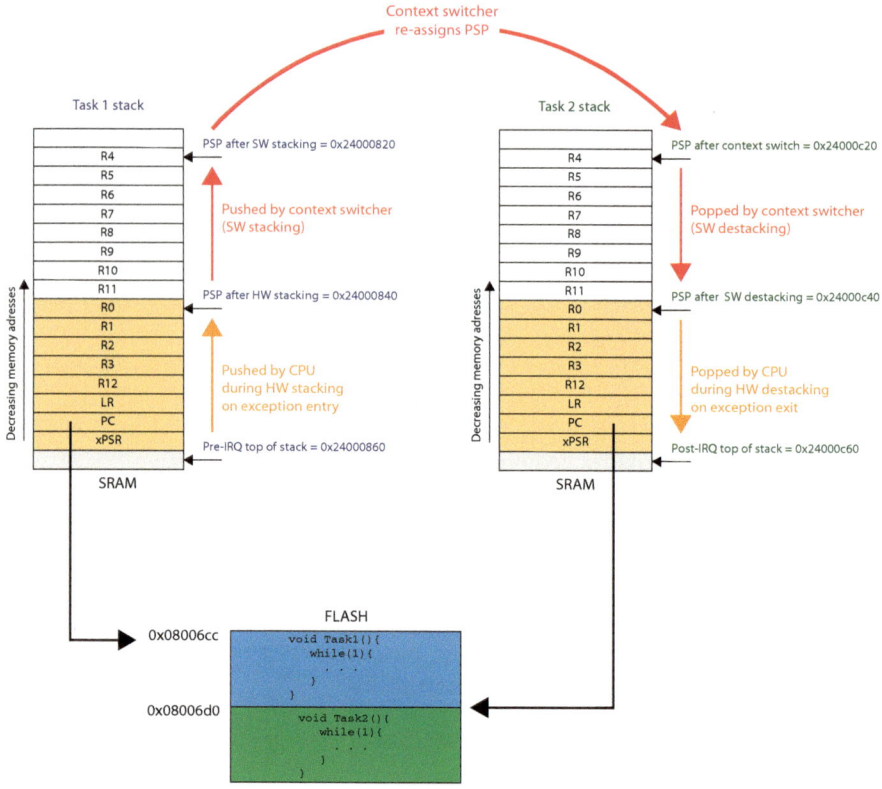

Fig. 2.16 The modification progress of the PSP stack pointer during context switching

Figure 2.16 shows the chronology of the stack pointer when a context switch happens between Task1 and Task2. The following code presents the function that implements the context switcher:

```
int ContextSwitch(int current_task, TCB_Type pTCB[]){
  volatile int new_task;

  pTCB[current_task].sp = (unsigned int*) __get_PSP();

  // select next task in round-robin fashion
  new_task = current_task + 1;
  if (new_task == NTASKS) new_task = 0;

  __set_PSP((unsigned int)pTCB[new_task].sp);

  return new_task;
}
```

Listing 2.13 The functions ContextSwitch() that implements context switching.

The parameters of the `ContextSwitch` functions are the index of the current task (`current_task`) and the pointer to the TCB table (`pTCB`). The function return the index of a new task.

2.4.7.9   SysTick Handler

Finally, we can implement the SysTick handler that will perform the task switch:

```
void SysTick_Handler(void)
{
  unsigned int tmp;
  // 1. Save context of the interrupted task:
  if (current_task != -1){
    __asm__  volatile ( "MRS %0, psp\n\t"
             "STMFD %0!, {r4-r11}\n\t"
             "MSR psp, %0\n\t" : "=r" (tmp) );

    // 2. Switch context:
    current_task = ContextSwitch(current_task, TCB);

    // 3. restore context of the new task:
    __asm__  volatile ( "MRS %0, psp\n\t"
                   "LDMFD %0!, {r4-r11}\n\t"
                   "MSR psp, %0\n\t" : "=r" (tmp) );
  }
}
```

Listing 2.14   The SysTick handler used to perform task switch.

The SysTick handler performs the following steps:

1. Saves the context (R4-R11) of the interrupted task on the task's stack using PSP.
2. Switch context (swap stack pointers) using the `switch_context()` function.
3. Restore the context (R4-R11) of the new task from its stack using PSP.
4. Return from interrupt and restore the exception frame of the new task from its stack.

2.4.7.10   Starting the Scheduler

Finally, we are ready to start our scheduler within the main function. To do so, we need to:

1. Initialize scheduler.
2. Switch to NOT PRIVILEGED mode with PSP as the stack pointer by setting the last two bits in the CONTROL register.
3. Call Task0.
4. Within Task0, wait for the first SysTick interrupt.

The following code shows how to start the scheduler:

```
unsigned int stackRegion[NTASKS * TASK_STACK_SIZE];
TCB_Type TCB[NTASKS];
```

```
void (*TaskFunctions[NTASKS])();
int current_task = -1;

void Task0(){
  while(1) {}
}
void Task1(){
  while(1) {}
}
void Task2(){
  while(1) {}
}
void Task3(){
  while(1) {}
}


int main(void)
{
  TaskFunctions[0] = Task0;
  TaskFunctions[1] = Task1;
  TaskFunctions[2] = Task2;
  TaskFunctions[3] = Task3;
  // Init scheduler:
  InitScheduler(stackRegion, TCB, TaskFunctions);
  current_task = 0;
  // Start SysTick timer with the highest priority:
  HAL_InitTick(0);
  // Switch to NOT PRIVILEDGED with PSP:
  __set_CONTROL(0x00000003);
  // Call the first task:
  Task0();  // never return!
  while(1){}
}
```

Listing 2.15 Starting the scheduler.

To write into the CONTROL register (which is not memory-mapped), the above code uses the _ _set_CONTROL function defined in the CMSIS library as:

```
/**
  \brief    Set Control Register
  \details Writes the given value to the Control Register.
  \param [in]    control  Control Register value to set
  */
__STATIC_FORCEINLINE void __set_CONTROL(uint32_t control)
{
  __ASM volatile ("MSR control, %0" : : "r" (control) : "memory");
}
```

Listing 2.16 The CMSIS definition of inline assembly function for writing into the CONTROL register.

2.4.7.11 Using PendSV for Context Switching

The approach with the SysTick handler used to perform the context switching would, however, not work with other interrupts (peripheral interrupts, for example). The SysTick handler would interrupt IRQ handlers as well, and stack registers affected by the peripheral IRQ handler and unstack task's registers, resulting in undefined

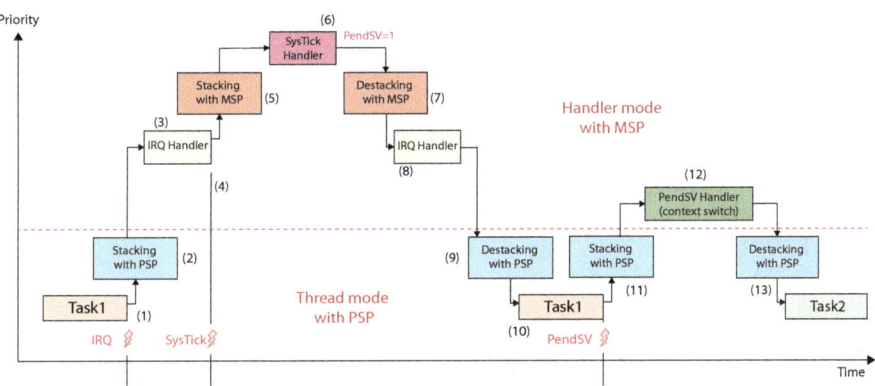

Fig. 2.17 A simple task scheduler based on PendSV interrupts

behaviour of both tasks and peripheral interrupt handler. This would undoubtedly result in the hard fault.

The PendSV (Pending Supervisor Call) interrupt is commonly used for context switching in ARM Cortex-M microcontrollers due to several advantages and characteristics that make it well-suited for this purpose. The PendSV interrupt has the lowest possible priority among all exceptions and interrupts. This makes it an ideal choice for context switching, as it doesn't interfere with other higher-priority interrupts or exceptions. The PendSV exception will interrupt only the non-priority tasks and certainly not any exception handler. The low-priority nature of PendSV ensures that it doesn't preempt other exceptions or interrupts, providing predictable and deterministic behaviour during context switches. This predictability is essential in real-time systems. PendSV can be triggered explicitly through software by setting the PendSV bit in the ICSR register within the System Control Block. This allows for precise control over when context switches occur. Typically, the PendSV interrupt is set pending from the SysTick handler.

Figure 2.17 shows the solution to this problem with the PendSV interrupt. Usually, the SysTick interrupt has the highest priority among all exceptions and interrupts with configurable priority. If an interrupt request (IRQ) takes place before the SysTick exception, the SysTick exception might preempt the IRQ handler. In this case, we should not carry out the context switching. The PendSV exception solves the problem by delaying the context-switching request until all other IRQ handlers have completed their processing. To do this, the PendSV is programmed as the lowest-priority exception. The Systick handler sets the pending status of the PendSV, and the context switching is carried out within the PendSV exception. Let us describe the solution in Fig. 2.17:

1. Task1 is preempted by an IRQ interrupt request.
2. Task1's hardware stack frame is stacked on the process stack using the PSP.
3. The IRQ handler executes.

4. The SysTick exception eventually preempts the IRQ handler.
5. The hardware stack frame of the IRQ handler is stacked on the main stack using the MSP register.
6. The SysTick handler sets the PendSV bit. Hence, PendSV interrupt is pending.
7. The SysTick exits, and the hardware stack frame of the IRQ handler is popped from the main stack.
8. The IRQ handler continues its execution.
9. The IRQ handler exits, and the hardware stack frame of the Task1 is popped from the process stack.
10. The Task1 continues its execution.
11. PendSV is fired and the Task1's hardware stack frame is stacked on the process stack using the PSP.
12. PendSV handler performs the context switching.
13. PendSV handler exits and the Task2's hardware stack frame is popped from the process stack using the PSP.
14. Task2 executes.

Hence, the solution to implement a scheduler based on the SysTick and PendSV exceptions is simple. Firstly, we move the code for context switching from the SysTick handler into the PendSV handler:

```
void PendSV_Handler(void)
{
  volatile unsigned int tmp1=0;
  volatile unsigned int tmp2=0;
  // 1. Save context of the interrupted task:
  __asm__ volatile ( "MRS %0, psp\n\t"
                     "STMFD %0!, {r4-r11}\n\t"
                     "MSR psp, %0\n\t" : "=r" (tmp1) );


  // 2. Switch context:
  current_task = ContextSwitch(current_task, TCB);

  // 3. restore context of the new task:
  __asm__ volatile ( "MRS %0, psp\n\t"
                     "LDMFD %0!, {r4-r11}\n\t"
                     "MSR psp, %0\n\t" : "=r" (tmp2) );
}
```

Listing 2.17 PendSV handler performs context switching.

Secondly, the SysTick handler only sets PendSV pending in the ICSR register:

```
void SysTick_Handler(void)
{
  // Set the PendSV Pending bit in  ICSR:
  SCB->ICSR |= (unsigned long)0x01 << 28;
}
```

Listing 2.18 The SysTick handler only sets PendSV pending.

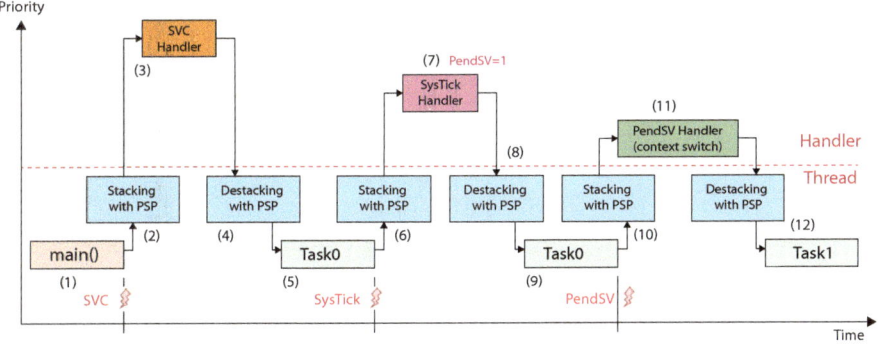

Fig. 2.18 Starting the scheduler with the SVC exception

2.4.7.12   Using the Supervisor Call (SVC) Exception to Start the Scheduler

Instead of directly calling the first task (Task0) from the main function, the first task should be initialized and started in the same way as the others. In other words, the scheduler should rely on the exception return to start the first task. For this purpose, we can use the Supervisor Call (SVC) exception. Recall that the SVC instruction triggers the SVC exception. Due to the interrupt priority behaviour of the Cortex-M processors, the SVC instruction can only be used in thread mode or exception handlers that have a lower priority than the SVC itself. Otherwise, a HardFault exception would be generated. The SVC instruction is a privileged operation that allows a task in an unprivileged mode to request a service from the operating system (or kernel) running in a privileged mode. This separation of privilege levels ensures that only trusted code can initiate scheduling or other system-related operations. Figure 2.18 shows the process of starting and running the scheduler using the SVC exception. Let us describe the solution in Fig. 2.18:

1. The `main()` function initializes the scheduler (i.e., initializes all tasks) and eventually executes the SVC instruction.
2. The `main()` function is preempted by the SVC exception, and its hardware stack frame is stacked on the process stack using the PSP
3. The SVC handler sets the PSP to point to the top of the Task0 stack, restores the context (R4-R11) of the Task0 and exits.
4. Upon exception exit, the hardware stack frame of Task0 is restored, therefore returning control to Task0.
5. Task0 executes until the end of its time slot.
6. The SysTick exception preempts Task0, saving its hardware stack frame onto its stack.
7. The SysTick handler sets the PendSV bit. Hence, the PendSV interrupt is pending.
8. The SysTick exits, and the hardware stack frame of Task0 is popped from the main stack.
9. Task0 continues its execution.

10. PendSV is fired, and the Task0 hardware stack frame is stacked on the process
    stack using the PSP.
11. PendSV handler performs the context switching.
12. PendSV handler exits, and Task1's hardware stack frame is popped from the
    process stack using the PSP.
13. Task1 now executes.

To initialize the scheduler that uses the SVC exception to start the first task, we
use the following function:

```
void InitSchedulerSVC(unsigned int* pStackRegion,
    TCB_Type pTCB[],
    void (*TaskFunctions[])()){
    unsigned int* pTaskStackBase;

    // 1. create all tasks:
    for(int i=0; i<NTASKS; i++){
        pTaskStackBase = pStackRegion + (i+1)*TASK_STACK_SIZE;
        TaskCreate(&pTCB[i], pTaskStackBase, TaskFunctions[i]);
    }

    // 2. initialize all tasks
    //     The main() and will be first interrupted by SVC.
    //     Task0 will be entered from SVC Handler
    for(int i=0; i<NTASKS; i++){
        TaskInit(&pTCB[i]);
    }

    // set PSP to Task0.SP:
    __set_PSP((unsigned int)pTCB[0].sp);
}
```

Listing 2.19 The function InitSchedulerSVC() creates all tasks and initializes the stack
frames for all tasks.

Contrary   to   the   function   InitScheduler(),   the   function
InitSchedulerSVC() initializes the stack and both stack frames of all tasks.
The SVC handler simply sets the PSP to point to the top of Task0's stack and restores
the context (R4-R11) of the first task:

```
void SVC_Handler(void)
{
    /* We are here, because main() called SVC. As we interrupted main(),
     * there is no need to save its context.
     * We should never return to main()!!
     * The SVC_Handler should start the first task - Task0
     * The Task 0 is started by restoring its SW context and
     * its HW context upon the exception return.
     */
    // set PSP to Task0.SP:
    __set_PSP((unsigned int)TCB[0].sp);
    current_task = 0;
    // Restore the context of the Task 0:
    __RESTORE_CONTEXT();
}
```

Listing 2.20 The SVC Handler.

The following code shows how to start the scheduler:

```
unsigned int stackRegion[NTASKS * TASK_STACK_SIZE];
TCB_Type TCB[NTASKS];
void (*TaskFunctions[NTASKS])();
int current_task = -1;

void Task0(){
  while(1) {}
}
void Task1(){
  while(1) {}
}
void Task2(){
  while(1) {}
}
void Task3(){
  while(1) {}
}


int main(void)
{
  TaskFunctions[0] = Task0;
  TaskFunctions[1] = Task1;
  TaskFunctions[2] = Task2;
  TaskFunctions[3] = Task3;
  // Init scheduler:
  InitSchedulerSVC(stackRegion, TCB, TaskFunctions);
  // Start SysTick timer:
  HAL_InitTick(0);
  // Switch to NOT PRIVILEDGED with PSP:
  __set_CONTROL(0x00000003);
  // Start the scheduler:
  __asm volatile("svc 0");
  while(1){}
}
```

Listing 2.21   Starting the scheduler using the SVC exception.

The code for the scheduler can be found here:
https://github.com/bulicp/ContextSwitchM7-book.git.

2.5   RISC-V Interrupts and Exceptions

RISC-V architecture defines different privilege modes that determine the level of access and control a program or process has over the system's resources. A privileged mode in a CPU refers to a specific operating mode in which the CPU has access to various system resources. Privileged modes are often used in modern computer architectures to ensure the proper operation, security, and control of the system. Privileged modes are crucial in separating user-level programs from system-level operations and for managing system security, isolation, and resource allocation. For example, a modern CPU restricts a user program from accessing system critical resources (e.g., special CPU registers, memory regions, special instructions, etc.), while the system programs may access all system resources. Privileged modes are the mechanism to achieve this differentiation between user-level and system-level programs. Modern CPUs usually have a separate set of control and status registers (CSRs) for each privileged mode and a special control register that tells which privileged mode the CPU is currently running. Depending on the status of this special control register (i.e., current privileged mode), the CPU can access the corresponding set of CSRs and execute only the instructions allowed in the current privileged level. For example, if the CPU is currently running in a user-privileged mode, it can execute only the standard instruction set. At the same time, executing some special instructions that can alter critical system resources is prohibited. Besides, programs running in user-privileged mode can never alter the content of this special control register and thus switch between privileged modes. But wait, how can we change a privileged mode once the CPU runs in user-privileged mode? Well, it depends on the current privileged mode:

1. If the CPU runs in user-level privileged mode, the only way to switch to a system-level privileged mode is through exceptions (traps or interrupts). Exceptions can trigger mode transitions. When an exception (a trap or an interrupt) occurs, the CPU automatically switches to system-level privileged mode, and the exception handling routine executes in the system-level privileged mode. Upon exiting the exception handler, the CPU automatically switches to the previous (e.g., user-level) privileged mode.
2. If the CPU runs in system-level privileged mode, the CPU can switch to a user-level privileged mode simply by executing a special instruction that alters the content of the special control register and, hence, changes the current system-level privileged mode to user-level privileged mode. CPUs have specific instructions that are used to initiate mode transitions. These instructions are often called privileged and can only be executed when the CPU is in a system-level privileged mode.

2.5.1 RISC-V Privileged Modes

In order to be able to understand interrupts and interrupts handling in RISC-V, we'll briefly describe and explain the privileged modes in RISC-V. Privileged modes are a fundamental part of RISC-V's flexibility, as they enable various operating systems, hypervisors, and security models to be implemented on the same instruction set architecture. Here is a brief description and explanation of three basic privileged modes in RISC-V:

1. **User Mode** (U): User mode is the lowest privilege mode in RISC-V. In this mode, a user-level application or program runs with restricted access to system resources. User mode provides the least privilege and is suitable for application-level code. In user mode, applications can execute most instructions but have limited access to privileged instructions and control registers. User mode can execute basic instructions, access memory, and perform arithmetic operations.

However, it cannot directly manipulate control and status registers (CSRs) related to exception handling or interrupt control.

2. **Supervisor Mode (S):** Supervisor mode is a privilege level above user mode. It is designed for operating system kernel code, which needs greater control over system resources and privilege to perform tasks like context switching and managing hardware devices. Supervisor mode has more access to control registers and instructions compared to user mode. It can perform operations related to exception handling, interrupt control, and system management. S-mode can execute privileged instructions that deal with system control and exception handling. It can access and modify most control and status registers (CSRs), including those related to interrupts and exceptions.

3. **Machine Mode (M):** Machine mode is the highest privilege mode in RISC-V. It provides complete control over the system, including access to all resources and system-wide configuration. M-mode has full access to all instructions, control registers, and hardware resources, making it suitable for tasks such as system initialization, low-level device control, and platform management. M-mode can execute all RISC-V instructions, including those reserved for privileged and system-level operations. It can access and modify all control and status registers (CSRs), and it has control over exceptions and interrupts across all privilege levels. Upon reset, RISC-V enters machine mode.

The E31 RISC-V core in FE-310 SoC supports only Machine and User privilege modes. The transition between privilege modes in E31 RISC-V is typically controlled by changing specific bits in control and status registers (CSRs). The machine mode handles these transitions, ensuring that the processor switches between user and machine modes appropriately. Additionally, exceptions and interrupts may trigger mode transitions, allowing the processor to respond to exceptional conditions or external events. As all exceptions (traps and interrupts) execute in Machine mode, we will restrict the description of exceptions only to this privilege mode.

2.5.2   RISC-V Machine Modes Exceptions

According to the RISC-V Privileged Architecture, the E31 RISC-V CPU comprises five control and status registers for Machine privilege mode:

1. **mstatus:** In RISC-V, the **mstatus** (Machine Status) register is a critical control and status register (CSR) used to manage and store various information related to the Machine privilege mode. The **mstatus** register plays a central role in controlling exception handling, interrupt handling, and the overall operation of the processor in machine mode. The **mstatus** register keeps track of and controls the CPU's current operating state, including whether or not interrupts are enabled. A summary of the **mstatus** bits related to interrupts in the E31 RISC-V CPU is provided in Fig. 2.19. Note that this is not a complete description of **mstatus** as it contains fields unrelated to interrupts. For the full description of **mstatus**, please

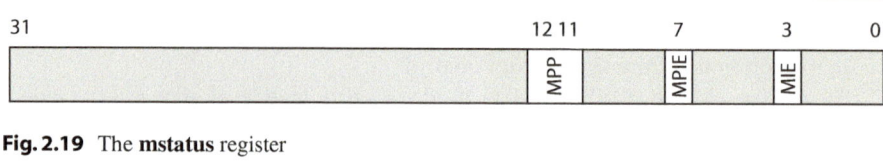

Fig. 2.19 The **mstatus** register

Fig. 2.20 The **mie** register

consult the RISC-V Instruction Set Manual, Volume II: Privileged Architecture. The **mstatus** register contains the following exception-related bits:

a. **MIE** (Machine Interrupt Enable): This bit controls whether machine-level interrupts are globally enabled or disabled. When MIE is set, the CPU can process machine-level interrupts; when it is cleared, machine-level interrupts are disabled.

b. **MPIE** (Machine Previous Interrupt Enable): This bit stores the previous state of MIE before it was modified due to an interrupt. It helps manage interrupt nesting by preserving the previous interrupt-enable state.

c. **MPP** (Machine Previous Privilege Mode): This two-bit field stores the previous privilege mode before the CPU entered machine mode due to an interrupt. It is used during return from interrupt to return to the appropriate privilege mode after processing an interrupt.

2. **mie:** The **mie** (Machine Interrupt Enable) register is responsible for enabling or disabling various types of interrupts that can interrupt the execution of the CPU in machine mode. Individual interrupts are enabled by setting the appropriate bit in the **mie** register. The **mie** register is depicted in Fig. 2.20. The **mie** register contains the following bits:

a. **MSIE** (Machine Software Interrupt Enable): This bit controls whether machine-level software interrupts are enabled or disabled. When MSIE is set, the CPU can process machine-level software interrupts; otherwise, machine-level software interrupts are disabled.

b. **MTIE** (Machine Timer Interrupt Enable): This bit controls whether machine-level timer interrupts are enabled or disabled. When MTIE is set, the CPU can process machine-level timer interrupts.

c. **MEIE** (Machine External Interrupt Enable): This bit controls whether machine-level external interrupts are enabled or disabled. When MEIE is set, the CPU can process machine-level external interrupts.

3. **mip:** The **mip** (Machine Interrupt Pending) register indicates which interrupts are currently pending. The **mip** register is depicted in Fig. 2.21. When an interrupt occurs, the corresponding bit in **mip** is set to 1. When the CPU takes an interrupt,

Fig. 2.21   The **mip** register

the corresponding bit in **mip** is cleared. The **mip** register contains the following bits:

a. **MSIP** (Machine Software Interrupt Pending): When MSIP is set, the Machine Software Interrupt is pending.
b. **MTIP** (Machine Timer Interrupt Pending): When MTIP is set, the Machine Timer Interrupt is pending.
c. **MEIP** (Machine External Interrupt Pending): When MEIP is set, the MachineExternal Interrupt is pending.

If more than one interrupt is pending, the RISC-V CPU prioritizes the interrupts as follows, in decreasing order of priority: Machine External Interrupts (highest priority), Machine Software Interrupts, and Machine Timer Interrupts (lowest priority).

4. **mcause:** In RISC-V architecture, the **mcause** register is a control and status register (CSR) that is used to provide information about the cause of an exception or interrupt that occurred in machine mode. A summary of the **mcause** bits related to interrupts in the E31 RISC-V CPU is provided in Fig. 2.22. When a trap is taken in machine mode, the most significant bit in **mcause** (bit INT) is 0, and the ten least-significant bits (EXCEPTION CODE field) are written with a code indicating the event that caused the trap. When an interrupt is taken, the most significant bit of **mcause** (bit INT) is set to 1, and the ten least-significant bits (EXCEPTION CODE field) contain the interrupt number, using the same encoding as the bit positions in the **mip** register. Table 2.3 lists exception codes and their description. For example, a Machine Timer Interrupt causes **mcause** to be set to 0x80000007.

5. **mtvec:** The **mtvec** register has two main functions. Firstly, it specifies the base address for the vector table, which contains the addresses of exception handlers. Secondly, it sets the mode by which the E31 CPU will process exceptions. The RISC-V CPU can process exceptions in two modes: direct and vectored. In **direct mode**, the **mtvec** register holds the address of a single global exception handler. The processor directly jumps to this global handler's address when a trap or interrupt occurs. In direct mode, we might use a single handler for all exceptions, simplifying the exception-handling process. However, it may not be suitable for systems requiring fine-grained control over exception handling. In **vectored mode**, the **mtvec** register holds the base address of the vector table. In this mode, the processor uses a 10-bit field in the **mcause** register to index the vector table and find the appropriate handler for the specific trap or interrupt that occurred. The vectored mode allows more flexibility in handling various exceptions and interrupts with different routines. In vectored mode, we can have multiple handlers for

Fig. 2.22 The **mcause** register

Table 2.3 **mcause** Exception codes and their description

INT	Exception code	Description
0	0	Instruction address misaligned
0	1	Instruction access fault
0	2	Illegal instruction
0	3	Breakpoint
0	4	Load address misaligned
0	5	Load access fault
0	6	Store address misaligned
0	7	Store access fault
0	8	Environment call from U-mode
0	11	Environment call from M-mode
1	3	Machine software interrupt
1	7	Machine timer interrupt
1	11	Machine external interrupt

different exceptions and interrupts, allowing us to handle each type of exception or interrupt differently. This is often the preferred way of interrupt handling. The **mtvec** register is depicted in Fig. 2.23. The **mtvec** register contains the following bit fields:

a. **MODE:** This 2-bit field sets the interrupt processing mode (00-Direct, 01-Vectored).
b. **BASE:** This 30-bit field contains the vector table base address. This field requires 64-byte alignment.

Table 2.4 describes how an address of the exception handler is computed in two different interrupt processing modes. **Direct mode** means all interrupts and exceptions trap to the same handler, and there is no vector table implemented. It is the handler's responsibility to execute code to figure out which interrupt occurred. The handler in direct mode should first read the bit 31 in **mcause** to determine if an interrupt or exception occurred. It should then execute appropriate code based on the EXCEPTION CODE field in **mcause**, which contains the respective interrupt or exception code. **Traps always use the direct mode**. Hence, all exceptions trap to the same handler. For example, suppose the BASE is set to 0x20011500. When an exception occurs, the PC is set to 0x20011500, and the first instruction of the exception handler should be at this address. **Vectored mode** allows for

Fig. 2.23 The **mtvec** register

Table 2.4 **mtvec** Modes and address of exception handler encoding

MODE	Interrupt processing mode	Address of exception handler
0	Direct	PC = BASE *NOTE: Exceptions are not vectored. All exceptions trap to the same handler. The handler executes code to figure out which exception occurred.*
1	Vectored	PC = BASE + 4 x mcause[EXCEPTION CODE] *NOTE: BASE must be 64-byte aligned. This is to avoid an adder in the above computation*

creating a vector table that hardware uses for lower interrupt handling latency. **Only interrupts can use the vectored mode**. When an interrupt occurs in vectored mode, the PC will get assigned by the hardware to the address of the vector table index corresponding to the interrupt ID. From the vector table index, a subsequent jump will occur from there to service the interrupt. The interrupt handler offset is calculated by PC = BASE + 4 x mcause[EXCEPTION CODE]. The vectored mode does not require the software overhead to determine which interrupt occurred. In this mode, when an interrupt occurs, the execution jumps directly to the vector table offset for the corresponding interrupt. For example, suppose the global and machine timer interrupts are enabled, and the BASE is set to 0x20011500. If the vectored mode is selected and a machine timer interrupt occurs, the EXCEPTION CODE in the **mcause** register will be 0x07. Then, the address of the interrupt handler that processes the machine timer interrupts will be 0x20011500 + 4 x 0x07 = 0x20011500 + 0x1C = 0x2001151C. Hence, when the interrupt is taken, the PC is set to 0x2001151C, and the first instruction of the interrupt handler should be at this address.

Configuring these five Control and Status Registers registers correctly is crucial for proper exception handling in RISC-V systems, as they dictate where the processor should jump when an exception occurs and how exceptions are managed. These CSRs are not memory-mapped and can only be accessed through special privileged instructions: **csrr** and **csrw** for read and write, respectively. Hence, to work with these CSRs, developers must use assembly language instructions to read and modify these registers as needed.

2.5.3 FE-310 Interrupts

The SiFive Freedom E310, also known as FE310, is a microcontroller based on the RISC-V architecture. It's designed for embedded and IoT applications and is notable for being one of the early implementations of the RISC-V ISA. Let us have a deeper view of interrupts supported in SiFive Freedom E310. The FE310 SoC supports two types of RISC-V interrupts: local and global. Local interrupts are signalled directly to the RISC-V E31 CPU with a dedicated interrupt line for each local interrupt. The RISC-V E31 CPU has three interrupt lines for external, software and timer interrupts (Fig. 2.24). Software and timer interrupts are local interrupts generated by the Core-Local Interruptor (CLINT). Besides the software and timer interrupts, various I/O devices (e.g., UART, GPIO, etc.) can use global interrupts to activate the external interrupt line and to interrupt the CPU. Global interrupts from I/O devices are routed through a Platform-Level Interrupt Controller (PLIC), which will be described later.

The CLINT is a mandatory component in RISC-V processor systems. It's responsible for managing timer-related and software-generated interrupts at the core level. The CLINT generates two interrupts:

1. Machine Timer Interrupts: The CLINT contains a timer called the Machine Timer, which can generate timer interrupts for various purposes, including timekeeping, scheduling, and triggering tasks at specific intervals.
2. Machine Software Interrupts: In RISC-V, the software can generate software interrupts to communicate with the operating system. In general, the program running in user mode is not allowed to call operating system procedures. Hence, the only way a user program makes a system call is by generating a software interrupt. The software interrupt handler running in machine mode then calls an

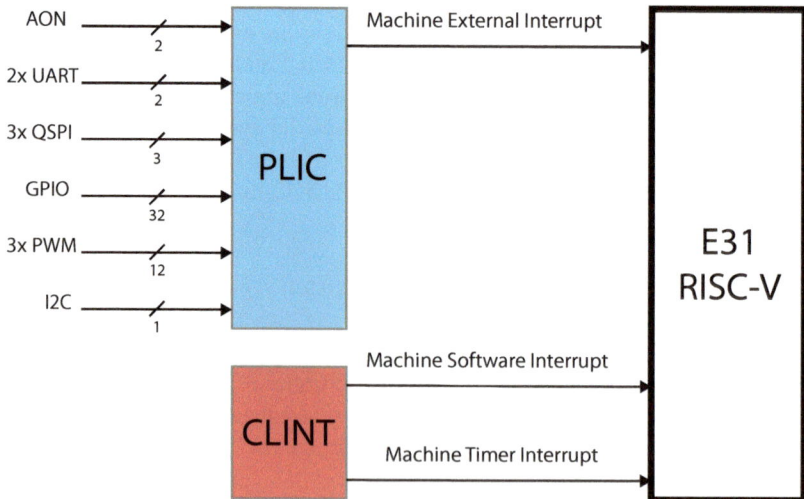

Fig. 2.24 FE310 interrupt architecture block diagram

Table 2.5 Memory map for CLINT registers on SiFive FE310 SoC

Address	Width	Register
0x02000000	4B	**msip**
0x02004000	8B	**mtimecmp**
0x0200BFF8	8B	**mtime**

operating system procedure. The CLINT can be used to handle these software-generated interrupts.

The CLINT comprises memory-mapped control and status registers related to software and timer interrupts. Table 2.5 shows the memory map for CLINT on SiFive FE310.

2.5.3.1 Machine Software Interrupts

A machine software interrupt is an interrupt generated by software running in machine mode to request attention from the processor for specific tasks or events. Machine software interrupts are generated by writing '1' to the **msip** register within CLINT. The **msip** register is a 32-bit memory-mapped register where the upper 31 bits are hardwired to zero. The least significant bit of the **msip** register is reflected in the MSIP bit of the **mip** register. On reset, the **msip** register is cleared to zero.

2.5.3.2 Machine Timer Interrupts

CLINT, which is a mandatory part of RISC-V architecture, provides a 64-bit real-time counter, which monotonically increases at a clock speed, and its content is visible as a memory-mapped register **mtime**. In the FE310 SoC, CLINT is responsible for providing the real-time counter. Machine timer interrupt is a local interrupt, which can be generated by using two architecturally defined timer registers: **mtime** and **mtimecmp**:

1. **mtime** register: The 64-bit **mtime** register stores the current value of the 64-bit timer counter. The software can read this register to determine the current time.
2. **mtimecmp** register: The **mtimecmp** register holds a value that is compared with the **mtime** register. When **mtime** reaches the value stored in **mtimecmp**, it triggers a timer interrupt. This register is used to set up timer interrupts for specific time intervals.

In summary, the machine timer generates timer interrupts when the **mtime** matches or exceeds the value stored in the **mtimecmp** register. This feature is crucial for implementing preemptive multitasking, where the processor can switch between tasks at predefined time intervals.

2.5.4   Interrupt Entry and Exit

Interrupt entry and exit refer to the processes by which a RISC-V processor handles interrupts. These processes involve transitioning from regular program execution to an interrupt handler and returning to regular program execution after the interrupt is serviced. In the following subsections, we describe and explain interrupt entry and exit in RISC-V.

2.5.4.1   Interrupt Entry
When a machine interrupt occurs:

1. The value of the MIE bit in **mstatus** is copied into the MPIE bit in **mstatus**, and then MIE is cleared, effectively disabling interrupts.
2. The privilege mode prior to the interrupt is saved in the MPP field in **mstatus**.
3. The cause of the interrupt is encoded into EXCEPTION CODE in **mcause**.
4. The current PC is copied into the **mepc** register, and then the PC is set to the value specified by **mtvec** as described in Table 2.4.

   At this point, control is handed over to software in the interrupt handler with interrupts disabled. Interrupts can be re-enabled by explicitly setting the MIE bit in **mstatus** or by executing the `mret` instruction to exit the handler.

2.5.4.2   Interrupt Exit
To exit from a machine interrupt, the `mret` instruction must be executed at the end of the interrupt handler. When a `mret` instruction is executed, the following occurs:

1. The privilege mode is set to the value encoded in the MPP field in **mstatus**.
2. In the **mstatus** register, the MIE bit is set to the value of MPIE.
3. The PC is set to the value of **mepc**, hence pointing to the instruction, which was interrupted.

   At this point, control is handed over to the previously interrupted program.

2.5.5   Implementing Vector Table and Handlers

Implementing a vector table and handlers in assembly language for RISC-V involves setting up a program structure to store the addresses of exception handlers and configuring the system to use this table when exceptions occur to jump to the interrupt-specific handler. Below are the steps to implement an exception table and handlers in RISC-V assembly:

1. **Define the Vector Table**: Create a program structure that serves as the vector table. As we have learned, the address of the first instruction of an interrupt handler is calculated using the BASE address of the vector table and the exception cause (Table 2.4). Each entry in the vector table occupies exactly 4 bytes, and there is

only room for one instruction per handler in the vector table. Therefore, the only instructions in the exception table should be the jump instructions that transfer control to an interrupt-specific handler. An example of the vector table is as follows:

```
 1  #  -----------------------------------
 2  #
 3  #   V  E  C  T  O  R     T  A  B  L  E
 4  #
 5  #    must be 64-byte aligned.
 6  #  -----------------------------------
 7
 8  .balign 64
 9  .global _vector_table
10  _vector_table:                         # BASE
11      j _default_handler
12      j _default_handler
13      j _default_handler
14      #  ---------------------------
15      j _msw_interrupt_handler     # 3
16      #  ---------------------------
17      j _default_handler
18      j _default_handler
19      j _default_handler
20      #  ---------------------------
21      j _mtim_interrupt_handler    # 7
22      #  ---------------------------
23      j _default_handler
24      j _default_handler
25      j _default_handler
26      #  ---------------------------
27      j _mext_interrupt_handler    # 11
28      #  ---------------------------
```

Listing 2.22 A vector table for E31 RISC-V.

The vector table is populated with jump instructions to transfer control to interrupt-specific handlers. For example, the jump instruction (j _mtim_interrupt_handler) that causes the jump to the timer interrupt handler is placed at the offset $7 \times 4 = 0x1C$ from the beginning of the vector table. So when a machine timer interrupt occurs, the PC is set to BASE + 0x1C and the CPU will execute the j _mtim_interrupt_handler instruction.

We can see from Listing 2.22 that besides the jump instructions to exception handlers for software, timer and external interrupts, there is also a jump instruction to _default_handler in all other entries in the vector table. We have already learned that there are only three interrupt sources in FE310 SOC (software, timer and external), so why do we need the fourth interrupt handler _default_handler? This is to ensure that in case of a trap (INT=0 in **mcause**), the CPU executes _default_handler.

2. **Register the Base Vector Table Address:** We should configure the **mtvec** register to point to the exception table. Also, we should set the preferred interrupt processing mode in mtvec. Listing 2.23 presents the RISC-V assembly code to register the base address and to select the vectored mode:

```
 1  #------------------------------------------
 2  #     Register the base address for vector table
 3  #         in mtvec
 4  #
 5  #@arguments:
 6  #     # a0 - interrupt vector table base address
 7  #     # a1 - interrupt processing mode
 8  #              (0x0 - direct, 0x1 - vectored)
 9  #------------------------------------------
10  .balign 4
11  .global register_handler
12  .type register_handler, @function
13  register_handler:
14      # prologue:
15      addi sp, sp, -16      # Allocate the routine
16                           #    stack frame
17      sw ra, 12(sp)        # Save the return address
18      sw fp, 8(sp)         # Save the frame pointer
19      sw s1, 4(sp)
20      sw s2, 0(sp)
21      addi fp, sp, 16      # Set the framepointer
22
23      or a0, a0, a1        # OR base address with mode
24      csrw mtvec, a0       # and save into mtvec
25
26      # epilogue:
27      lw s2, 0(sp)
28      lw s1, 4(sp)
29      lw fp, 8(sp)         # restore the frame pointer
30      lw ra, 12(sp)        # restore the return address
31      addi sp, sp, 16      # de-allocate the routine
32                           #    stack frame
33      ret
```

Listing 2.23 Assembly function for registering the vector table base addreess.

3. **Define Exception/Interrupt Handler:** Write the exception/interrupt handler routines in assembly language. Each handler should be a separate section of code that corresponds to a specific exception type and ends with the `mret` instruction. The prologue of an interrupt handler usually begins with saving the registers onto the stack to avoid overwriting the contents of the saved registers (s0–s11). After the body of the exception handler executes, the epilogue of an interrupt handler restores the saved registers from the stack. Finally, the handler returns with `mret`, an instruction unique to machine mode. The `mret` instruction restores the PC from **mepc**, the previous interrupt-enable setting, and the privilege mode as described in Sect. 2.5.4.2. For example, the following code (Listing 2.24) presents the RISC-V assembly code for a machine timer interrupt handler:

```
 1  #------------------------------------------
 2  #    Machine Timer Interrupt Handler
 3  #------------------------------------------
 4  .balign 4
 5  .global _mtim_interrupt_handler
 6  _mtim_interrupt_handler:
 7
 8  # Prologue :
 9  #     save 16 ABI caller registers
10  #     (ra, t0-t6, a0-a7)
```

```
11  addi sp, sp, -16*4     # Allocate the routine stack frame
12  sw t0,  0*4(sp)
13  sw t1,  1*4(sp)
14  sw t2,  2*4(sp)
15  sw t3,  3*4(sp)
16  sw t4,  4*4(sp)
17  sw t5,  5*4(sp)
18  sw t6,  6*4(sp)
19  sw a0,  7*4(sp)
20  sw a1,  8*4(sp)
21  sw a2,  9*4(sp)
22  sw a3, 10*4(sp)
23  sw a4, 11*4(sp)
24  sw a5, 12*4(sp)
25  sw a6, 13*4(sp)
26  sw a7, 14*4(sp)
27  sw ra, 15*4(sp)
28
29  # Decode interrupt cause
30  csrr t0, mcause       # read exception cause
31  bgez t0, 1f           # exit if not an interrupt
32
33  # Increment timer compare by 1000 cycles
34  li t0, 0x0200BFF8     # load the mtime address
35  lw t1, 0(t0)          # load mtime (LO)
36  lw t2, 4(t0)          # load mtime (HI)
37  li t3, 1000           # load 1000 cycles
38  add t3, t1, t3        # increment lower bits by 1000
39  sltu t1, t3, t1       # generate carry-out
40  add t2, t2, t1        # increment upper bits with carry
41
42  li t0, 0x02004000     # load the mtimecmp address
43  sw t3, 0(t0)          # update mtimecmp (LO)
44  sw t2, 4(t0)          # update mtimecmp (HI)
45
46  1:
47  # Epilogue: restore ABI caller regs
48  lw t0,  0*4(sp)
49  lw t1,  1*4(sp)
50  lw t2,  2*4(sp)
51  lw t3,  3*4(sp)
52  lw t4,  4*4(sp)
53  lw t5,  5*4(sp)
54  lw t6,  6*4(sp)
55  lw a0,  7*4(sp)
56  lw a1,  8*4(sp)
57  lw a2,  9*4(sp)
58  lw a3, 10*4(sp)
59  lw a4, 11*4(sp)
60  lw a5, 12*4(sp)
61  lw a6, 13*4(sp)
62  lw a7, 14*4(sp)
63  lw ra, 15*4(sp)
64  addi sp, sp, 16*4     # de-allocate the routine stack frame
65  mret
```

Listing 2.24 Assembly code for the machine timer interrupt.

The code in Listing 2.24 assumes that interrupts are globally enabled in **mstatus** (MIE = 1), that timer interrupts have been enabled in **mie**, and that **mtvec** has been set to the base address of the vector table with the interrupt processing mode set to vectored. The prologue preserves 16 registers according to RISC-V ABI (Application Binary Interface). You may find this a little odd—why waste

16 instructions and 64 bytes in memory to save these registers? Well, it turns out there is a very good reason we do this. When writing an interrupt handler in RISC-V assembly language, it's essential to save and restore the necessary registers to ensure the proper operation of the interrupted program. The specific registers that should be saved onto the stack can vary depending on the RISC-V privilege mode, the interrupt source, and the calling conventions of the platform. However, here's a general guideline for which registers we should consider saving:

a. **ra** register stores the return address for function calls. Saving and restoring this register ensures that control can return correctly to the interrupted program.

b. **Caller-Saved Registers t0-t6** can be freely modified by the caller (interrupted program) without the caller being responsible for saving their original values. If the interrupt handler modifies any of these registers, we should save and restore them to maintain the integrity of the interrupted program.

c. **Stack Pointer** when the interrupt handler needs additional stack space. In such a case, we need to save and restore the stack pointer to ensure that stack operations do not interfere with the interrupted program's stack.

d. **Other Registers Used by the Interrupt Handler**. Depending on the specific needs of the interrupt handler, we may use additional registers for temporary storage or calculations or for passing arguments. If these registers are modified, we should save and restore them.

After the prologue, the handler decodes the exception cause by examining **mcause**: interrupt if **mcause** < 0, trap otherwise. Then, it simply increments the time comparator so that the next timer interrupt occurs about 1000 timer cycles in the future. The handler is not preemptible, as it keeps interrupts disabled throughout the handler. Finally, the epilogue restores saved registers and returns with mret.

We can also write interrupt handlers in C. To write an interrupt handler in C for a RISC-V-based system, we typically need to use a combination of assembly language and C code. For example, reading and writing CSRs (e.g., **mcause**) is only possible with the special csrr, csrw instructions; hence, we are forced to use assembly language for such operations. The exact details of how to implement interrupt handlers in C can vary depending on your platform and compiler, but we will give a general outline of how to write an interrupt handler in C for a RISC-V system:

a. **Mark the Function as an Interrupt Handler**: Usually, we use compiler-specific attributes or pragmas to mark the function as an interrupt handler. This attribute is crucial for the compiler to generate prologue and epilogue sequences for an interrupt handler and to put the mret instruction at the end of the generated code. The following C code presents how to mark a function as an interrupt handler:

```
/*
 * Use "interrupt" attribute to indicate that the specified
 * function is an interrupt handler.
 * The compiler generates function entry and exit
 * sequences suitable for use in an interrupt handler
 * when this attribute is present.
 */

__attribute__((interrupt)) void interrupt_handler(void) {
    // Interrupt handling code
}
```

Listing 2.25 Interrupt handler function in C.

b. **Use inline assembly for accessing CSRs:** To read/write the CSRs registers in RISC-V, we should use inline assembly. The exact details of how to use inline assembly depend on the compiler, so we should always consult the compiler manual. Here is an example of how to write inline assembly to read the **mcause** register in C:

```
unsigned int mcause_value;

// Inline assembly to read mcause
asm volatile(
    "csrr %0, mcause"  // Read mcause into %0
    : "=r" (mcause_value)  // Output : mcause_value
);
```

Listing 2.26 Inline assembly to read **mcause**.

The volatile qualifier is necessary as GCC optimizers sometimes discard asm statements if they determine there is no need for the output variables. Using the volatile qualifier disables these optimizations.

Listing 2.27 presents the machine timer interrupt handler.

```
unsigned int *pMTime    = (unsigned int *)0x0200bff8;
unsigned int *pMTimeCmp = (unsigned int *)0x02004000;

__attribute__ ((interrupt)) void mtime__handler (void) {

    unsigneg int mcause_value;
    // Decode interrupt cause:
    // Non memory-mapped CSR registers can only be accessed
    // using special CSR instructions. Hence, we should use
    // inline assembly:
    __asm__ volatile ("csrr %0, mcause"
                      : "=r" (mcause_value) /* output */
                      : /* input : none */
                      : /* clobbers: none */
    );

    if (mcause_value & 0x8000007) { // mtime interrupt!
        // Increment timer compare by 500 ms:
        *pMTimeCmp = *pMTime + 16384;
    }
}
```

Listing 2.27 Machine timer interrupt handler in C.

4. **Enable Global Interrupts:** To enable machine-level interrupts, we should set
   the MIE bit in the **mstatus** register. The following code (Listing 2.28) presents
   the RISC-V assembly code to enable global machine-level interrupts :

```
.equ MSTATUS_MIE_BIT_MASK ,   0x00000008   # bit 3

#------------------------------------------
#     Enable global interrupts in mstatus
#------------------------------------------
.balign 4
.global enable_global_interrupts
.type enable_global_interrupts , @function
enable_global_interrupts:
    # prologue:
    addi sp, sp, -16      # Allocate the routine
                          #      stack frame
    sw ra, 12(sp)         # Save the return address
    sw fp, 8(sp)          # Save the frame pointer
    sw s1, 4(sp)
    sw s2, 0(sp)
    addi fp, sp, 16       # Set the framepointer

    li t0, MSTATUS_MIE_BIT_MASK
    csrs mstatus, t0      #  set the MIE bit in mstatus

    # epilogue:
    lw s2, 0(sp)
    lw s1, 4(sp)
    lw fp, 8(sp)          # restore the frame pointer
    lw ra, 12(sp)         # restore the return address
    addi sp, sp, 16       # de-allocate the routine
                          #     stack frame
    ret
```

Listing 2.28 Assembly function for enabling global interrupts in the **mstatus** register.

5. **Enable Particular Interrupt:** Depending on what particular interrupt (software,
   timer or external) we would like to enable, we should set an appropriate bit in
   the **mie** register. Listing 2.29 presents the RISC-V assembly code to enable the
   machine timer interrupt:

```
.equ MIE_MTIE_BIT_MASK ,       0x00000080   # bit 7

#------------------------------------------
#     Enable machine timer interrupt in mie
#------------------------------------------

.balign 4
.global enable_mtimer_interrupt
.type enable_mtimer_interrupt , @function
enable_mtimer_interrupt:
    # prologue:
    addi sp, sp, -16      # Allocate the routine
                          #      stack frame
    sw ra, 12(sp)         # Save the return address
    sw fp, 8(sp)          # Save the frame pointer
    sw s1, 4(sp)
    sw s2, 0(sp)
```

```
19    addi fp, sp, 16        # Set the framepointer
20
21    li t0, MIE_MTIE_BIT_MASK
22    csrs mie, t0           # set MTIE in mie
23
24    # epilogue :
25    lw s2, 0(sp)
26    lw s1, 4(sp)
27    lw fp, 8(sp)           # restore the frame pointer
28    lw ra, 12(sp)          # restore the return address
29    addi sp, sp, 16        # de-allocate the routine
30                           #    stack frame
31    ret
```

Listing 2.29   Assembly function for enabling the machine timer interrup in the **mie** register.

2.5.6   Case Study: A Simple Task Scheduler on RISC-V Based FE310

Having delved into the intricacies of context switching on the ARM Cortex-M7 architecture, we now embark on a compelling journey to explore the analogous process on
the RISC-V architecture. Our exploration of context switching on ARM Cortex-M7
processors has given us insights into the nuanced dance of saving and restoring task
states, managing interrupts, and orchestrating seamless transitions between tasks.
Now, with the backdrop of ARM's methodologies, we set our sights on RISC-V—a
modular and versatile architecture that captivates developers and researchers alike
with its openness and adaptability. In our previous foray into ARM Cortex-M7, we
uncovered the distinctive features of the ARM Cortex-M7 architecture. We navi-
gated the intricacies of saving and restoring register states, mitigating interruptions,
and steering the efficient flow of tasks. Armed with this knowledge, we are ready
to apply these principles to the RISC-V. By understanding the parallels and dis-
tinctions between ARM Cortex-M7 and RISC-V, we are poised to master the art
of crafting efficient and tailored context-switching routines for diverse computing
environments.

Recall that context switching, a pivotal aspect of modern computing, is a process that allows a system to seamlessly transition between multiple tasks, ensuring
responsiveness and the efficient use of computational resources. Exploring context
switching in the RISC-V architecture unveils a journey through the intricacies of
multitasking and efficient resource utilization on RISC-V, characterized by its simplicity, modularity, and open design philosophy, which provides a flexible canvas for
implementing context-switching mechanisms.

Contrasting RISC-V's approach to context switching with the ARM Cortex-M7,
notable differences emerge. We have learned that the ARM Cortex-M7 employs a
specific set of registers (e.g., banked stack pointer) and a dedicated interrupt-handling
mechanism to facilitate context switching. Its unique stack frame format and the
presence of two different stack pointers, one for tasks (PSP) and one for interrupt
handlers (MSP), contribute to efficient task switching. Another key distinction lies
in the register sets used during context switching. While both architectures involve

saving and restoring register values, the specific registers and their organization differ. Understanding the intricacies of register usage in each architecture is crucial for crafting efficient context-switching routines.

This case study explores the intricacies of context switching in the RISC-V architecture, delving into its underlying mechanisms and the challenges involved in orchestrating efficient task transitions.

2.5.6.1 Background

A simple round-robin task scheduler (Fig. 2.25) on RISC-V-based FE310 processors effectively manages multiple tasks or threads in a cooperative multitasking environment. In this scheduler, each task is given a fixed time slice (quantum) during which it can execute. When its time slice expires, the scheduler switches to the next task in the queue. **The task scheduler relies on the interrupts and stacks to achieve context switching**. The machine timer (**mtime**) interrupts will be used for context switching.

When switching contexts, the scheduler needs a way to keep track of which tasks are doing what using a task table. Recall from the previous sections that the ARM Cortex-M7 processor has two separate stack pointers, allowing stack separation between the kernel and tasks, which in turn simplifies the context switch procedure. RISC-V-based FE-310 has only one stack pointer, which slightly complicates the context switching and forces us to carefully manage the stack within the interrupt handler. Besides, both tasks and kernel will run in machine mode.

Figure 2.26 shows the scheduler operations during a context switch in more detail. When a Machine Timer interrupt occurs, the execution switches to the machine timer interrupt handler. Once in the machine timer interrupt handler, the scheduler pushes the interrupted Task1 registers **x1** (return address), **x5-x31**, **epc**, and **mstatus** onto the task's stack and saves its SP in the task's TCB. Contrary to ARM-Cortex M7, RISC-V does not automatically save critical registers (i.e., it does not implement hardware stacking). Upon interrupt entry, RISC-V only saves the return address into **epc** and the status of the interrupted procedure into **mstatus**. Hence, the interrupt handler is responsible for saving the complete context of the interrupted task: registers

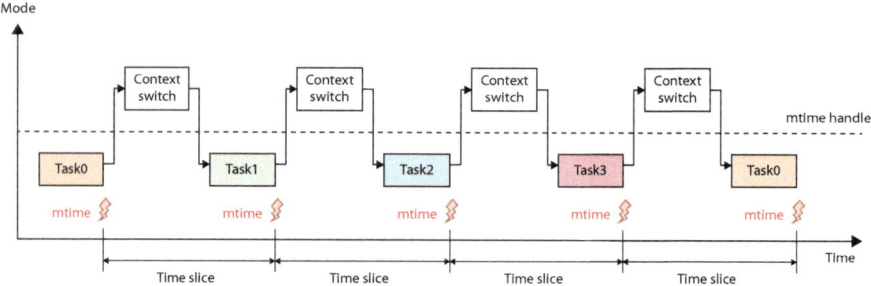

Fig. 2.25 A simple task scheduler on RISC-V based FE310

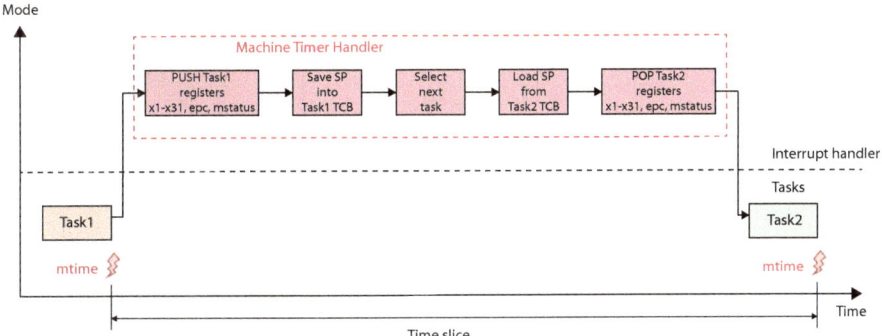

Fig. 2.26 Operations during a context switch

x1, **x5-x31**, return address contained in **epc** and the processor status before the interrupt occurred (contained in **mstatus**).

Then, the scheduler selects the next task (Task2) in a round-robin fashion. Before returning from the machine timer interrupt handler, the scheduler is responsible for loading the Task2 SP and restoring the Task2 context (**x1**, **x5-x31**, **epc**, and **mstatus**) from the Task2 stack. Finally, upon interrupt exit, **mepc** is copied to **pc** and the execution returns to the new task, Task2.

As we have already learned, three routines are required to implement and run the scheduler: create new tasks, initialize tasks, and perform the context switch. Besides, several data structures are required to implement and manage the stack for each task and represent each task's state. The stack region used to implement the tasks' stacks, and the task control block are the same as in the Sect. 2.4.7. The following subsections provide a step-by-step description of implementing a simple round-robin scheduler on a RISC-V-based FE310 processor.

2.5.6.2   Task Creation

The `TaskCreate()` function saves the address of the task's stack and the address of the task's function into the task's TCB.

The following code presents the function used to create a new task:

```
void TaskCreate(TCB_Type* pTCB,
                unsigned int* pTaskStackBase,
                void (*TaskFunction)()){

  pTCB->sp          = (unsigned int*) pTaskStackBase;
  pTCB->pTaskFunction = TaskFunction;
}
```

Listing 2.30 The function `TaskCreate()` that creates a new task.

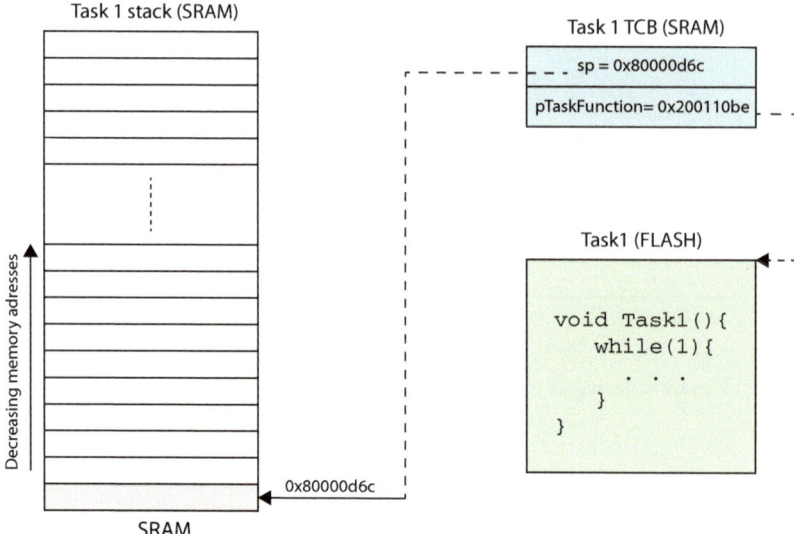

Fig. 2.27 Memory layout and content after calling the `TaskCreate()` function

The parameters of the above `TaskCreate()` function are:

- pTCB—a pointer to a task's TCB,
- pStackBase—pointer task's stack block,
- TaskFunction—address of a task's function.

Figure 2.27 illustrates the memory layout and the contents of the task's stack and TCB after creating Task1 using the `TaskCreate()` function.

2.5.6.3 Task Initialisation
The following code presents the function used to initialize a new task:

```
void TaskInit(TCB_Type* pTCB){
  Context_TypeDef* pStackFrame;

  // Make the room for the stack frame and
  // set pointer to the top of stack frame:
  pStackFrame = (Context_TypeDef*)((void*)pTCB->sp -
                    sizeof(Context_TypeDef));

  // populate Stack Frame
  pStackFrame->mepc    = (unsigned int) (pTCB->pTaskFunction);
  pStackFrame->x1      = 0xFFFFFFFF; // (ra: task never exits)
  pStackFrame->mstatus = (0x03 << 11) | (0x01 << 7); // value 0x1880

  // Set task's stack pointer in the TCB to point at the top of the
  // task's SW stack frame
  pTCB->sp      = (unsigned int*) pStackFrame;
}
```

Listing 2.31 The function `TaskInit()` that initializes a new task.

The only parameter of the above `TaskInit()` function is a pointer to a task's TCB. The `TaskInit()` function performs the following steps:

1. Initialise the pointer to the stack frame. The stack frame will hold the task's context. We need to prepare the stack frame for each new task so that when the task switch occurs, the frame will be ready for de-stacking and, hence, entering a new task. To make this task easier, we will abstract the stack frame with the following C structure:

```
/*
 * The RISC-V context is saved in the following stack frame,
 * where the global(tp) and thread (tp) pointers
 * are currently assumed to be constant so are not saved:
 */
typedef struct {
  unsigned int mepc;     // (sp+0)
  unsigned int x1;       // (sp+1)
  unsigned int t5;       // (sp+2)
  unsigned int x6;       // (sp+3)
  unsigned int x7;       // (sp+4)
  unsigned int x8;       // (sp+5)
  unsigned int x9;       // (sp+6)
  unsigned int x10;      // (sp+7)
  unsigned int x11;      // (sp+8)
  unsigned int x12;      // (sp+9)
  unsigned int x13;      // (sp+10)
  unsigned int x14;      // (sp+11)
  unsigned int x15;      // (sp+12)
  unsigned int x16;      // (sp+13)
  unsigned int x17;      // (sp+14)
  unsigned int x18;      // (sp+15)
  unsigned int x19;      // (sp+16)
  unsigned int x20;      // (sp+17)
  unsigned int x21;      // (sp+18)
  unsigned int x22;      // (sp+19)
  unsigned int x23;      // (sp+20)
  unsigned int x24;      // (sp+21)
  unsigned int x25;      // (sp+22)
  unsigned int x26;      // (sp+23)
  unsigned int x27;      // (sp+24)
  unsigned int x28;      // (sp+25)
  unsigned int x29;      // (sp+26)
  unsigned int x30;      // (sp+27)
  unsigned int x31;      // (sp+28)
  unsigned int mstatus;  // (sp+29)
  unsigned int unused1;  // (sp+30)
  unsigned int unused2;  // (sp+31)
} Context_TypeDef;
```

Listing 2.32 A C structure used to abstract the task's stack frame.

2. Now, as the pointer to the stack frame, `pStackFrame`, is set, we can populate the frames with initial values. The stack frame is populated as follows:

- **mstatus** = 0x00001880. We set the task's previous privilege level (field MPP in **mstatus**) to '11' (machine mode), and the task's previous interrupt flag MPIE to '1'.

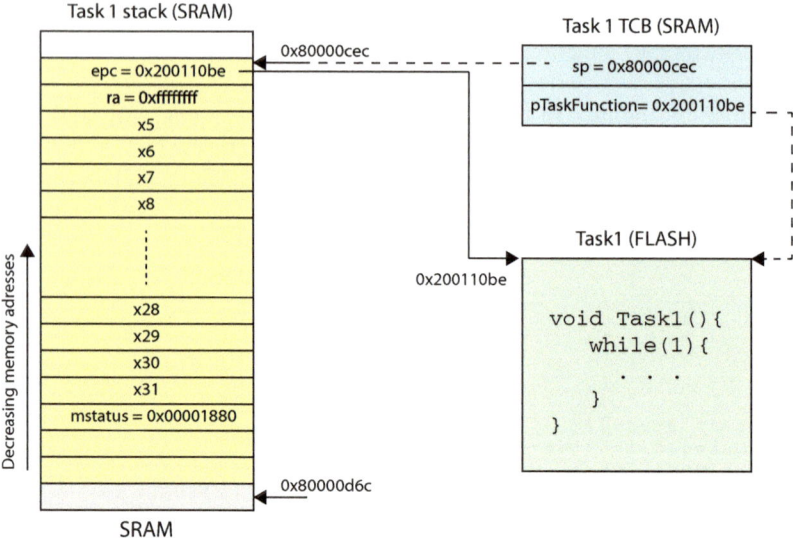

Fig. 2.28 Memory layout and content after calling the `TaskInit()` function

- **mepc** = the address of the task,
- **x1 (ra)** = 0xFFFFFFFF—in our case, tasks never finish, so the return address is set to 0xFFFFFFFF.

3. Finally, it saves the address of the top of the stack frame into the task's SP entry in the task's TCB.

After these steps, a new task is ready to be executed for the first time when the task switch occurs, and the task is selected for execution. Figure 2.28 illustrates the memory layout and the contents of the task's stack and TCB after creating Task1 using the `TaskInit()` function.

2.5.6.4 Scheduler Initialisation
The following code presents the function used to initialize all tasks:

```
void InitScheduler(unsigned int* pStackRegion, TCB_Type pTCB[],
                   void (*TaskFunctions[])()){
  unsigned int* pTaskStackBase;

  // 1. create all tasks:
  for(int i=0; i<NTASKS; i++){
    pTaskStackBase = pStackRegion +
                     (i+1)*TASK_STACK_SIZE;
    TaskCreate(&pTCB[i], pTaskStackBase,
               TaskFunctions[i]);
  }

  // 2. initialize all tasks:
```

```
     for(int i=0; i<NTASKS; i++){
       TaskInit(&pTCB[i]);
     }
 }
```

Listing 2.33 The function `InitScheduler()` creates and initializes all tasks.

The function `InitScheduler()` performs the following steps:

1. Creates all tasks.
2. Initializes all tasks

After these steps, everything is set up for the first context switch. Figure 2.29 illustrates the memory layout and the task's stack after initializing the scheduler using the `InitScheduler()` function.

2.5.6.5 Machine Timer Interrupt Handler

Finally, we can implement the machine timer interrupt handler that will perform the task switch. Contrary to ARM Cortex-M7, RISC-V does not have two separate stack pointers for handlers and user programs. Hence, interrupt handlers on RISC-V-based FE310 use the same stack pointer as interrupted tasks. If we implemented the machine timer interrupt handler in C, the stack pointer would become corrupted by the interrupt handler itself because the C compiler would generate the prologue code according to the calling convention. On the other hand, assembly language enables precise management of the stack pointer, allowing for the preservation of the current task's context and the restoration of the new task's context. Therefore, an

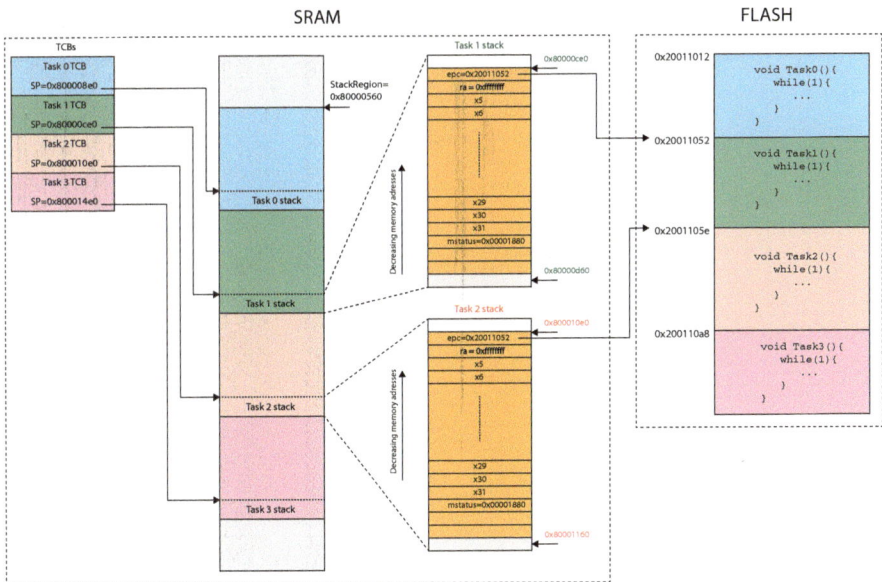

Fig. 2.29 Memory layout and content after initializing four tasks during the scheduler initialization

interrupt handler used for context switching on a RISC-V-based processor should be
written in assembly language. Here is the machine interrupt handler used for context
switching:

```
/*-----------------------------------------
     Machine Timer Interrupt Handler
-----------------------------------------*/
.balign 4, 0
.global _mtim_interrupt_handler
_mtim_interrupt_handler:
    # Save context:
    __macro_SAVE_CONTEXT

    # Increment time slice (tick)
    __macro_INCREMENT_TICK

    # switch context
    __macro_SWITCH_CONTEXT

2:
    # Restore context
    __macro_RESTORE_CONTEXT

    mret
```

Listing 2.34 The machine timer interrupt handler used to perform task switch.

The machine timer interrupt handler performs the following steps:

1. Saves the context of the interrupted task on the task's stack using the
   __macro_SAVE_CONTEXT macro, defined as:

```
.macro  __macro_SAVE_CONTEXT
    addi sp, sp, -CONTEXT_SIZE
    sw x1,      1*WORD_SIZE(sp)
    sw x5,      2*WORD_SIZE(sp)
    sw x6,      3*WORD_SIZE(sp)
    sw x7,      4*WORD_SIZE(sp)
    sw x8,      5*WORD_SIZE(sp)
    sw x9,      6*WORD_SIZE(sp)
    sw x10,     7*WORD_SIZE(sp)
    sw x11,     8*WORD_SIZE(sp)
    sw x12,     9*WORD_SIZE(sp)
    sw x13,     10*WORD_SIZE(sp)
    sw x14,     11*WORD_SIZE(sp)
    sw x15,     12*WORD_SIZE(sp)
    sw x16,     13*WORD_SIZE(sp)
    sw x17,     14*WORD_SIZE(sp)
    sw x18,     15*WORD_SIZE(sp)
    sw x19,     16*WORD_SIZE(sp)
    sw x20,     17*WORD_SIZE(sp)
    sw x21,     18*WORD_SIZE(sp)
    sw x22,     19*WORD_SIZE(sp)
    sw x23,     20*WORD_SIZE(sp)
    sw x24,     21*WORD_SIZE(sp)
    sw x25,     22*WORD_SIZE(sp)
    sw x26,     23*WORD_SIZE(sp)
    sw x27,     24*WORD_SIZE(sp)
    sw x28,     25*WORD_SIZE(sp)
    sw x29,     26*WORD_SIZE(sp)
```

```
     sw  x30,       27*WORD_SIZE(sp)
     sw  x31,       28*WORD_SIZE(sp)

     csrr t0, mepc
     sw  t0,         0*WORD_SIZE(sp)
     csrr t0, mstatus
     sw  t0,        29*WORD_SIZE(sp)
.endm
```

Listing 2.35 __macro_SAVE_CONTEXT macro.

2. Increments the time slice using the __macro_INCREMENT_TICK macro:

```
.macro __macro_INCREMENT_TICK
    # Increment timer compare by TIME_SLICE cycles
    la t0, CLINT_MTIME        # load the mtime address
    lw t1, 0(t0)              # load mtime (LO)
    lw t2, 4(t0)              # load mtime (HI)
    li t3, TIME_SLICE
    add t3, t1, t3            # increment lower bits
                             # by TIME_SLICE cycles
    sltu t1, t3, t1          # generate carry-out
    add t2, t2, t1           # add carry to upper bits
    la t0, CLINT_MTIME_CMP
    sw t3, 0(t0)             # update mtimecmp (LO)
    sw t2, 4(t0)             # update mtimecmp (HI)
.endm
```

Listing 2.36 __macro_INCREMENT_TICK macro.

3. Switch context (swap stack pointers) using the __macro_SWITCH_CONTEXT:

```
.macro __macro_SWITCH_CONTEXT
    la t0, current_task
    lw t1, 0(t0)             # t1 holds current_task
    sll t4, t1, 3           # t4 = t1 * 8
    la t5, TCB              # t5 <- &TCB[0]
    add t5, t5, t4          # t5 <- &TCB[current_task]
    sw sp, 0(t5)            # save sp of the current
                           #   task

    # select a new task in round-robin:
    addi t1, t1, 1
    li t2, NTASKS
    bne t1, t2, 1f
    li t1, 0
1:  sw t1, 0(t0)

    sll t4, t1, 3          # t4 = t1 * 8
    la t5, TCB             # t5 <- &TCB[0]
    add t5, t5, t4         # t5 <- &TCB[current_task]
    lw sp, 0(t5)           # load sp of the current task
.endm
```

Listing 2.37 __macro_SWITCH_CONTEXT macro.

4. Restores the context of the new task from its stack using the __macro_RESTORE_
CONTEXT macro:

```
.macro    __macro_RESTORE_CONTEXT
     lw t0,        0*WORD_SIZE(sp)
     csrw mepc,  t0
     lw t0,       29*WORD_SIZE(sp)
     csrw mstatus,  t0

     lw x1,        1*WORD_SIZE(sp)
     lw x5,        2*WORD_SIZE(sp)
     lw x6,        3*WORD_SIZE(sp)
     lw x7,        4*WORD_SIZE(sp)
     lw x8,        5*WORD_SIZE(sp)
     lw x9,        6*WORD_SIZE(sp)
     lw x10,       7*WORD_SIZE(sp)
     lw x11,       8*WORD_SIZE(sp)
     lw x12,       9*WORD_SIZE(sp)
     lw x13,      10*WORD_SIZE(sp)
     lw x14,      11*WORD_SIZE(sp)
     lw x15,      12*WORD_SIZE(sp)
     lw x16,      13*WORD_SIZE(sp)
     lw x17,      14*WORD_SIZE(sp)
     lw x18,      15*WORD_SIZE(sp)
     lw x19,      16*WORD_SIZE(sp)
     lw x20,      17*WORD_SIZE(sp)
     lw x21,      18*WORD_SIZE(sp)
     lw x22,      19*WORD_SIZE(sp)
     lw x23,      20*WORD_SIZE(sp)
     lw x24,      21*WORD_SIZE(sp)
     lw x25,      22*WORD_SIZE(sp)
     lw x26,      23*WORD_SIZE(sp)
     lw x27,      24*WORD_SIZE(sp)
     lw x28,      25*WORD_SIZE(sp)
     lw x29,      26*WORD_SIZE(sp)
     lw x30,      27*WORD_SIZE(sp)
     lw x31,      28*WORD_SIZE(sp)
     addi sp,  sp,  CONTEXT_SIZE
.endm
```

Listing 2.38 __macro_RESTORE_CONTEXT macro.

2.5.6.6 Using the Environment Call (Ecall) Exception to Start the Scheduler

We have already learned that instead of directly calling the first task (Task0) from the main function, the scheduler should rely on the exception return to start the first task. For this purpose, we can use the environment call exception. In the RISC-V architecture, environment calls, often abbreviated as "ecalls," are a mechanism by which a user-level program can request services or system functions from the kernel. The ecall instruction initiates an environment call. This instruction triggers an exception, which is trapped at the exception handler on the address stored in the BASE field of the **mtvec** register. Figure 2.30 shows the process of starting and running the scheduler using the environment call exception.

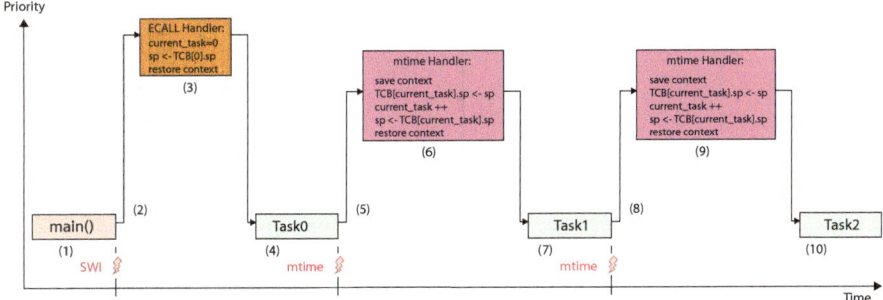

Fig. 2.30 Starting the scheduler with the SVC exception

The environment call handler simply sets the SP to point to the top of Task0's stack, initializes the first tick, enables the machine timer interrupt and restores the context of the first task:

```
.balign 4, 0
.global _exception_handler
_exception_handler:
    # Decode exception cause:
    csrr t0, mcause     # read exception cause
    bltz t0, 2f         # exit if not an exception

    # Check if ECALL:
1:  li t1, 0xB          # ecall from M-mode
    bne t1, t0, 2f

    /*
      ECALL:
      1. load the SP of the first task (SP <- TCB[0].sp)
      2. increment tick to set the first timer interrupt
      3. enable MTIME interrupt
      4. restore context of the first task
      5. Upon return, the first task is executed and
         the scheduler is running
    */
    /* 1. Load SP of the first task: */
    la t1, TCB          // load the address of TCB[0]
    lw sp, 0(t1)        // load sp from TCB[0].sp

    # 2. Increment time slice (tick)
    __macro_INCREMENT_TICK

    /* 3. Enable MTIME interrupt */
    li t0, 0x00000080
    csrs mie, t0

    # 4. Restore the context of Task0
    __macro_RESTORE_CONTEXT
2:
    mret
```

Listing 2.39 The environemt call exception handler.

Finally, here is the main function which initializes and starts the scheduler:

```
int main() {

    // Set the task functions:
    TaskFunctions[0] = Task0;
    TaskFunctions[1] = Task1;
    TaskFunctions[2] = Task2;
    TaskFunctions[3] = Task3;

    // Init scheduler:
    InitScheduler(stackRegion, TCB, TaskFunctions);
    current_task = 0;

    // Set up vectored interrupts and enable CPU's interrupts
    _register_handler(_vector_table, INT_MODE_VECTORED);
    _enable_global_interrupts();

    //Environment call - start the scheduler:
    __asm__ volatile("ecall");

    // We should never return here...
    while(1){}
    return 0;
}
```

Listing 2.40   Initializing and starting the the scheduler.

2.6   ARM 9 Exceptions and Interrupts

The ARM9 supports the following six types of interrupts and exceptions:

- Fast interrupt Request,
- Interrupt Request,
- Data and Prefetched abort exceptions,
- Undefined instruction exception, and
- Software interrupt, and
- Reset.

The interrupt instruction SWI raises the software interrupts. The software interrupts allow a program running in the user mode to request privileged operations such as OS functions. The Prefetch abort exception occurs when the CPU fetches an instruction from an illegal address. The Data abort exception occurs when a data transfer instruction attempts to load or store data at an illegal address. The Undefined instruction exception occurs when the processor cannot recognize the currently fetched instruction. The Interrupt request occurs when the processor's external interrupt request pin (IRQ) is asserted (LOW), and the interrupt mask bit (I) in the current program status register (CPSR) is cleared (interrupts enabled). The Fast interrupt request occurs when the processor's external fast interrupt request pin (FIQ) is asserted (LOW), and the interrupt mask bit (F) in the current program status register (CPSR) is cleared (fast interrupts enabled). The Reset interrupt occurs when the processor's reset pin is asserted.

2.6.1   Vector Table and Interrupt Priorities

ARM9 processors use the vectored interrupt handling method. Each interrupt/exception has its own entry in the vector table. Each entry in the vector table has only 32 bits, which is insufficient to contain the full code for a handler; hence, each entry commonly contains a branch instruction or load PC instruction to the actual handler. Table 2.6 shows the interrupt/exception, its address in the vector table, and its priority. As interrupts/exceptions can coincide, the CPU has to use a priority mechanism to handle the most important interrupt/exception. For example, the Reset interrupt has the highest priority and takes precedence over all other interrupts/exceptions. All interrupts/exceptions disable further interrupts/exceptions by setting the I bit in the CPSR register. The Reset and Fast Interrupt Request also set the F bit in the CPSR register, thus masking the Fast interrupt request. Listing 2.41 shows a typical method of implementing a vector table for ARM9 processors.

```
          .org 0x00000000
Vector_Table:
          b Reset_Handler
          b Undefined_Handler
          b SWI_Handler
          b Prefetch_Handler
          b Abort_Handler
          nop                           // never used
          b IRQ_Handler
          b FIQ_HAndler


Reset_Handler:
          <handler instructions>
Undefined_Handler:
          <handler instructions>
SWI_Handler:
          <handler instructions>
Prefetch_Handler:
          <handler instructions>
Abort_Handler:
          <handler instructions>
IRQ_Handler:
          <handler instructions>
FIQ_Handler:
          <handler instructions>
```

Listing 2.41 ARM vector table and interrupt handlers.

Listing 2.41 shows a typical method of implementing a vector table for ARM9 processors. The vector table starts at the address 0x00000000. Each entry in the vector table is 32 bits long and contains a branch instruction (B) to the interrupt handler. When, for example, a Data Abort exception occurs, the CPU stops the execution of the current running program, saves the program context, and moves the vector 0x00000010 into the program counter. This way, the b Abort_Handler instruction is fetched, and the CPU jumps to Abort_Handler.

As we already said, the Reset interrupt is the highest priority interrupt and is always taken whenever the Reset pin is asserted. The reset handler is responsible for initializing the system and other interrupt sources and setting the stack pointer.

Table 2.6 ARM9 vector table

Interrupt/exception	Vector table address	Priority (1-High, 6-Low)
Reset	0x00000000	1
Undefined instruction	0x00000004	6
Software interrupt	0x00000008	6
Prefetch abort	0x0000000C	5
Data abort	0x00000010	2
Interrupt request	0x00000018	4
Fast interrupt request	0x0000001C	3

So, the Reset interrupt masks automatically all other interrupts before their sources are initialized. Only then can the reset handler enable other interrupts. Hence, during the first few instructions of the reset handler, we should avoid SWI, undefined instructions, and memory accesses that can cause the Data and Prefetch aborts.

The Fast Interrupt Request (FIQ) occurs when a peripheral asserts the processor's FIQ pin. The peripheral device must hold the FIQ input low until the processor acknowledges the interrupt request. As a response to FIQ, the CPU disables both Interrupt and Fast Interrupt requests. Hence, no external device can interrupt the CPU unless the IRQ and FIQ interrupts are re-enabled by software. The Fast Interrupt Request reduces the execution time of the exception handler relative to a normal interrupt by removing the requirement for register saving (minimizing the overhead of context switching).

The Interrupt Request (IRQ) is a normal interrupt that occurs when a peripheral device asserts the IRQ pin. The peripheral device must hold the IRQ input pin low until the processor acknowledges the interrupt request. An IRQ has a lower priority than the FIQ and Data Abort and is masked on entry to an FIQ or Data Abort sequence. On entry to the IRQ handler, the further IRQ interrupts are disabled and should remain disabled until the current interrupt source has been acknowledged and the IRQ pin has been de-asserted.

We can notice from Table 2.6 that both Software Interrupt and Undefined Instruction have the same level of priority since they cannot occur simultaneously.

2.6.2   ARM9 Interrupt Handling

ARM9 processors are 5-stage pipelined machines with Instruction Fetch (IF), Instruction Decode (ID), Execution (EX), Memory (MEM), and Write-Back (WB) stages. In a pipelined machine, an instruction is executed step by step and is not completed for several clock cycles. An external interrupt can occur at any time during the execution of an instruction. Also, other instructions in the pipeline can raise exceptions that may force the machine to abort the instructions in the pipeline before they have been completed. One of the problems with interrupts in the pipelined CPUs is when to halt instruction in the pipeline. In the case of external interrupts, one possible solution would be to execute all fetched instructions before handling the interrupt request. However, the problem with this approach would be a long interrupt latency. The other solution would be to halt the execution of all fetched instructions and fetch them again upon returning from the interrupt handler. This way, we would have minimal interrupt latency. Obviously, this is not a good idea because some instructions, such as STORE instructions, can modify the content in memory and should not be stopped and executed again. Also, arithmetic instructions might have already changed the content of the status register (usually in the Execution stage) and should not be dismissed. The most common solution to the problem is to execute all instructions that have been issued into the execution stage. In the case of an external interrupt in ARM9, the CPU executes all instructions in the stages EX, MEM, and WB while dismissing two instructions in the stages IF and ID.

To resume, in the case of an external interrupt, the CPU has to let all instructions that were issued for execution complete and flush all succeeding instructions from the pipeline. In the case of an exception caused by an instruction, the CPU should stop executing the offending instruction, let all preceding instructions complete, and flush all succeeding instructions from the pipeline. Only then can the CPU start saving the context and fetching the instruction pointed by the interrupt vector (the first instruction in the interrupt handler).

Let us now look at how ARM9 handles the IRQ interrupts. When an IRQ interrupt occurs, the ARM 9 processor executes the three instructions issued for execution and will flush the last two fetched instructions. The last two fetched instructions are from the addresses PC (the instruction that is currently in the IF stage) and PC-4 (the instruction that is currently in the ID stage). The instruction in the EX stage is from the address PC-8. This is important to notice because the last executed instruction before entering the interrupt handler was from the address PC-8, but the program counter contains the address PC. The first instruction to execute upon returning from the interrupt handler is one that was in the ID stage when the interrupt request occurred. Hence, the instruction address that should be fetched upon returning from the interrupt handler is PC-4.

When an IRQ interrupt occurs, the ARM9 processor executes the instructions that are issued for execution. Then, the following hardware procedure is executed:

- the CPU saves the Current Program Status (CPSR) register into the Saved Program Status (SPSR) register; hence, the processor automatically saves the status of the interrupted program. The CPSR register is a special-purpose register in ARM9 processors that contains arithmetic flags and interrupt masks,
- the CPU automatically disables interrupts by setting the I bit in the CPSR register,
- the CPU saves the current program counter (PC) into the link register (LR). This way, the LR register holds the return address. It is important to note that the CPU saves the address of the last fetched instruction and does not automatically correct this value to point to the instruction that was in the ID stage when the interrupt

occurred. Hence, it is the programmer's responsibility to adjust the value in PC upon returning from the interrupt handler and

- the CPU fetches the instruction from the interrupt vector 0x00000018.

Now, the interrupt handler starts. The above procedure is hard-wired in the CPU and involves no instruction fetch and execution. When an interrupt handler has completed, it must move both the return value in the LR register minus 4 to the PC and the SPSR to the CPSR. This action restores both the PC and the CPSR and returns to the interrupted program. Listing 2.42 shows a typical return method from an IRQ interrupt handler.

```
IRQ_Handler:
        <handler instructions>
        ...
        ...
        subs pc, lr, #4       //  pc <- lr-4
```

Listing 2.42 A typical IRQ interrupt handler

Many instructions in ARM9 can have an "s" suffix. The "s" suffix ensures that when the program counter is the destination register, the CPSR register is automatically restored from the SPSR register. The same holds for the subs instruction in Listing 2.42. Hence, the instruction subs pc,lr,#4 firstly saves the LR-4 into the program counter (remember that the programmer is responsible for correctly restoring the return address into the program counter upon returning from the handler) and then restores CPSR from SPSR.

It is important to stress that not all interrupt/exception handlers use the same instruction to return. For example, the Data abort exception occurs in the MEM stage. Hence, only the instruction in the WB stage is executed, while the instructions from the IF, ID, and EX stages are flushed. When the Data abort exception occurs, the instruction in the EX stage is from the address PC-8. Thus, the Data abort handler uses subs pc,lr,#8 to return:

```
IRQ_Handler:
        <handler instructions>
        ...
        ...
        subs pc, lr, #8       //  pc <- lr-8   !!!!
```

Listing 2.43 A typical Data abort exception handler

2.6.3 Interrupt Handlers in C

Interrupt handlers can be written in an assembly language or a high-level language like C. Usually, we want to avoid the assembly language as much as possible and program in our favorite high-level language. Remember that the CPU executes an interrupt handler directly, and the protocol for executing an interrupt handler differs from calling and executing a standard C function. Most importantly, an ISR has to end with some "interrupt return" opcode, whereas usual C functions end with ordinary "return" opcode. We have seen previously that the ARM interrupt handlers should return with SUBS opcode, which is used to restore the PC from LR-4 and CPSR from SPSR. In the case of an ordinary subroutine, the return opcode for ARM would be MOV PC, LR (restores PC from LR). A programmer could be tempted to write an interrupt handler like this:

```
/* How NOT to write an interrupt handler */
void my_interrupt_handler(void)
{
    /* do something */
}
```

Listing 2.44 How not to write an interrupt handler.

This simply cannot work. The compiler needs to understand that this is to be an interrupt handler and that the SUBS PC,LR,#4 instruction should be the last instruction used to return. The compiler will simply use the MOV PC, LR instruction to return.

Some compilers, such as GCC, Clang, and ARMCC, to name a few, have directives like #pragma or special function attributes, allowing you to declare a routine interrupt. For example, the *interrupt* function attribute in GCC indicates that the specified function is an interrupt handler. The compiler then generates function entry and exit sequences suitable for use in an interrupt handler when this attribute is present.

The correct (GCC) way of implementing an interrupt handler in C is:

```
/* GCC style interrupt handler */
__attribute__((interrupt)) void my_interrupt_handler()
{
    /* do something */
}
```

Listing 2.45 GCC style interrupt handler.

The ARMCC compiler offers the __irq function declaration keyword to write C interrupt handlers. The __irq keyword preserves all registers used by the interrupt handler and exits the handler by setting the PC to (LR–4) and restoring the CPSR to its original value from SPSR. Also, if the kernel calls a subroutine, __irq preserves the link register (LR), which is corrupted by the subroutine call.

```
/* ARMCC style interrupt handler */
__irq void my_interrupt_handler()
{
    /* do something */
}
```

Listing 2.46 ARMCC style interrupt handler.

However, not only the directive or function qualifier designates the interrupt handlers. Compilers often require that the handler declaration contains a special function argument specifying the kind of interrupt (for example, IRQ or Abort). The compiler uses this special argument to restore the PC from LR correctly (for example, LR-4 for IRQ or LR-8 for Data abort). All these attributes and arguments defined and used by a particular compiler prevent the handler code from being portable.

2.7 Intel Interrupts

Intel processors have two external pins for external interrupts:

- INTR pin—it is used to signal for normal (maskable) interrupts.
- NMI pin—it is used to signal nonmaskable interrupts.

Besides interrupts, Intel processors can detect exceptions from two sources:

- Processor exception—triggered from the processor as a result of some exceptional conditions within the processor (e.g., divide by zero). These exceptions are further classified as faults, traps, and aborts.
- Software interrupts—triggered with the processor instruction INT.

Exceptions are classified as:

- Faults are either detected before the instruction begins to execute or during the execution of the instruction. A fault is an exception that can generally be corrected and that, once corrected, allows the program to be restarted with no loss of continuity. The return address for the fault handler points to the faulting instruction rather than to the instruction following the faulting instruction.
- A trap is an exception reported immediately following the execution of the instruction INT. Traps allow the execution of a program or task to be continued without losing program continuity. The return address for the trap handler points to the instruction to be executed after the trapping instruction.
- An abort is an exception that does not allow a restart of the program or task that caused the exception. Aborts are used to report severe errors.

The Intel processor services interrupts and exceptions only between the end of one instruction and the beginning of the next. This is referred to as the instruction boundary. Certain conditions and flag settings cause the processor to inhibit certain interrupts and exceptions at instruction boundaries. The IF (interrupt-enable flag) bit in the FLAGS register (this is the status register in Intel x86 microprocessors that contains the current state of the processor) controls the acceptance of external interrupts signaled via the INTR pin. When IF = 0, INTR interrupts are masked; when IF = 1, INTR interrupts are enabled. The Intel processor instructions CLI (Clear Interrupt-Enable Flag) and STI (Set Interrupt-Enable Flag) are used to clear/set the IF flag.

Table 2.7 Intel Exceptions and Interrupts. Only a few exceptions and interrupts are shown

Vector number	Description	Type
0	Division by zero	Fault
1	Debug	Fault
2	NMI	Interrupt
3	Breakpoint	Trap
...	...	...
	...	...
14	Page Fault	Fault
...	...	...
32–255	External interrupts on INTR	Interrupt

If more than one interrupt or exception is pending at an instruction boundary, the processor services one of them at a time according to their priority. In general, aborts have the highest priority, followed by traps, NMI, and INTR. The faults have the lowest priority.

Each architecturally defined exception and interrupt in Intel processors is assigned a unique identification number, called a **vector number**. The processor uses the vector number assigned to an interrupt as an index in the interrupt vector table. The allowable range for vector numbers is 0–255. The Intel architecture reserves vector numbers in the range 0 through 31 for architecture-defined exceptions and interrupts. Vector numbers in the range 32–255 are designated as user-defined interrupts and are assigned to external I/O devices to enable those devices to send interrupts. One characteristic of Intel processors, distinguishing them from ARM processors is that the peripheral device that caused an interrupt must provide the vector number to the CPU. Table 2.7 shows vector number assignments and exception types for architecturally defined exceptions and interrupts.

In the older Intel processors (before 80386), the interrupt table is called IVT (interrupt vector table). The IVT is an array of 32-bit interrupt vectors stored consecutively in memory and indexed by an interrupt vector. The IVT permanently resides at the same location in memory, ranging from 0x0000 to 0x03ff, and consists of 256 four-byte interrupt vectors (i.e., pointers to the interrupt/exception handlers). When responding to an exception or interrupt, the processor multiplies the vector number by four to form the address of the entry in the IVT.

In modern Intel processors, the interrupt table is called IDT (interrupt descriptor table). The IDT is an array of 8-byte descriptors stored consecutively in memory and indexed by an interrupt vector. Each descriptor contains information describing how to access the interrupt/exception handler. The IDT may reside anywhere in physical memory. The processor has a special register (IDTR) to store both the physical base address and the length in bytes of the IDT. When an interrupt occurs, the processor multiplies the interrupt vector by eight and adds the result to the IDT base address. With the help of the IDT length, the resulting memory address is then verified to be within the table; if it is too large, an exception is generated. If everything is okay, the 8-byte descriptor stored at the calculated memory location is loaded, and actions are taken according to the descriptor's contents. As said, the interrupt descriptor table (IDT) associates each vector number with a descriptor for the instructions that service the associated event. Because there are only 256 vector numbers, the IDT contains up to 256 descriptors. It can contain fewer than 256 entries; entries are required only for vector numbers that are actually used.

The interrupt handling procedure in the Intel processor is rather complicated. Here, we omit all the details and give only the basic concepts. When responding to an exception or interrupt, the processor first saves the current state of the interrupted program or task (the status FLAGS register and the program counter) on the stack. Each entry in the IDT (or IVT) holds the start address of the interrupt handler. The processor thus reads the start address of the handler from the IDT (or IVT) into the program counter and starts the handler's execution. To return from an exception- or interrupt-handler handler, the handler uses the IRET instruction. The IRET instruction is similar to the RET instruction used to return from normal procedures, except it restores the saved status register FLAGS.

2.8   Interrupt Controllers

We have seen that the interrupt line from a peripheral device should be connected to the CPU's interrupt input signal. In such a way, a peripheral device can interrupt the CPU and require its attention. The CPU will sense this interrupt input signal at every instruction fetch and know that the peripheral device needs attention. But what should we do if there is more than one peripheral device that would like to interrupt the CPU? What if there are tens of external peripheral devices, which is often the case in real computer systems? Should we add an interrupt input pin to the CPU for every external peripheral device? A large number of interrupt input pins on the CPU for every external device would complicate the CPU interfacing and increase the error probability.

One possible solution to solve this problem would be to have one level-sensitive interrupt input pin (IRQ) on the CPU shared by all external peripheral devices. This solution is illustrated in Fig. 2.31. Whenever an interrupt is asserted, the CPU branches to the interrupt handler associated with the IRQ pin. This interrupt handler would poll each and every I/O peripheral device to determine which device asserted the interrupt line. So, the CPU will handle the interrupt request on the IRQ pin as vectored interrupts but will also, within the interrupt handler, use the polled interrupts method to check the interrupt's cause. Every modern I/O peripheral device has an addressable (memory-mapped) status register. There is usually one bit in this status register, referred to as an interrupt-pending bit, which is set internally by the interrupting device when the device asserts the interrupt line. The CPU can read this interrupt-pending bit in the status register to determine the interrupting device. The processor branches to a specific device-service routine if the interrupt-pending bit is set. The interrupt handling bit is cleared within this device-specific routine, and the

Fig. 2.31 Example system with several I/O devices sharing one interrupt input signal

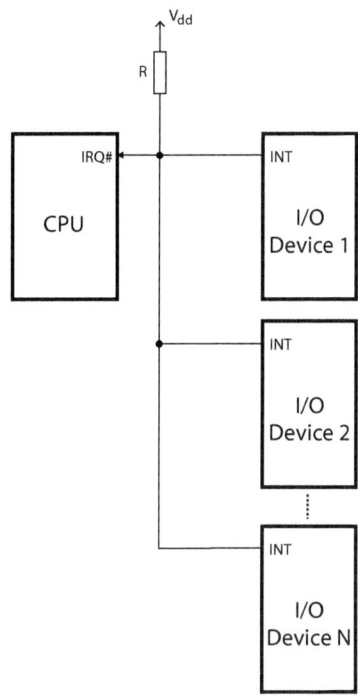

interrupt request is serviced. Listing 2.47 shows pseudocode for the IRQ interrupt handler that uses polling to determine which device has requested the interrupt. The downside to this technique is that it is time-consuming.

```
/*
 * Polling Handler
 */
__attribute__((interrupt)) void polling_IRQ_handler () {

    /* Check the interrupt pending bit in the I/O device 1 */
    if (IO1_status_reg & (1<<INT_PEND_BIT)) {
    /* I/O Device 1 Code */
    IO1_status_reg &= ~(1<<INT_PEND_BIT); // clear int pending bit

    /* Do something */

    }

    /* Check the interrupt pending bit in the I/O device 2 */
    if (IO2_status_reg & (1<<INT_PEND_BIT)) {
    /* I/O Device 2 Code */
    IO2_status_reg &= ~(1<<INT_PEND_BIT); // clear int pending bit

    /* Do something */

    }

    ...
```

```
25    /* Check the interrupt pending bit in the I/O device N */
27    if (ION_status_reg & (1<<INT_PEND_BIT)) {
      /* I/O Device N Code */
29    ION_status_reg &= ~(1<<INT_PEND_BIT); // clear int pending bit

31    /* Do something */

33    }
    }
```

Listing 2.47 IRQ interrupt handler with polling.

A better solution to this problem would be to use a special device—an **interrupt controller**. The interrupt controller is a special device which:

- combines all external interrupt requests onto one CPU IRQ line,
- prioritizes them (decides which interrupt request will be routed to the CPU when more than one I/O device has requested the interrupt),
- routes the selected interrupt request to the CPU's IRQ input signal and
- most importantly, it provides the CPU with the information on which device has requested the interrupt. Commonly, it provides the CPU with the interrupt vector; thus, the CPU does not have to poll I/O peripheral devices.

Figure 2.32 illustrates the structure of a system that uses an interrupt controller. All potential external interrupt sources are routed through the interrupt controller. In the case of one or more interrupt requests, the interrupt controller prioritizes the interrupt inputs. It then transfers the interrupt request with the highest priority to the CPU, along with the interrupt vector. This sequence is performed by hardware in the interrupt controller and not by software in the CPU. Hence, the interrupt controller provides a much faster response to an interrupt request.

In summary, interrupt controllers are essential in modern computer systems as they facilitate the handling of diverse interrupt requests generated by various hardware components. Interrupt controllers optimize CPU resources by allowing the CPU to respond to events as they occur (in contrast to continuous polling of hardware), ensuring that the processor only performs work when necessary. In systems where multiple interrupts can occur simultaneously, interrupt controllers manage the nesting of interrupts. They ensure that an interrupt can be interrupted by a higher-priority one while maintaining the correct execution order. They also manage interrupt priorities, ensure timely response to critical events, and optimize system resources by allowing the CPU to handle events as they occur, all of which are fundamental for efficient and responsive system operation.

Although the operations in an interrupt controller are performed by hardware, interrupt controllers are programmable. It means that they typically have a common set of addressable (memory-mapped) registers, which enable the system programmer to set the priorities and interrupt vectors for each interrupt source before the interrupt controller is used. The following sections will cover a few real-world interrupt controllers used with ARM and Intel processors.

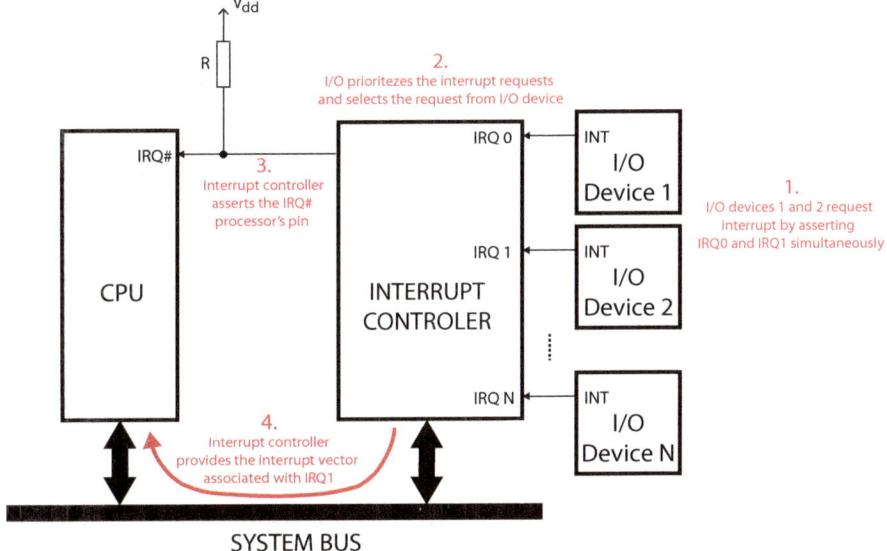

Fig. 2.32 A system with an interrupt controller

2.8.1   ARM Advanced Interrupt Controller

The Advanced Interrupt Controller (AIC) is an 8-level priority vectored interrupt controller, providing handling of up to thirty-two interrupt sources. It is used with ARM9 processors. Figure 2.33 illustrates the block diagram of an ARM9-based system with AIC. The AIC drives the FIQ# (fast interrupt request) and the IRQ# (standard interrupt request) inputs of an ARM9 processor. Inputs of the AIC are external interrupts coming from the peripheral I/O devices.

The Interrupt Source 0 (IS 0) is always connected to the FIQ processor's input. The interrupt sources 1 to 31 (IS 1 to IS 31) can be connected to the interrupt outputs of peripheral devices. An 8-level priority controller drives the IRQ line of the processor. Each interrupt source has a programmable priority level of 7 (the highest priority) to 0 (the lowest priority).

As soon as an interrupt request occurs on an interrupt source, the IRQ# line is asserted. If several interrupt sources have asserted the interrupt request, the priority controller determines the interrupt source with the highest priority, which will be serviced. If several interrupt sources of equal priority are pending, the interrupt with the lowest interrupt source number is serviced first. If an interrupt request happens during the interrupt service in progress, it is delayed until the software indicates to the AIC the end of the current service. Figure 2.33 illustrates the simplified internal structure of AIC. AIC employs an interrupt vectoring scheme. The interrupt handler addresses (interrupt vectors) corresponding to each interrupt source can be stored in the AIC's registers SVR1 to SVR31 (Source Vector Register 1–31). When one or more interrupt requests occur, the content of the SVR corresponding to the interrupt source with the highest priority is automatically transferred to the Interrupt Vector

Fig. 2.33 Simplified internal structure of AIC

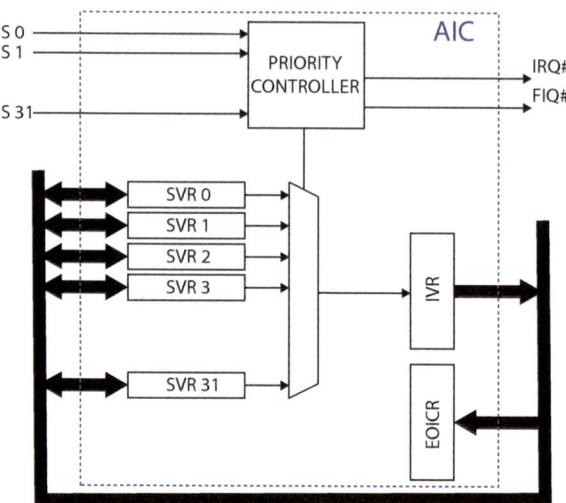

Register (IVR). To obtain the start address of the interrupt handler, the CPU must read the IVR register. In the ARM9-based systems, the IVR register is always mapped at the absolute address 0xFFFFF100. Remember that the interrupt vector for IRQ interrupt is 0x00000018. Hence, the IVR is accessible from the ARM interrupt vector at address 0x00000018 through the following instruction:

```
ldr pc,[pc,#-0XF20]
```

When the processor executes this instruction, it loads the value in IVR into its program counter, thus branching the execution on the correct interrupt handler. Besides, reading the IVR also de-asserts the IRQ# line on the processor. But from where does the value -0xF20 come in the above instruction? Recall that the instructions are executed in the EX stage. By the time the above instruction is issued into the EX stage, the PC has already been increased by 8 and is equal to 0x00000020. This is because the CPU has fetched two more instructions. Hence, we have to subtract 0x2F0 from 0x00000020 to obtain 0xFFFF F100.

Before returning, the interrupt handler must indicate to the AIC the end of the current service by dummy writing to the EOICR register (End Of Interrupt Command Register). This will re-enable the further interrupts in AIC. The return from the interrupt handler is, as we have already learned, performed by the subs pc,lr,#4 instruction. This has the effect of returning from the interrupt to whatever was being executed before and of restoring the CPSR from the SPSR.

An example of the procedure for obtaining the interrupt vector is in Fig. 2.34. Let us assume a peripheral device asserts the interrupt request at the IS 1 line of AIC (step 1). Assuming that no other IS line has been asserted and that the CPU services no interrupt, the priority controller in AIC immediately asserts the CPU's IRQ# signal (step 1). Then, the priority controller selects the SVR1 register, and its

Fig. 2.34 Simplified internal operation of AIC

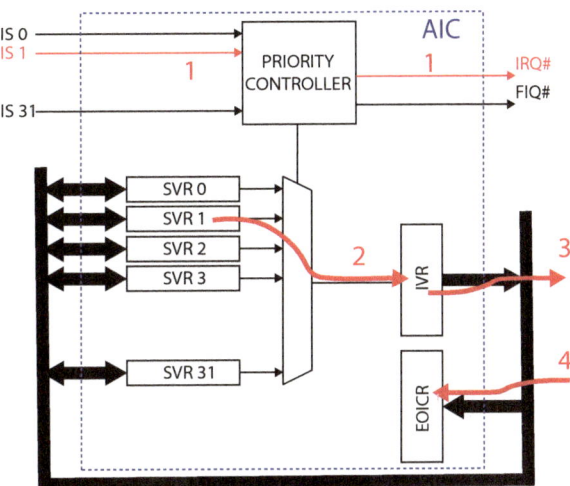

content is transferred into the IVR register (step 2). The CPU detects that IRQ# has been asserted, stops the instruction execution, and saves the context of the interrupted program. It then fetches the instruction `ldr pc,[pc,#-0XF20]` from the IRQ interrupt vector (0x00000018). This instruction moves the content of the IVR register into the program counter (step 3) and CPU branches on the IRQ handler. Before returning from the IRQ handler, the CPU dummy writes into the EOICR (step 4).

As we have seen, when AIC is used to route external interrupt requests from peripheral devices to the CPU, the instruction at the interrupt vector 0x00000018 is not a branch instruction (B) to the interrupt handler, but the instruction that loads the IVR into PC (which also acts as a branch). The same holds for the FIQ vector. Hence, accordingly, we should change the interrupt vector table from Listing 2.41. Also, the interrupt handlers for interrupt sources IS1 to IS31 should dummy write to EOICR before returning. Listing 2.48 shows the updated interrupt vector table and pseudocode for an ISx interrupt handler.

```
            .org 0x00000000
Vector_Table:
            b Reset_Handler
            b Undefined_Handler
            b SWI_Handler
            b Prefetch_Handler
            b Abort_Handler
            nop
            lr pc, [pc, #-0xF20]      // load IVR into PC
            lr pc, [pc, #-0xF20]      // load IVR into PC


ISx_Handler:
            <handler instructions>
            ...
            <write to EOICR>
            subs pc,lr,#4
```

Listing 2.48 ARM vector table and ISx handler when AIC is present in the system.

2.8.2 RISC-V Platform-Level Interrupt Controller in FE310

In Sect. 2.5.3, we learned that SiFive FE310 SoC contains two interrupt controllers: The Core Local Interruptor (CLINT) and the Platform Level Interrupt Controller. The Core Local Interruptor (CLINT) is a mandatory component in RISC-V-based systems, which provides two local interrupts (software and timer) to the RISC-V core. The PLIC is another interrupt controller in the SiFive FE310s. It is responsible for managing global interrupts from various IO devices in the system and distributing them to the RISC-V through the Machine External Interrupt line.

The FE310 SoC has multiple peripherals (timers, GPIO pins, UARTs, etc.) that can generate interrupts. These peripheral devices generate (drive) 52 interrupt sources. The PLIC aggregates these interrupt sources and generates the interrupt request over the Machine External Interrupt line. Table 2.8 lists peripheral devices and associated interrupt sources. For example, each GPIO pin can generate one interrupt source; hence, the GPIO interface generates 32 interrupt sources.

The PLIC supports multiple priority levels for interrupts, allowing us to prioritize critical events over less critical ones. Priority levels are configurable. If two or more interrupt sources generate interrupt requests, the PLIC will select the source with the highest priority level. Each PLIC interrupt source can be assigned a priority by writing to its 32-bit memory-mapped priority register **priority**. The memory addresses of 52 **priority** registers are `0x0C000000 + 4 x SourceID`. For example, the address of the UART0's **priority** register is `0x0C00000C`. The FE310-G003 supports seven (7) levels of priority. A priority value of 0 means "never interrupt" and disables the interrupt for the source. Priority 1 is the lowest active priority, and priority 7 is the highest. Besides, global interrupts with the lowest source ID have the highest priority. In such a way, if two or more global interrupts with the same priority level are triggered, the PLIC will service first the one with the lowest source ID.

Table 2.8 Peripheral devices and their associated interrupt sources in FE310 PLIC

Device	Interrupt source IDs
WDT	1
RTC	2
UART0	3
UART1	4
QSPI0	5
SPI1	6
SPI2	7
GPIO	8–39
PWM0	40–43
PWM1	44–47
PWM2	48–51
I2C	52

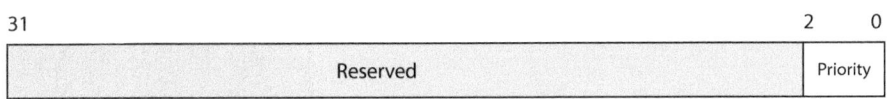

Fig. 2.35 The **priority** register

The **priority** register is depicted in Fig. 2.35. The three least significant bits in the **priority** encode the priority level.

Besides priority levels, PLIC enables Per-Source Interrupt Control. Each global interrupt source connected to the PLIC can be individually enabled or disabled by setting the corresponding bit in two registers: **enable1** and **enable2**. This feature allows fine-grained control over which sources can generate interrupts. The **enable1** and **enable2** are memory-mapped and can be accessed as a contiguous array of two memory words at addresses 0x0C002000 (**enable1**) and 0x0C002004 (**enable2**). The enable bit for interrupt source ID is stored in the bit (ID mod 32) of the word (ID/32). For example, the enable bit of the interrupt source 3 (UART0) is stored in the bit (3 mod 32)=3 of the word (3/32)=0, which is accessible as the register **enable1**. Similarly, the enable bit of the interrupt source 39 (GPIO pin 31) is stored in the bit (39 mod 32)=7 of the word (39/32)=1, which is accessible as the register **enable2**. Bit 0 of **enable1** represents the non-existent interrupt source ID 0 and is hardwired to 0.

When one or more interrupt sources trigger the interrupt request to PLIC, PLIC will select the interrupt source with the highest priority and trigger the interrupt on the Machine External Interrupt line of the RISC-V CPU. At the same time, PLIC will write the ID of the highest priority interrupt source into its 32-bit **claim** register, memory-mapped at 0x0C200004. The RISC-V will execute the Machine Internal Interrupt handler. This handler should then read the **claim** register. This read will return the ID of the highest-priority pending interrupt or zero if there is no pending interrupt. In such a way, the CPU will recognize which interrupt source has triggered the interrupt request. This step informs the PLIC that we're handling the interrupt and prevents it from reasserting the same interrupt while we're servicing it. After appropriately servicing the interrupt source, the Machine Internal Interrupt handler should write the interrupt ID it received from the **claim** register back to the **claim** register.

2.8.2.1   Implementing PLIC Vector Table and Handlers

Here, we will try to provide a complete code example for using the Platform-Level Interrupt Controller (PLIC) in the SiFive FE310 microcontroller. The code snippets in this subsection will hopefully demonstrate to you how to set up the vector table for PLIC interrupt sources, initialize the PLIC, handle a specific interrupt source, and acknowledge (complete) the interrupt.

Implementing a vector table, interrupt handlers and basic routines for the Platform-Level Interrupt Controller (PLIC) in the SiFive FE310 microcontroller in assembly and C language involves defining the vector table, writing assembly code for each interrupt handler, and writing other routines for PLIC in C. Below, we provide a step-by-step guide to implement this:

1. **Define the Vector Table**. In assembly, we define the interrupt vector table for PLIC as a table (an array) of jump instructions to interrupt handlers. Each jump instruction in the vector table corresponds to a specific interrupt source. We can place the vector table at an arbitrary memory location, providing it is correctly aligned:

```
# -------------------------------------------------
#
#   P L I C    V E C T O R     T A B L E
#
# -------------------------------------------------
.balign 4
.global _plic_ext_vector_table
_plic_ext_vector_table:
        j _panic_handler                # PLIC src 0
        j _aon_wdt_handler              # PLIC src 1
        j _aon_rtc_handler              # PLIC src 2
        j _uart0_handler                # PLIC src 3
        j _uart1_handler                # PLIC src 4
        j _qspi0_handler                # PLIC src 5
        j _spi1_handler                 # PLIC src 6
        j _spi2_handler                 # PLIC src 7
        j _gpio0_handler                # PLIC src 8
        j _gpio1_handler                # PLIC src 9
        j _gpio2_handler                # PLIC src 10
        j _gpio3_handler                # PLIC src 11
        j _gpio4_handler                # PLIC src 12
        j _gpio5_handler                # PLIC src 13
        j _gpio6_handler                # PLIC src 14
        j _gpio7_handler                # PLIC src 15
        j _gpio8_handler                # PLIC src 16
        j _gpio9_handler                # PLIC src 17
        j _gpio10_handler               # PLIC src 18
        j _gpio11_handler               # PLIC src 19
        j _gpio12_handler               # PLIC src 20
        j _gpio13_handler               # PLIC src 21
        j _gpio14_handler               # PLIC src 22
        j _gpio15_handler               # PLIC src 23
        j _gpio16_handler               # PLIC src 24
        j _gpio17_handler               # PLIC src 25
        j _gpio18_handler               # PLIC src 26
        j _gpio19_handler               # PLIC src 27
        j _gpio20_handler               # PLIC src 28
        j _gpio21_handler               # PLIC src 29
        j _gpio22_handler               # PLIC src 30
        j _gpio23_handler               # PLIC src 31
        j _gpio24_handler               # PLIC src 32
        j _gpio25_handler               # PLIC src 33
        j _gpio26_handler               # PLIC src 34
        j _gpio27_handler               # PLIC src 35
        j _gpio28_handler               # PLIC src 36
        j _gpio29_handler               # PLIC src 37
        j _gpio30_handler               # PLIC src 38
        j _gpio31_handler               # PLIC src 39
        j _pwm0_handler                 # PLIC src 40
        j _pwm0_handler                 # PLIC src 41
        j _pwm0_handler                 # PLIC src 42
        j _pwm0_handler                 # PLIC src 43
        j _pwm1_handler                 # PLIC src 44
        j _pwm1_handler                 # PLIC src 45
        j _pwm1_handler                 # PLIC src 46
        j _pwm1_handler                 # PLIC src 47
```

```
57    j _pwm2_handler              # PLIC src 48
58    j _pwm2_handler              # PLIC src 49
59    j _pwm2_handler              # PLIC src 50
60    j _pwm2_handler              # PLIC src 51
61    j _i2c_handler               # PLIC src 52
```

Listing 2.49 The PLIC interrupt vector table.

2. **Define Interrupt Handler Routines**. Write the assembly code for each interrupt handler. These routines should handle the specific interrupt and include any necessary operations. Here, we provide only the the basic code for GPIO13 interrupt handler:

```
1  .balign 4
2  .weak _gpio13_handler
3  _gpio13_handler:
4      # Your code goes here:
5      ...
6      ret
```

Listing 2.50 Assembly code for the GPIO13 (PLIC source 21) interrupt handler.

3. **Write the Machine External Interrupt handler**. This handler is invoked when the external interrupt is asserted:

```
1  /*--------------------------------------
2      Machine External Interrupt Handler
3  --------------------------------------*/
4  .balign 4
5  .global _mext_interrupt_handler
6  .type _mext_interrupt_handler, @function
7  _mext_interrupt_handler:
8      # Prologue : save 16 ABI caller registers
9      ...
10
11     # Decode interrupt cause:
12     csrr t0, mcause       # read exception cause
13     bgez t0, 1f           # exit if not an interrupt
14
15     # Claim the interrupt - read CLAIM
16     #   A non-zero read contains the ID of
17     #   the highest pending interrupt.
18     la t0, PLIC_CLAIM     # load the address of CLAIM reg
19     lw t1, 0(t0)          # read CLAIM
20     slli t2, t1, 2        # id*4 to obtain the offset
21
22     # load the address of the PLIC
23     #   external interrupt vector table
24     la t3, _plic_ext_vector_table
25     add t3, t3, t2        # ext_vector_table + 4*id
26     jalr t3               # call interrupt handler
27
28  1:
29     # epilogue: restore ABI caller regs
30     ...
31
32     mret
```

Listing 2.51 Assembly code for the machine external interrupt handler.

The machine external interrupt handler:

1. decodes the interrupt cause (same as in the machine time handler),
2. reads the interrupt source ID from the **claim** register in PLIC,
3. calculates the address of the interrupt handler by adding 4xID to the base address of the PLIC vector table, and
4. calls the interrupt handler.

4. **Set PLIC priorities**. Write the C function to set the interrupt priorities if needed:

```
#define PLIC_INT_PRIORITY_BASE        0x0C000000

/* Set interrupt priority
 *
 * Interrupt source id: 1-52
 *  * Interrupt priority levels 7
 *  * Bits 2:0
 *  * 0 - never interrupt/disables interrupt
 *  * 1 - lowest active priority
 *  * 7 - highest priority */

void plic_set_priority(unsigned int source, unsigned int priority){

    *((unsigned int *) PLIC_INT_PRIORITY_BASE + source) = priority;
}
```

Listing 2.52 C function for setting PLIC interrupt priority for a given source.

5. **Enable PLIC source**. Write the C function to enable the specific interrupt source:

```
#define PLIC_INT_ENABLE1              0x0C002000

/*
 * Enable interrupt source in enable registers
 */
void plic_enable_source(unsigned int source){
    unsigned int bit_position = source | 32;
    unsigned int enable_reg = source / 32;

    *((unsigned int *) PLIC_INT_ENABLE1 + enable_reg) |= (1 << ↵
        bit_position);
}
```

Listing 2.53 C function for enabling a PLIC interrupt source.

2.8.3 ARM Cortex-M Nested Vectored Interrupt Controller

The Nested Vectored Interrupt Controller (NVIC) is a crucial component of ARM Cortex-M microcontrollers, including the Cortex-M7. It serves as the central hub for managing and controlling interrupts and exceptions in these processors. Figure 2.36 shows the relation between the NVIC unit, the Processor Core, and the peripherals. The NVIC supports up to 240 interrupts (IRQ1 to IRQ240), each with up to 256

Fig. 2.36 The NVIC
controller in the Cortex-M7
core

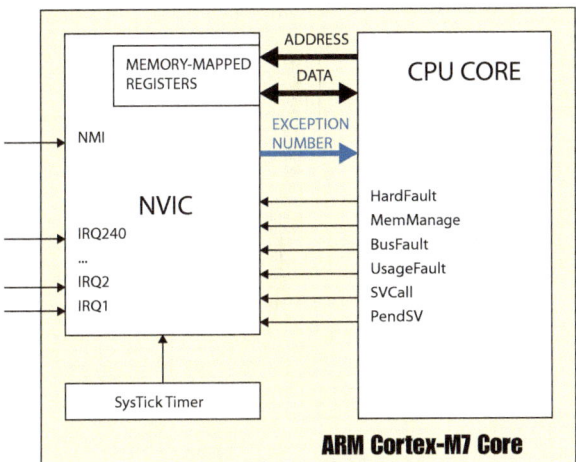

priority levels (0–255), with a higher level corresponding to a lower priority. The interrupts/exceptions can originate from various sources, such as external peripherals, internal hardware, or system events. The NVIC manages the prioritization of interrupts, allowing the system to handle multiple interrupt requests simultaneously and determine the order in which these interrupts are serviced based on their assigned priority levels.

One of the unique features of the NVIC is its ability to handle nested interrupts. It allows higher-priority interrupts to preempt the processing of lower-priority interrupts, maintaining the integrity of the system's operation. Moreover, it provides control over enabling and disabling interrupts, allowing the software to manage which interrupt sources are active or inactive. This capability is crucial for managing critical sections of code and ensuring the system's responsiveness.

The NVIC manages the routing of interrupts, determining which interrupt handler (function) should be executed when a specific interrupt occurs. The processor knows where exception handlers are located in memory thanks to exception vectors (i.e., addresses in memory of exception handlers) inside the vector table. The NVIC is responsible for sending exception numbers, which are used as indices of the exception vectors in the vector table. It enables the CPU to find the interrupt vector and immediately jump to the associated interrupt handler rather than polling interrupt sources to determine which one requested the interrupt. By allowing the CPU to respond to events as they occur (rather than continuously polling the hardware), the NVIC optimizes CPU resources and reduces power consumption.

The processor core interacts with the NVIC through a set of memory-mapped registers, which provide control over enabling and disabling interrupts, allowing the software to manage which interrupt sources are active or inactive.

Developers interact with the NVIC through the CMSIS (Cortex Microcontroller Software Interface Standard). CMSIS is a software interface that allows programmers to configure interrupts and manage interrupt handling in an efficient and standardized manner. For instance, using the CMSIS software interface, developers can easily assign different priorities to interrupts, enable or disable interrupts, and configure interrupt vectors.

As mentioned, NVIC comprises several registers that facilitate interrupt configuration, prioritization, and handling. Below are some key registers commonly found in the NVIC of ARM Cortex-M microcontrollers:

1. Eight **Interrupt Set Enable Registers (NVIC_ISER0 - NVIC_ISER7)**, which enable interrupts and show which interrupts are enabled. Each bit in these registers corresponds to a specific interrupt source, allowing individual interrupt control. If a pending interrupt is enabled, the NVIC activates the interrupt based on its priority. If an interrupt is not enabled, asserting its interrupt signal changes the interrupt state to pending, but the NVIC never activates the interrupt, regardless of its priority.
2. Eight **Interrupt Clear Enable Registers (NVIC_ICER0 - NVIC_ICER7)** registers disable interrupts and show which interrupts are disabled. Each bit in these registers corresponds to a specific interrupt source, allowing individual interrupt control.
3. Eight **Interrupt Set Pending Registers (NVIC_ISPR0 - NVIC_ISPR7)** registers force interrupts into the pending state and show which interrupts are pending. They control whether an interrupt is marked as pending or not. Each bit in these registers corresponds to a specific interrupt source, allowing individual interrupt control.
4. Eight **Interrupt Clear Pending Registers (NVIC_ICPR0 - NCVIC_ICPR7)** registers remove the pending status of an interrupt. Each bit in these registers corresponds to a specific interrupt source, allowing individual interrupt control.
5. Eight **Interrupt Active Bit Registers (NVIC_IABR0 - NVIC_IABR7)** registers indicate which interrupts are active. A bit is read as one if the status of the corresponding interrupt is active or active and pending.
6. 60 **Interrupt Priority Registers (NVIC_IPR0 - NVIC_IPR59)** registers provide an 8-bit priority field for each interrupt. These registers are byte-accessible, and there is a total of 240 8-bit priority fields in 60 IPRs.

Table 2.9 lists the memory-mapped NVIC registers and their addresses. The availability and configuration of these registers may vary slightly across different Cortex-M microcontroller variants, so the specific functionalities and register names may differ in certain models.

2.8.3.1 Interrupt Priority Levels

In Cortex-M7 cores, the priority of each interrupt is defined through the corresponding 8-bit Priority field in the IPR register. This 8-bit field allows up to 255 different priority levels. However, in practice, only the four upper bits of this field are used to decrease the complexity of NVIC and lower the power consumption. Figure 2.37 shows how the content of IPR is interpreted. This means that we have only sixteen maximum priority levels. The lower this number is, the higher the priority is. The

Table 2.9 NVIC registers

Address	Name	Description
0xE000E100-0xE000E11C	NVIC_ISER0 - NVIC_ISER7	Interrupt Set-Enable Registers
0xE000E180-0xE000E19C	NVIC_ICER0 - NVIC_ICER7	Interrupt Clear-Enable Registers
0xE000E200-0xE000E21C	NVIC_ISPR0 - NVIC_ISPR7	Interrupt Set-Pending Registers
0xE000E280-0xE000E29C	NVIC_ICPR0 - NVIC_ICPR7	Interrupt Clear-Pending Registers
0xE000E300-0xE000E31C	NVIC_IABR0 - NVIC_IABR7	Interrupt Active Bit Register
0xE000E400-0xE000E4EC	NVIC_IPR0 - NVIC_IPR59	Interrupt Priority Register

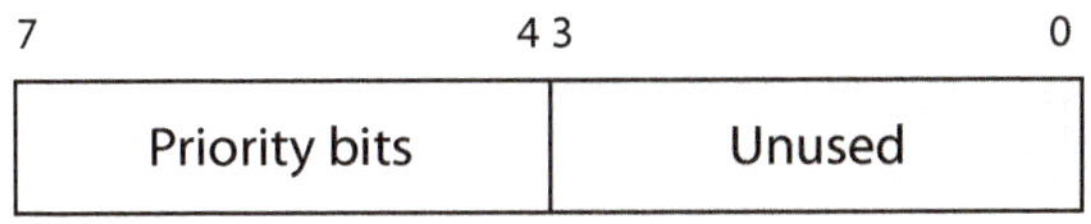

Fig. 2.37 The 8-bit priority field in an interrupt priority register

four priority bits can be further logically subdivided into two parts: a series of bits defining the **preemption priority** and a series of bits defining the **sub-priority**. The preemption priority level rules the preemption priorities between exceptions. If an exception with a priority higher than another one fires, it will preempt the execution of the lower-priority exception. The sub-priority determines which exception handler will be executed first in case of multiple pending exceptions with the same preempt priority, and it will not act on preemption. The way the 4-bit priority field is logically subdivided is called a **priority grouping** and is defined in the ARM Cortex-M7 System Control Block. Once defined, a priority grouping is common to all interrupts used in the system.

Figure 2.38 shows five possible priority groupings in ARM Cortex-M7 processors. In each priority grouping scheme, the most significant bits within the overall priority level represent the preemption priority, which determines the priority between different exceptions and their ability to preempt each other. The least significant bits within the priority level represent the sub-priority. They manage the order of handling interrupts with the same preemption priority, allowing fine-grained control over which interrupt is serviced first among those with the same preempt bits. The choice of grouping determines the balance between high-priority preemption and the finer-grained management of interrupts with the same high-priority level. For instance, a system configured with three preemption priority bits and one sub-priority bit (Priority Grouping 3) allows for eight levels of preemption and two sub-priority levels within each preemption level. The priority grouping concept provides a means to tailor the interrupt handling scheme to the specific requirements of an application, allowing for more precise control over the order in which interrupts are processed and handled within the Cortex-M7 architecture.

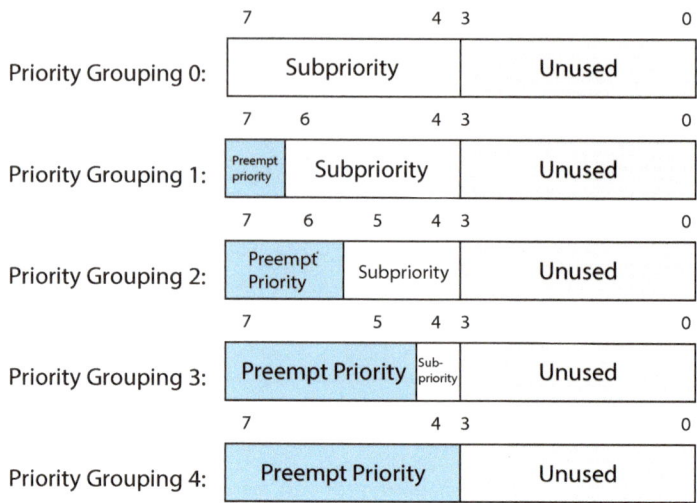

Fig. 2.38 Priority grouping in ARM Cortex-M7

2.8.4 Case Study: External Interrupts in STM32H7xx Microcontrollers

Figure 2.39 shows the block diagram of the interrupt circuitry block connected to the processor core in an STM32H7xx microcontroller. The interrupt circuitry consists of

1. The NVIC unit tightly coupled to the processor core within the ARM-Cortex M7. The internal peripherals within STM32H7xx (e.g., UARTs, timers, etc.) are connected to IRQ lines of the NVIC.

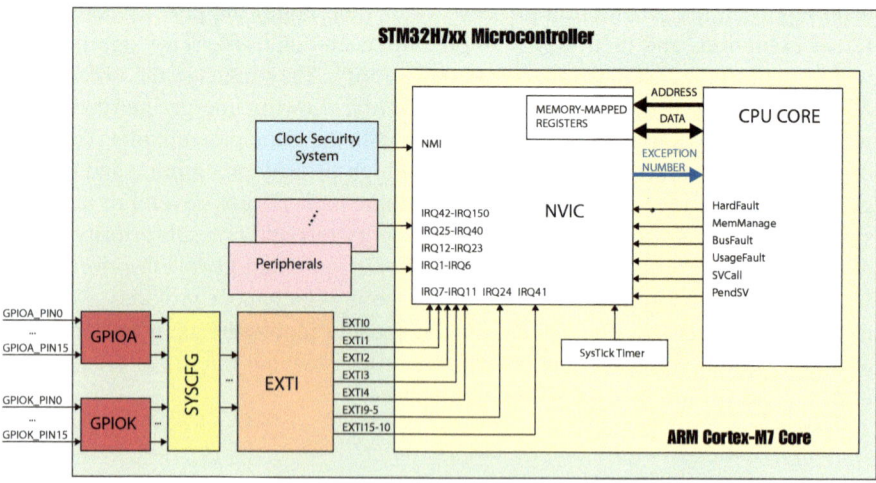

Fig. 2.39 The NVIC and EXTI controllers in the STM32H7xx microcontrollers

2. An additional dedicated interrupt controller, named Extended Interrupt and Event
Controller (EXTI), responsible for the interconnection between the external I/O
interrupt signals and the NVIC controller, as we will see next.

The Clock Security System (CSS) in STM32H7xx microcontrollers is a feature
designed to enhance the reliability and robustness of clock sources used within
the microcontroller. It provides a safeguard against potential failures in the clock
system to ensure the proper functioning of the microcontroller in various operating conditions. The CSS is connected to the NMI input of the NVIC controller.
The STM32H7xx microcontroller contains 11 16-bit GPIOs named GPIOA through
GPIOK. In total, 176 GPIO pins can be used to generate external interrupt requests.
In STM32X7xx microcontrollers, the EXTI is used to generate interrupts from GPIO
pins. The EXTI is a peripheral that enables interrupt requests based on specific GPIO
pin events, such as a rising or falling edge, allowing external events to trigger interrupt
requests.

2.8.4.1   Extended Interrupt and Event Controller (EXTI)

Figure 2.40 shows the block diagram of the EXTI controller. The main features of the
EXTI controller are an independent trigger and mask on each interrupt/event line and
a dedicated pending (status) bit for each interrupt line. The EXTI controller manages
25 input interrupt lines in total. The EXTI includes memory-mapped registers that
allow the programmer to set interrupt trigger conditions (rising edge, falling edge,
or both) and enable interrupts on specific EXTI lines. Some of the EXTI registers,
which we are interested in in this subsection, are:

1. **Interrupt mask register (EXTI_IMR)** manages the interrupt mask status for
each EXTI line. Setting a bit in this register enables the interrupt from that line.
2. **Rising trigger selection register (EXTI_RTSR)** enables/disables the rising trigger for each EXTI input line. When enabled, the EXTI generates an interrupt
request when a rising edge is detected on an input line.
3. **Falling trigger selection register (EXTI_FTSR)** enables/disables falling trigger
for each EXTI input line. When enabled, the EXTI generates an interrupt request
when a falling edge is detected on an input line.
4. **Pending register (EXTI_PR)** indicates the pending status of the interrupt for
each EXTI line. Reading a bit in this register shows if an interrupt request is

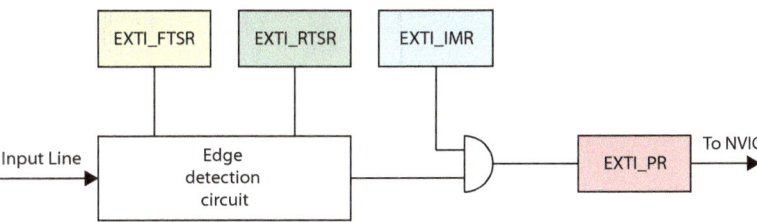

Fig. 2.40 EXTI block diagram

pending on that line. We should write '1' to the bit in the interrupt handler to clear the pending state of the corresponding interrupt.

The EXTI controller has internal interrupt control logic that monitors the GPIO pins' status and triggers interrupt requests when the configured events occur. To generate the interrupt, the interrupt line should be configured and enabled. This is done by programming the two trigger registers (EXTI_RTSR and EXTI_FTSR) with the desired edge detection and by enabling the interrupt request by writing a '1' to the corresponding bit in the interrupt mask register (EXTI_IMR). When the selected edge occurs on the external interrupt line, and the interrupt on that line is enabled in the interrupt mask register (EXTI_IMR), the pending bit corresponding to the interrupt line is set in the pending register (EXTI_PR). Also, EXTI generates an interrupt request to the NVIC. Once the interrupt handler is executed, this request must be reset by writing a '1' in the pending register (EXTI_PR) from the interrupt handler. EXTI interrupts can be individually prioritized using the NVIC, allowing different external events to have different levels of priority. To configure a line as an interrupt source, we use the following procedure:

1. Set the corresponding mask bit (EXTI_IMR)
2. Configure the trigger selection bits of the interrupt lines (in the EXTI_RTSR and EXTI_FTSR registers)
3. Configure the enable and mask bits that control the NVIC IRQ channel mapped to the external interrupt controller (EXTI) so that an interrupt coming from one of the 25 lines can be correctly acknowledged.

2.8.4.2   System Configuration Controller (SYSCFG)

As we already said, there are 176 GPIO pins in STM32H7xx microcontrollers, which can generate external interrupt requests. All these pins are routed to 16 input lines in the EXTI controller using the System Configuration controller. In STM32 micro-controllers, including the STM32F7 series, the System Configuration (SYSCFG) controller is crucial in routing GPIO pins to the External Interrupt (EXTI) lines. This routing process involves four memory-mapped specific registers within the SYSCFG (SYSCFG_EXTICR1 to SYSCFG_EXTICR4) and 16 multiplexors in the SYSCFG. Figure 2.41 shows the EXTI configuration registers available in the SYSCFG module. Four SYSCFG registers, SYSCFG_EXTICR1 to SYSCFG_EXTICR4, contain 16 groups of four select bits for 16 multiplexers, allowing the configuration of which of 16 GPIO pins is connected to a specific EXTI line. The EXTI line multiplexing functionality provided by SYSCFG enables the routing of GPIO pins to EXTI lines, as presented in Fig. 2.42.

The 16 EXTI controller lines, EXTI0 to EXTI15, are connected to the NVIC controller using only 7 IRQ inputs. The EXTI0, EXTI1, EXTI2, EXTI3 and EXTI4 lines are connected to their dedicated NVIC IRQ inputs IRQ7 to IRQ11 (see Fig. 2.39). The EXTI lines, EXTI5 to EXTI9, share the NVIC IRQ24 input, and the EXTI lines EXTI10 to EXTI5 share the NVIC IRQ41 input (see Fig. 2.39). Figure 2.43

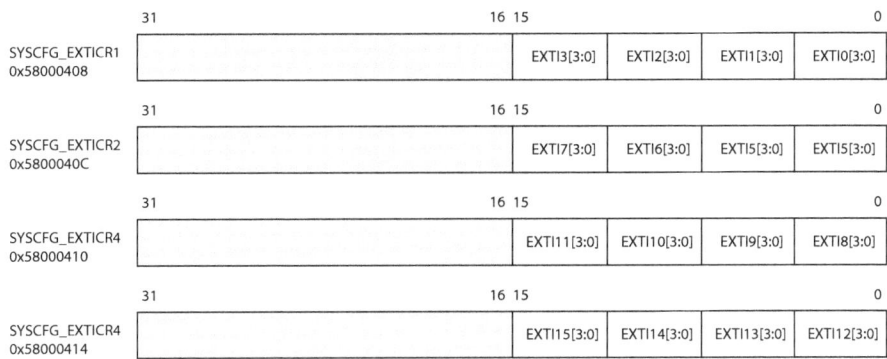

Fig. 2.41 The SYSCFG external interrupt configuration registers and their addresses

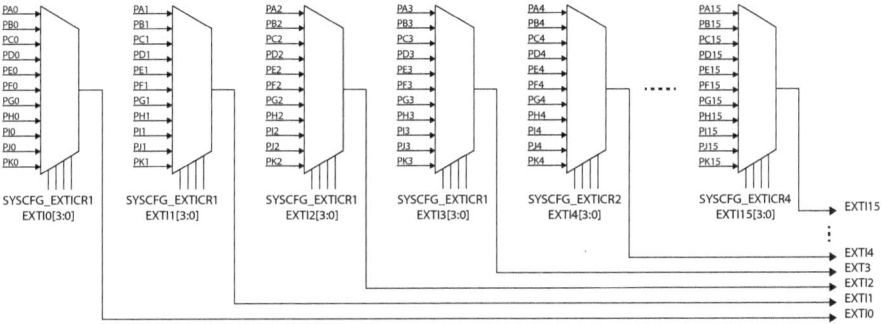

Fig. 2.42 The mapping of GPIO pins to EXTI lines in the System Configuration (SYSCFG) module in an STM32H7 MCU

illustrates how EXTI lines, EXTI0 to EXTI15, are mapped to exception handlers in ARM Cortex-M7 cores.

2.8.4.3   Triggering Interrupts on the GPIO Pins

Suppose we want to trigger interrupts when the rising edge is detected on GPIOC Pin 13. The process of setting up the STM32H7xx system involves the following steps:

1. **GPIO Pin Configuration:** Configure GPIOC Pin 13 as input.
2. **SYSCFG Configuration:** After configuring the GPIOC Pin 13, we need to map the pin 13 to the EXTI13 line. This involves configuring the associated multiplexor in the SYSCFG controller. To configure the multiplexor which maps GPIOC Pin 13 to EXTI13, we need to write '0010' to the 4-bit field EXTI13[3:0] in the SYSCFG_EXTICR4 register.
3. **EXTI Configuration:** Set the triggering conditions for the EXTI line 13 associated with GPIOC Pin 13. As we want the interrupt to be triggered by a rising edge, we should set the bit associated with EXTI13 in the Rising trigger selection reg-

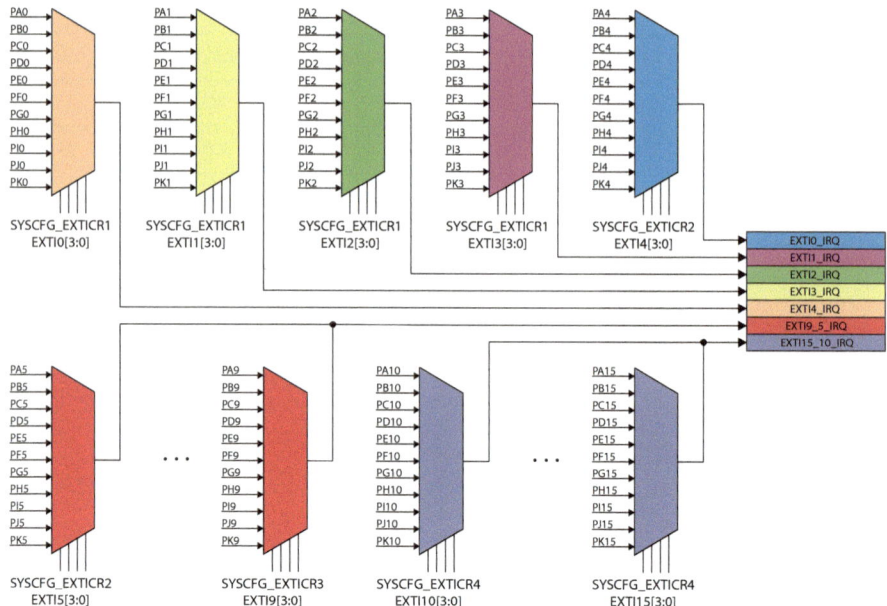

Fig. 2.43 The mapping of the EXTI lines to exception numbers (handlers)

ister (EXTI_RTSR). We should also enable the interrupt associated with EXTI13 by setting the appropriate bit in the interrupt mask register (EXTI_IMR).

4. **NVIC Configuration:** Configure the NVIC controller. This involves enabling the IRQ41 in the Interrupt Set Enable Register NVIC_ISER1 and setting the priority level for IRQ41 in the priority register NVIC_IPR11.

5. **Exception Handler Implementation:** Implement the exception handler `EXTI15_10_IRQHandler()` for the NVIC IRQ line associated with EXTI13. When the configured event occurs on the GPIOC Pin13, the EXTI generates an interrupt request on NVIC IRQ41, and the corresponding handler is called. Within the ISR, perform the necessary actions in response to the external event. The mandatory part of the handler is to clear the peripheral pending bit (in the EXTI_PR register). The peripheral pending bit will be held high until it is cleared by the application code. If the peripheral pending bit is not cleared, the interrupt will be fired again, and the handler will run again.

Figure 2.44 illustrates how the interrupt request on the GPIOC Pin13 is routed through SYSFIG, EXTI and NVIC to the CPU core after performing the above-listed configuration steps.

To enable an external interrupt on GPIOC Pin13 in the STM32H7xx microcontroller, we typically use the STM32Cube HAL (Hardware Abstraction Layer) library provided by STMicroelectronics, which simplifies configuring and using the micro-

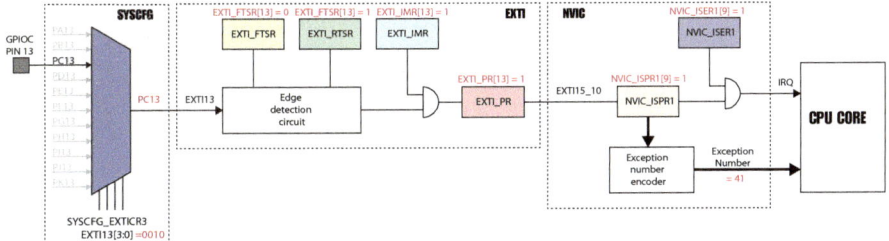

Fig. 2.44 Routing the interrupt request on the GPIOC Pin13 through SYSFIG, EXTI and NVIC to the CPU core

controller's peripherals. Below are the general steps to enable an external interrupt on GPIOC Pin13:

1. Configure GPIOC Pin13, SYSCFG and EXTI using `HAL_GPIO_Init()` function:

```
GPIO_InitTypeDef GPIO_InitStruct = {0};

__HAL_RCC_GPIOC_CLK_ENABLE();

GPIO_InitStruct.Pin = GPIO_PIN_13;
GPIO_InitStruct.Mode = GPIO_MODE_IT_RISING;
GPIO_InitStruct.Pull = GPIO_NOPULL;
HAL_GPIO_Init(GPIOC, &GPIO_InitStruct);
```

Listing 2.54 GPIO, SYSCFG and EXTI configuration.

The `HAL_GPIO_Init()` function sets the GPIO Pin13 as input, configures the SYSCFG controller to route GPIOC Pin13 to the EXTI13 line, and configures the EXTI controller to fire an interrupt when the rising edge occurs on the EXTI13 line.

2. Configure NVIC controller:

```
HAL_NVIC_SetPriority(EXTI15_10_IRQn, 0, 0);
HAL_NVIC_EnableIRQ(EXTI15_10_IRQn);
```

Listing 2.55 HAL functions used to configure NVIC.

3. Implement the `EXTI15_10_IRQHandler()` handler:

```
void EXTI15_10_IRQHandler(void)
{
    // Check if GPIO_PIN_13 triggered the interrupt:
    if (__HAL_GPIO_EXTI_GET_IT(GPIO_PIN_13) != 0x00U)
    {
        // Your code to handle the GPIO_PIN_13 interrupt goes here

        // Clear the EXTI13 pending bit in EXTI pending register
        __HAL_GPIO_EXTI_CLEAR_IT(GPIO_PIN_13);
    }
}
```

Listing 2.56 EXTI15_10_IRQ Handler.

2.8.5 Intel 8259A Programmable Interrupt Controler

Intel processors also have only a single interrupt input. As a personal computer has several peripheral devices that can raise interrupts, the Intel Programmable Interrupt Controller (PIC) 8259A is used to manage them. The 8259A PIC is a special interrupt controller designed particularly for Intel processors. It is connected between the interrupt-requesting peripheral device and the Intel processor. This means that the interrupt requests from peripheral devices are first transferred to the PIC, which in turn asserts the processor's interrupt input. Figure 2.45 illustrates the system with the 8259A PIC.

Fig. 2.45 A system with the 8259A PIC

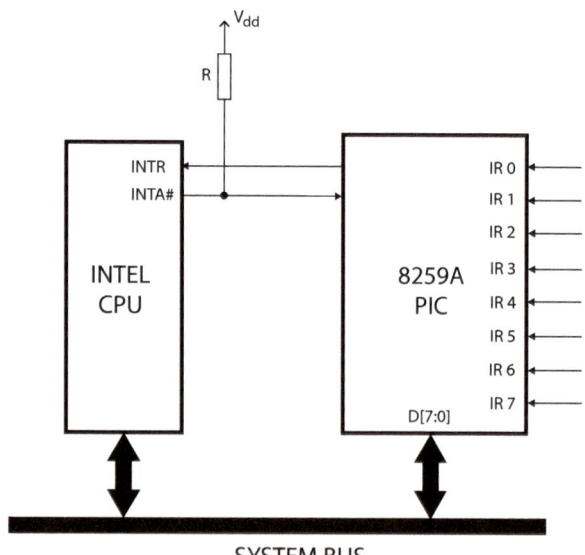

Fig. 2.46 Simplified internal
structure of the 8259A PIC

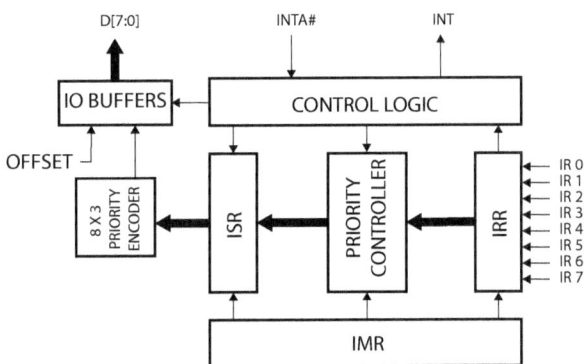

The 8259A was introduced in the early 1980s and was used in personal computers until the 1990s. It is still used in some Intel-based embedded systems. While not a separate chip anymore, the 8259A interface is still provided by the chipset on modern x86 motherboards. Although someone could say it is obsolete, its functioning will help us to understand the evolution of interrupt controllers in Intel-based computer systems.

The Intel 8259A Programmable Interrupt Controller handles up to eight vectored priority interrupts for the CPU. It is cascadable for up to 64 vectored priority interrupts without additional circuitry. The interrupt inputs have fixed priority based on their number, and the interrupts on their inputs may be either edge-triggered or level-triggered.

The 8592A PIC has the following set of registers: Interrupt Request Register (IRR), In-Service Register (ISR), and Interrupt Mask Register (IMR). The IRR register specifies which interrupts are pending. The ISR register specifies which interrupts have been acknowledged, and the IMR specifies which interrupts are to be ignored and not acknowledged. Figure 2.46 illustrates the simplified internal structure of the 8259A PIC.

The peripheral device that wishes to request an interrupt asserts one of the pins IR0 to IR7. If the interrupt is not masked in the IMR register, the 8259A PIC will set the corresponding bit in the interrupt request register (IRR). The IRR register remembers all the pending interrupt requests. As more peripheral devices can issue the interrupt request simultaneously, several bits may be set in the IRR register at the same time. At the same time, the 8259A sends an INT to the CPU. When the 8259A PIC asserts the interrupt request on the processor's INTR input, the processor recognizes this request on the next instruction fetch. It then stops the instruction fetch and automatically saves the program context onto the stack. The CPU then starts the so-called **interrupt-acknowledge cycle**.

Interrupt-acknowledge cycles are special bus cycles that enable the PIC to output an interrupt vector onto the data bus. This vector is fetched by the CPU and transferred into the program counter during the interrupt-acknowledge cycle. The value read during the interrupt-acknowledge cycle is then multiplied by 4 and used to load

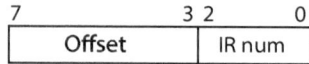

Fig. 2.47 The 8-bit vector number returned by the 8259A PIC. The lower three bits are the binary-coded number of the bit that was set in the ISR. The higher five bits are the offset that can be programmed during the 8259A PIC initialization

an interrupt vector from this address in memory. The Intel processors perform two back-to-back interrupt-acknowledge cycles in response to an active INTR input:

1. Firstly, the CPU responds by asserting the first INTA pulse. Upon receiving an INTA from the CPU, the priority controller in the 8259 passes the highest priority bit from IRR to the In-Service Register (ISR), and the corresponding IRR bit is reset. The set bit in the ISR indicates which interrupt request is being serviced.
2. Secondly, the processor asserts the second INTA pulse to instruct the 8259A to release an 8-bit interrupt number onto the Data Bus (D0-D7). This ends the interrupt-acknowledge sequence.

The CPU now reads the 8-bit interrupt number (n) and multiplies it by 4. This value represents the memory location address that holds the interrupt handler's start address. Hence, the CPU executes in hardware the following operation:

```
PC <-  Mem[n x 4]
```

The structure of the 8-bit vector number returned from the 8259A PIC is shown in Fig. 2.47. The lower three bits are the binary-coded number of the bit that was set in the ISR. The higher five bits are the offset that can be programmed during the 8259A PIC initialization. Recall that Intel stores its IVT table at the address 0x0000 and that the interrupt numbers 32 through 255 are reserved for external interrupts signaled on the INTR pin. If we want to map the IRQ interrupts from 8259A PIC at the address 0x0080 (=32x4) in the IVT, the offset returned in the 8-bit vector number should be 00100. In the case of the IR0 interrupt, the returned vector number is 0x20, which maps to 0x0080; in the case of the IR1 interrupt, the returned vector number is 0x21, which maps to 0x0084, etc.

To reset the bit in the ISR register, the interrupt handler should issue an End-Of-Interrupt (EOI) command to the 8259A PIC. The set bit is, therefore, deleted manually. The 8259A is now ready to process the next pending hardware interrupt request in IRR. The priority controller passes the highest priority bit from IRR to ISR, and the above sequence is repeated. Figure 2.48 illustrates the operation of the 8259A PIC when IR2 and IR4 are issued simultaneously, and IR2 has a higher priority than IR4.

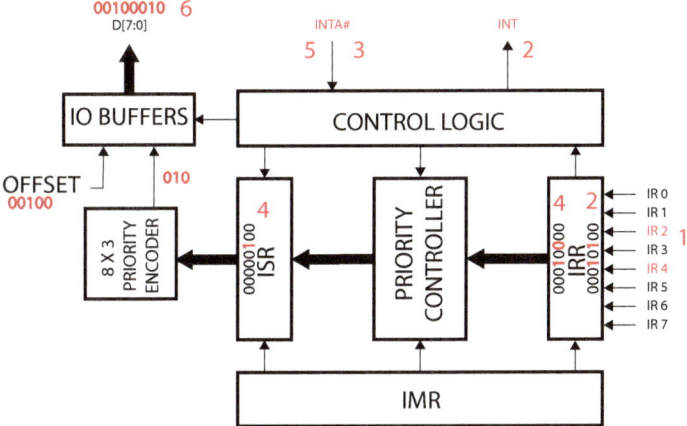

Fig. 2.48 The operation of the 8259A PIC when IR2 and IR4 are issued simultaneously, and IR2 has a higher priority than IR4. (1) Two peripheral devices assert the pins IR2 and IR4. (2) Assuming the interrupts are not masked in the IMR register, the 8259A PIC will set the corresponding bits in the interrupt request register (IRR). At the same time, the 8259A sends an INT to the CPU. (3) The CPU responds by asserting the first INTA pulse. (4) Upon receiving an INTA from the CPU, the priority controller in the 8259 passes the highest priority bit from IRR to ISR, and the corresponding IRR bit is reset. (5) The processor asserts the second INTA pulse. (6) The 8259A controller releases an 8-bit interrupt number onto the data bus (D0-D7)

Summary: AIC versus 8259A

Besides being designed for different processors, the main difference between the ARM AIC and Intel 8259A PIC is how the interrupt vector is obtained. In ARM AIC, the CPU reads the interrupt vector from an AIC's memory-mapped register using a LOAD instruction, while in 8259A PIC, the CPU reads the vector from the data bus without executing any instruction.

The former is considered faster (recall that instructions in ARM9 are executed in 5 clock cycles) but requires additional signaling between the interrupt controller and CPU (INTA) and a special interrupt-acknowledge cycle.

2.8.6 8259A PIC Cascading

With one 8259A PIC, eight interrupt sources could be managed. But soon, eight interrupt lines weren't enough. The 8259A PIC has the capability for **two-level cascading**. The first level is made up of one **master** PIC, and the second level is formed from up to eight **slave** PICs. Such a configuration can manage up to 64 peripheral interrupt requests. But in practice, only two PCs are used: one is the master, and the other is the slave. The PCs included two 8259A PICs chained together, and

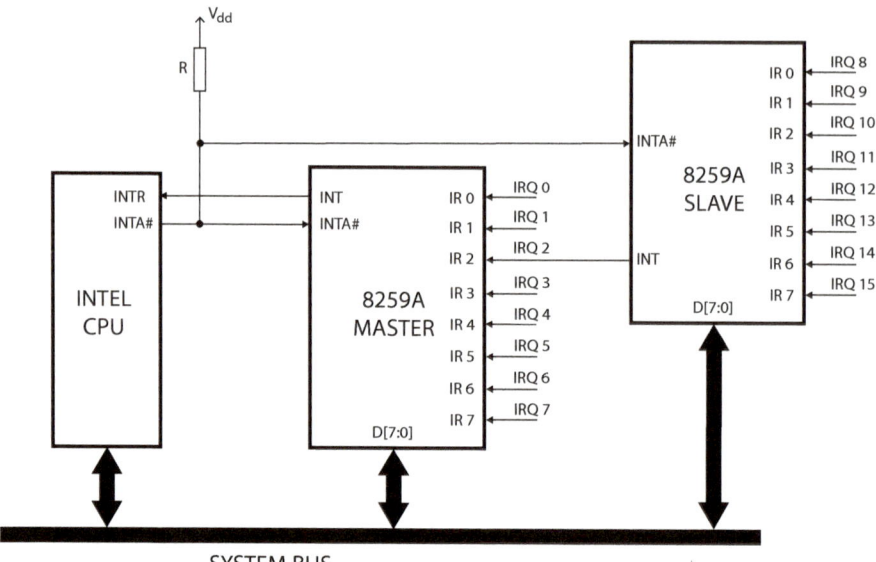

Fig. 2.49 A dual-PIC system

this setup became the de facto standard for the x86 platform. This scheme is referred to as the dual-PIC system. Figure 2.49 illustrates the dual-PIC system.

In the dual-PIC system, the slave's INT output is connected to the master's IR2 input. PC documentation established the following naming convention: IRQs 0 through IRQ7 are processed with the first Intel 8259 PIC (master), and IRQs from 8 to 15 are processed with the second Intel 8259 PIC (slave). Therefore, the slave's INT output is connected to the master's IRQ2 input. Only the master's INT output is connected to the CPU's INTR input and can signal about the incoming interrupts. INTA and D0 through D7 signals of both PICs are connected to the CPU and data bus as in the single-PIC configuration. Note that IRQ 2 is not available for device interrupts, and there are only fifteen interrupt inputs available for peripheral device interrupts.

The way in which an interrupt request is processed depends on whether the request is asserted on slave's or master's IRQ inputs. If the request is asserted on IRQs from 0 to 7 (master), it is processed in the same way as in the single-PIC configuration. Otherwise, the following steps are required:

1. when an interrupt request is placed on lines IRQ 8 through 15, the corresponding bit is set in the slave's IRR register. The slave asserts its INT output and signals the IRQ interrupt to the master PIC,
2. when master PIC receives the interrupt request on IRQ2, it sets the bit 2 in its IRR and asserts its INT output to signal the INTR request to the processor,
3. the CPE starts the interrupt-acknowledge sequence and sends the first INTA pulse.

4. upon receipt of the first INTA pulse, the highest priority bits in the master's and slave's IRRs are cleared, and the corresponding bits in both ISRs are set,
5. the CPE outputs the second INTA pulse and causes the slave PIC to output its 8-bit vector number.

The interrupt handler must send two EOI commands to clear both ISR bits.

But wait! How can we set two different vector numbers for interrupts signaled at the master's IR3 (IRQ3) input and at the slave's IR3 input (IRQ11)? Recall that the 8-bit vector number returned from the 8259A PIC contains a programmable offset in its higher five bits. Hence, the master and slave PICs should be initialized with different offsets. For example, we can set the master's offset to 00100 and the slave's offset to 00110. In this case, the interrupts from the master PIC will be mapped to the IVT addresses 0x0080-x009F, and the slave's interrupts will be mapped to the IVT addresses 0x00C0-0x00DF.

The original PCs used the ISA bus for their I/O devices. The interrupts on the ISA bus are edge-triggered. An I/O device asserts an interrupt by raising the signal from low to high. Edge-triggered interrupts inhibit the sharing of ISA interrupts by multiple devices, so each ISA device requires a dedicated interrupt input on the 8259As. A typical interrupt configuration at that time is presented in Table 2.10.

The Intel 8259 PIC has several limitations to interrupt servicing in modern computer systems:

Table 2.10 ISA interrupt assignments

IRQ	Assignment
0	System timer
1	Keyboard controller
2	Interrupt from slave controller
3	Serial ports COM 2/COM 4
4	Serial ports COM 1/COM 3
5	Sound card
6	Floppy disk controller
7	Parallel port 1 (Printer)
8	Real-time clock
9	ACPI
12	PS/2 mouse controller
13	Math (floating point) co-processor
14	ATA channel 1 (Primary IDE)
15	ATA channel 1 (Secondary IDE)

1. a limited number of interrupt lines necessitates the sharing of interrupts. Shared interrupts require the OS to poll multiple IO devices to determine who actually generated the interrupt,
2. interrupt priority is fixed based on IR number,
3. PIC does not support multiple CPUs.

At the time when the main bus for external devices was the ISA bus, this 8929A-based architecture was sufficient. It was only necessary that different peripheral devices did not connect to the same IRQ inputs since ISA uses edge-triggered interrupts, which are not shareable. However, the PCI bus later replaced the ISA bus, and interruptions in the PCI bus can be shared. Also, the PCI bus has been replaced by the serial message-based PCI Express (PCIe) bus, and more CPU cores are added to the computer system. The following sections will cover the evolution of the interrupt controller used in Intel-based computer systems, and we will learn how to handle the interrupts on the PCI bus where the number of peripheral devices exceeds the number 15, how to share the interrupt lines on the PCI bus, and finally, how to handle interrupts in modern multi-core PCIe-based systems.

2.8.7   Intel Advanced Programmable Interrupt Controler

By nature, the 8259A PIC can only send interrupts to one CPU, and in a multiprocessor system, it is desired to load CPUs in a balanced way. The solution to this problem was the new APIC (Advanced PIC) architecture. This architecture addressed many of the limitations of the older PIC-based architecture. The most apparent is the support for multiple CPUs.

At the system level, APIC consists of two parts (Fig. 2.50). One part resides in the I/O subsystem and is called the I/O APIC. It is responsible for routing interrupts from external devices to the other part, the Local APIC (LAPIC), which resides in each CPU. The local APIC and the I/O APIC communicate over a dedicated 3-wire serial APIC bus. The IOAPIC bus interface consists of two bi-directional data signals (APICD[1:0]) and a clock input (APICCLK). The modern systems may use a standard system bus instead of a separate APIC bus for this task. It is worth noting that it is possible to have several I/O APIC controllers in the system. For example, one for 24 interrupts in a southbridge, and the other one for 32 interrupts in a northbridge. The CPU's Local APIC contains the necessary intelligence to determine whether or not its processor should accept interrupts broadcast on the APIC bus. The Local APIC also provides local pending interrupts and handles all interactions with its local processor (e.g., the interrupt acknowledge sequence). Additionally, each I/O APIC has 24 interrupt lines and allows the priority of each interrupt to be set independently. The I/O APIC sends an interrupt vector to the local APIC, and, as a result, the OS does not have to interact with the I/O APIC until it sends the end-of-interrupt notification.

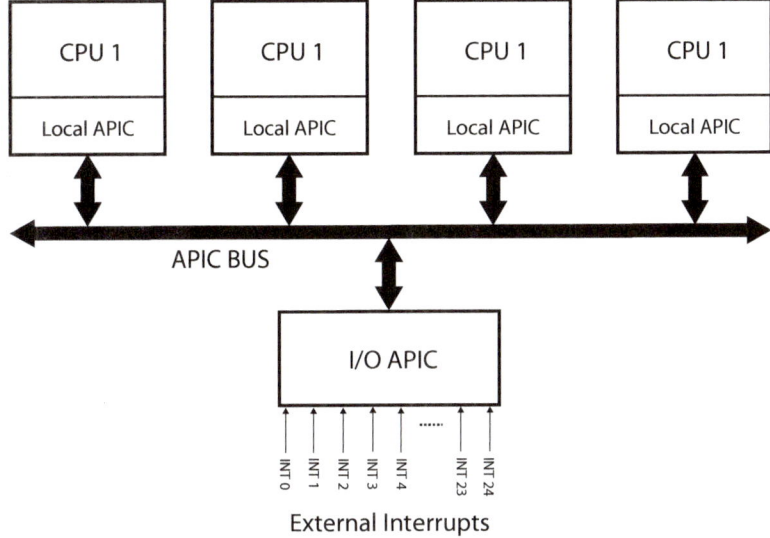

Fig. 2.50 An APIC based computer system

Summary: APIC

In the APIC-based systems, each CPU includes a local APIC that receives interrupt messages and uses them to assert interrupts on the CPU. The chipset includes one or more I/O APICs, which are responsible for converting device interrupt signals into messages that are delivered to one or more local APICs.

2.8.7.1 Local APIC

The Local Advanced Programmable Interrupt Controller (LAPIC) was introduced into the Pentium processor and is included in more recent Intel processor families. The local APIC performs two primary functions for the processor. It receives interrupts from the processor's interrupt pins and an external I/O APIC. It sends these to the processor core for handling. In multiple-processor systems, it sends and receives interprocessor interrupt (IPI) messages to and from other processors on the system bus. IPI messages are used to distribute interrupts among the processors in the system. When a local APIC sends an interrupt to its processor core (by asserting the processor's INTR line) for handling, the processor uses the interrupt and exception handling mechanism described in Sect. 2.7.

The LAPIC receives interrupts from several sources:

- Locally connected I/O devices: these interrupts are asserted by an I/O device connected directly to the processor's local interrupt pins (LINT0 and LINT1).

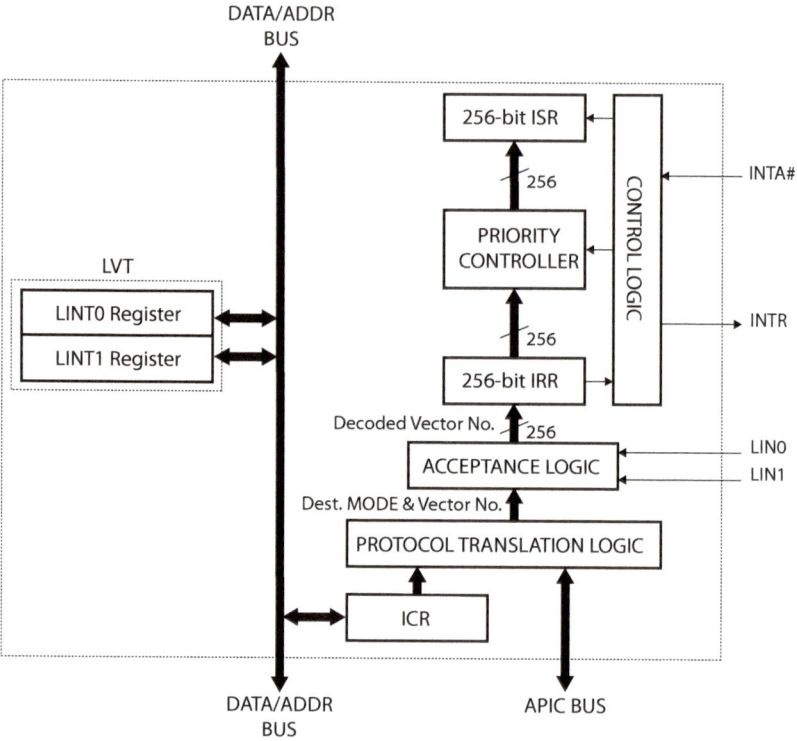

Fig. 2.51 Simplified internal structure of a Local APIC

- Inter-processor interrupts (IPIs): an Intel processor can use the IPI messages to interrupt another processor or group of processors on the system bus.
- Externally connected I/O devices: these interrupts are asserted by an I/O device connected to the interrupt input pins of an I/O APIC. Interrupts are sent as IPI messages from the I/O APIC to one or more LAPICs in the system.

Figure 2.51 illustrates a simplified internal structure of a Local APIC. The heart of the LAPIC is very similar to the 8259A: it contains the IRR and ISR registers and a priority controller. Besides, it contains two registers that form the local vector table LVT and the interrupt command register (ICR). In fact, the LAPIC is a rather complicated device containing a large set of addressable registers, timers, and other control logic. Figure 2.51 shows only the vital parts that are necessary to understand its interrupt handling. The Protocol Transition Logic block receives the IRI messages from the APIC bus. If the LAPIC is the destination, the Protocol Transition Logic block forwards the destination mode and the vector number from the IRI message to the Acceptance Logic block, which decodes the 8-bit vector number and forwards the bit from the decoded 256-bit word into the IRR register. The rest of the internal logic is described for each interrupt source in the text below.

Interrupts from locally connected I/O devices. Upon receiving a signal from the processor's LINT0 and LINT1 pins, the local APIC delivers the interrupt to the processor core using a group of APIC registers called the **local vector table**. A separate entry (i.e., a separate register) is provided in the local vector table for each local interrupt pin (LINT0 and LINT1). For example, if the LINT1 pin is going to be used as an NMI pin, the LINT1 entry in the local vector table can be set up to deliver an interrupt with vector number 2 (NMI interrupt in Table 2.7) to the processor core. The LVT consists of two 32-bit registers: LINT0 Register (specifies the interrupt number when an interrupt is signaled at the LINT0 pin) and LINT1 Register (specifies the interrupt number when an interrupt is signaled at the LINT1 pin). An interrupt number is an 8-bit number stored in the bits 0 through 7 in each LINT register.

Inter-processor interrupts (IPIs). A processor generates IPIs by writing to a special LAPIC register called the interrupt command register (ICR) in its local APIC. Writing to the ICR causes an IPI message to be generated and issued on the system bus or the APIC bus. An IPI message includes the processor destination number, the vector number, and the trigger mode (edge or level). When the target processor receives an IPI message, its local APIC handles the interrupt request automatically using information included in the message, such as vector number and trigger mode. The IPI mechanism is used in multi-processor systems to send or forward interrupts for a specific vector number. For example, a local APIC can use an IPI to forward an interrupt to another processor for servicing. Also, the IPI mechanism is used by I/O APIC to send an interrupt for a specific vector number that originates from an I/O device connected to I/O APIC. The interrupt command register (ICR) is a 64-bit local APIC register that allows software running on the processor to specify and send interprocessor interrupts (IPIs) to other processors in the system. The act of writing to the low 32 bits of the ICR causes the IPI message to be sent. Figure 2.52 illustrates the ICR register (only the bits that are important for understanding are specified/shown). The 8-bit Destination field specifies the target processor. The Destination Mode bit

Fig. 2.52 The LAPIC ICR register

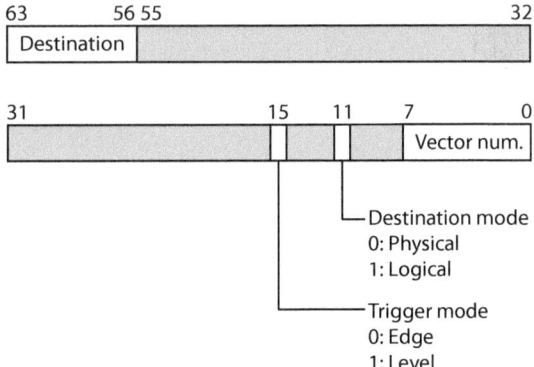

further specifies whether the destination is a physical (0) or logical (1) processor. The Trigger Mode bit selects the trigger mode: edge (0) or level (1).

Externally connected I/O devices. The local APIC can also receive interrupts from externally connected devices through the I/O APIC (see Fig. 2.50). The I/O APIC is responsible for receiving interrupts generated by system hardware and I/O devices and forwarding them to the local APIC as IPI messages. Each individual pin on the I/O APIC can be programmed to generate a specific interrupt vector when asserted. This vector is then sent to LAPIC as a part of an IPI message.

The local APIC handles the interrupts as follows:

1. if it receives a message on the APIC bus, it determines if it is the specified destination. If it is the specified destination, it accepts the message; otherwise, it discards the message,
2. if the local APIC determines that it is the designated destination for the interrupt, the local APIC sets the appropriate bit in the IRR,
3. when interrupts are pending in the IRR register, the local APIC sends them to the processor one at a time, based on their priority, similarly as in the 8259A. The processor responds with the interrupt acknowledge sequence. During the first INTA pulse, the LAPIC moves the highest priority bit from the IRR to the ISR. During the second INTA pulse, the LAPIC puts the interrupt vector on the data bus. If the interrupt request comes from a locally connected I/O device (at the LINT0 or LINT1 pins), the interrupt number is stored in the corresponding LVT entry in the LAPIC. If the interrupt request comes from a message, the interrupt number is contained in the message.

Completing the handler routine is indicated by instruction in the interrupt handler code that writes to the end-of-interrupt (EOI) register in the local APIC. Writing to the EOI register causes the local APIC to delete the interrupt from its ISR.

2.8.7.2 I/O APIC

The I/O Advanced Programmable Interrupt Controller (IOAPIC) manages multiprocessor interrupt and incorporates interrupt distribution across all processors. In systems with multiple I/O subsystems, each subsystem can have its own set of interrupts. Each interrupt pin is individually programmable as either edge or level triggered. The interrupt vector can be specified per interrupt input.

The I/O APIC (Fig. 2.53) consists of 24 interrupt input lines, a 24-entry Interrupt Redirection Table (IRT) with 64-bit entries, programmable registers, and a message unit for sending and receiving messages over the APIC bus. I/O devices signal interrupt requests by asserting one of the interrupt lines to the I/O APIC. The I/O APIC selects the corresponding entry in the IRT and uses the information in that entry to format an interrupt request message. Each entry in the IRT contains:

- a bit that indicates edge/level sensitive interrupt,
- the interrupt vector and priority, and
- the destination processor.

Fig. 2.53 The I/O APIC

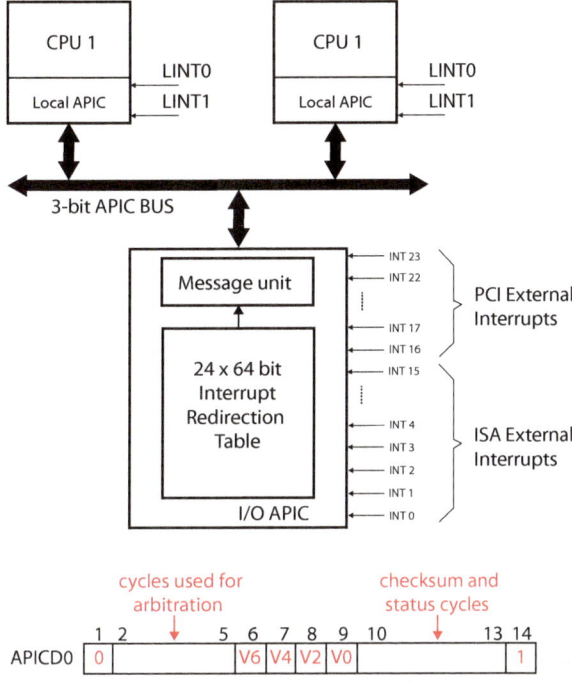

Fig. 2.54 The format of the
EOI message

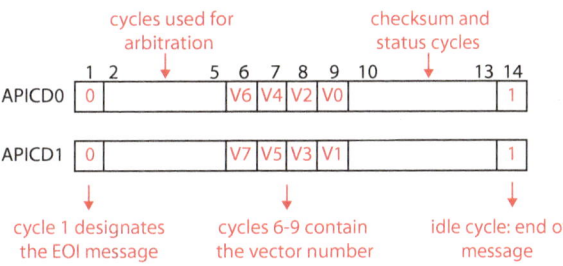

The information in the IRT entry is used to form and transmit an IRI message to other LAPICs via the APIC bus.

When an external interrupt request is signaled on the I/O APIC interrupt input, the I/O APIC controller will send an interrupt message to the LAPIC of one of the system CPUs. In this way, the I/O APIC controller helps balance interrupt load between processors.

APIC messages come in several formats and different lengths. Here, we present only two types of APIC messages: the EOI Message and the so-called Short Message. Local APICs use EOI messages to send an end-of-interrupt (EOI) occurring for a level-triggered interrupt to an I/O APIC. This message is needed so the I/O APIC knows when an interrupt has been serviced. In this way, the I/O APIC can differentiate between a new interrupt on the interrupt line versus the same interrupt on the interrupt line. I/O APICs use Short Messages for the delivery of external interrupts to local APICS.

The format of the EOI message is presented in Fig. 2.54. All EOI messages are 14 bits long and take 14 cycles on the APIC bus to transmit. Local APICs send 14-cycle EOI messages to the I/O APIC to indicate that the processor has accepted

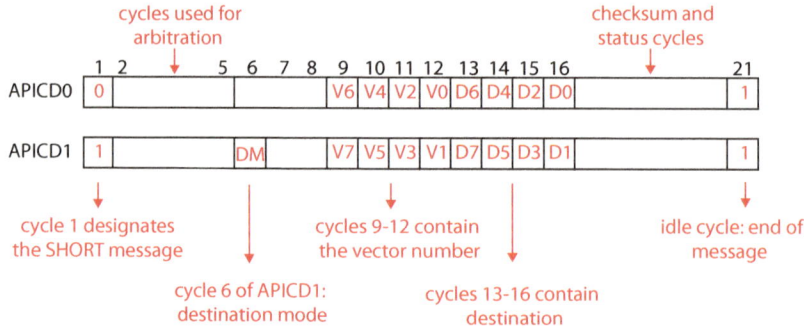

Fig. 2.55 The format of the SHORT message

a level-triggered interrupt. This message is a result of software writing into the EOI register of the local APIC. The first cycle is used to designate an EOI message. The vector number is sent in cycles 9 through 12. The local APIC gives the target of the EOI message by transmitting the interrupt vector number (V7 through V0). When this message is received, the I/O APIC resets the IRR bit for that interrupt. If the interrupt signal is still active after the IRR bit is reset, the I/O APIC treats it as a new interrupt. The last cycle, in which both data lines are set high, is used to signal the end of the message.

The format of a SHORT message is presented in Fig. 2.55. All SHORT messages are 21 bits long and take 21 cycles on the APIC bus to transmit. I/O APICS uses short messages to send external interrupts to local APICs. The first cycle is used to designate a SHORT message. The vector number is sent in cycles 9 through 12. If Destination Mode (DM) is 0, cycles 15 and 16 are the local APIC ID, and cycles 13 and 14 are sent as 1. In this case, the message is sent to a physical processor. If DM is 1, cycles 13 through 16 are the 8-bit Destination field that selects the logical processor. The last cycle, in which both data lines are set high, is used to signal the end of the message.

The I/O APIC with 24 input interrupt lines was used in the systems with the PCI bus. The APIC architecture could support up to 16 CPUs. The I/O APIC provided backward compatibility with the older 8259A PIC-based systems. Interrupts 0-15 were used for old ISA interrupts for compatibility with older systems, and interrupts 16-23 were meant for all the PCI devices. With this delimitation, all conflicts between ISA edge-triggered and PCI level-triggered interrupts could be easily avoided. This assignment of interrupts 0–15 provided only eight additional interrupts, which forced the sharing of PCI interrupts—two or more devices on the PCI bus were forced to share the same I/O APIC's input interrupt line. We will cover the PCI interrupt sharing and routine in the following sections.

One of the biggest differences between the 8259A PICs and I/O APICs is that the pins on I/O APICs are completely independent. With the 8259A PICs, the eight input pins are mapped to eight consecutive vectors in IDT (or IVT), and all of the interrupts are sent to the same CPU. In I/O APICs, on the other hand, each pin is programmed independently. The operating system assigns each pin its own vector and can be mapped to one or more CPUs.

2.9 PCI Interrupts

The Peripheral Component Interconnect (PCI) bus was added to the PCs in the mid-1990s. In the first years, PCI and ISA buses coexisted in the systems. In PCI, the term **device** refers to a piece of hardware plugged into the PCI slot and contains from one to eight **functions**. A multi-function PCI device is a physical PCI expansion board that embodies between two and eight PCI functions. For example, a single PCI device may include several USB controllers as functions. Another example would be a PCI card with one high-speed communications port and one parallel port. Hence, a PCI expansion card inserted in the PCI expansion slot is a PCI device, and a single expansion card may contain up to eight functions. However, from the operating system's perspective, each function on a PCI device is a logical operating device.

PCI allows devices to assert interrupts in two different ways:

1. The first way uses dedicated interrupt signals (lines) and is known as **Legacy INTx interrupts**.
2. The second way uses special memory writes sent over the data bus, just like APIC messages and is known as **Message Signaled Interrupts (MSI)**. First, we will cover Legacy INTx interrupts. Later, we are going to cover the MSI interrupts.

2.9.1 PCI Legacy Interrupts

A PCI card in a slot may have up to eight functions on it, but there are only 4 PCI interrupt pins: INTA#, INTB#, INTC#, and INTD#. PCI legacy interrupts are level-triggered; hence, they may be shared by multiple functions. Each function within the device is only permitted to use one of these interrupt pins to generate requests. A device containing only one function that uses only one interrupt pin must be bonded to INTA#. If a device includes more than one function, all functions within a device may be bonded to the same pin, INTA#, or each may be bonded to a dedicated pin (this would be true for a device with up to four functions). Also, a group of functions within the package may share the same interrupt pin. In the most simple (and most common) case, a PCI device has only one function with its interrupt going to the lane INTA#.

Figure 2.56 illustrates a simple interrupt model with two peripheral devices on the PCI bus; hence, both are bonded to INTA#. Each peripheral device embodies only one function that generates PCI interrupts. The first device is an ethernet card in the PCI slot 0 that generates interrupts on INTA# line, while the second device is a sound card in the PCI slot 3 that also generates interrupts on the INT#A line. Both devices share the same interrupt request signal trace on the system board, routed to the IRQ5 input on the Intel 8259A programmable interrupt controller (PIC). Indeed,

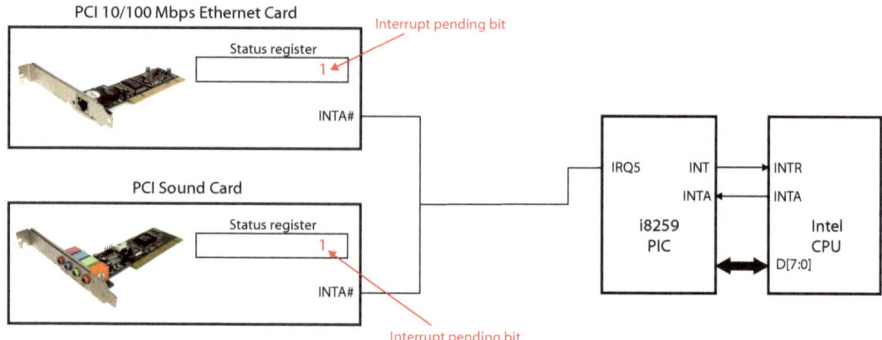

Fig. 2.56 Example system for shared interrupts on the PCI bus

the PCI standard does not limit the interrupt controller used to route PCI interrupts to the CPU as long it supports level-triggered interrupts; hence, in this example, we assume that the 8259A PIC is used for this purpose.

Let us assume that both devices assert the interrupt request and that the sound card asserted the interrupt first. The interrupts are asserted by driving the interrupt line LOW. In addition to asserting the interrupt request, both devices set an interrupt pending bit in their memory-mapped status registers so that interrupt handlers can access both status registers. Let us also assume that no other device in the system had asserted an interrupt request before the sound card and ethernet card.

When the Intel 8259A PIC detects the interrupt request on its IRQ5 input, it asserts the interrupt request on the processor's INTR input. The processor will recognize this request on the next instruction fetch. It then stops the instruction fetch and automatically saves the program context onto the stack. The CPU then starts the interrupt acknowledge sequence:

1. the CPU responds by asserting the first INTA pulse,
2. the 8259 PIC prioritizes the pending interrupt requests by setting the bit 5 in its ISR register and clearing the bit 5 in its IRR register,
3. the CPU outputs the second INTA pulse to instruct the 8259A PIC to release the 8-bit pointer onto the data bus (D0 to D7) where the CPU reads it.

Now, the processor has received the interrupt vector number associated with IRQ5. Let's assume that the interrupt vector number is $0x07$. The processor multiplies this value by 4, which yields the address of the interrupt vector entry, $0x0000001C$. The processor now reads the content of the memory location $0x0000001C$ to obtain the start address of the interrupt handler.

As both devices, the ethernet card and the sound card, share the same interrupt line IRQ5, the interrupt handler should contain the code to handle the interrupt requests from both devices. Let us assume that the interrupt handler contains both codes, the "ethernet" handler and the "sound card" handler. The simplified structure of the IRQ5 interrupt handler is presented in Listing 2.57. The interrupt handler first checks

which device has asserted the interrupt request by checking the interrupt pending bit
in the corresponding status register.

```
/*
 * IRQ5 Handler
 */
__attribute__((interrupt)) void irq5handler () {

    /* Check the interrupt pending bit in the Ethernet status reg */
    if (eth_status_reg & (1<<INT_PEND_BIT)) {
    /* Ethernet Handler Code */
    eth_status_reg &= ~(1<<INT_PEND_BIT); // clear int pending bit
    ...
    ...
    ...
    }

    /* Check the interrupt pending bit in the Sound card status reg */
    if (snd_status_reg & (1<<INT_PEND_BIT)) {
    /* Sound Card Handler Code */
    snd_status_reg &= ~(1<<INT_PEND_BIT); // clear int pending bit
    ...
    ...
    ...
    }

    /* return from interrupt */
}
```

Listing 2.57 IRQ5 interrupt handler

Listing 2.57 shows that the interrupt handler first checks the pending bit in the
ethernet card's status register. As this bit is asserted, the ethernet handler is executed
first. The ethernet handler clears the pending bit in the status register and processes
the interrupt request. Then, the IRQ5 handler proceeds with the sound card handler.
This handler checks the sound card's interrupt pending bit to determine if it requires
servicing. Since the pending bit is set, the main body of the sound card handler is
executed. It clears the interrupt pending bit and services the interrupt request. As
both devices have their pending bit clear, the interrupt line is de-asserted.

Hence, the system in Fig. 2.56 relies on the vectored interrupt handling to determine which interrupt request input in the PIC 8259A has been asserted but uses the
interrupt polling to determine which PCI device has asserted the interrupt request.
The sequence of polling determines the interrupt priority. This is why the ethernet
card is serviced first, although the sound card had asserted the interrupt request a
few moments before the ethernet card.

2.9.2 PCI Interrupts Routing

Each PCI device (function) that needs an interrupt comes with a fixed PCI interrupt
that can't be changed. But this PCI legacy interrupt signal (lane) can be mapped
(routed or redirected) to any APIC interrupt input. Thus, at one end, PCI legacy
interrupt lines (INTA# through INTD#) are signaled by a PCI function that needs
attention. At the other end, we have a CPU receiving an IDT vector. In the middle is

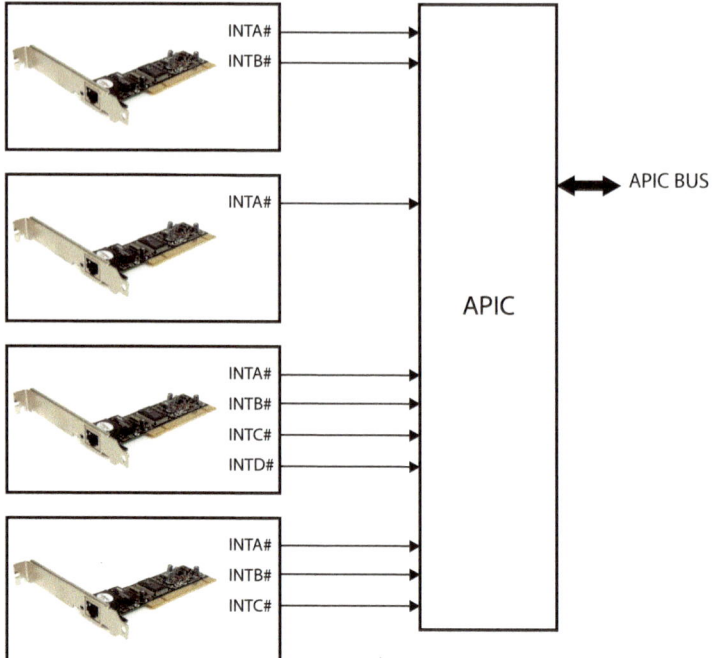

Fig. 2.57 Ideal routing of PCI interrupt lanes

an interrupt controller (most commonly, this would be an APIC pair: I/O APIC and LAPIC). Whenever any of the PCI legacy interrupt pins is asserted, the I/O APIC module supplies the vector associated with that input to the processor's embedded local APIC module. We have already learned that IR16-IR23 input I/O APIC pins are devoted to PCI. The upper four (IR20-IR23) are dedicated to PCI functions embedded in the chipset, while the lower four are dedicated to PCI legacy interrupt pins. The I/O APIC inputs IR20 through IR23 are called PIRQA through PIRQH when used for PCI interrupts. The acronym PIRQ stands for PCI interrupt request. We have also learned that several PCI devices or functions can use the same PCI legacy interrupt signal to assert interrupts. So, the question is, how the PCI legacy interrupt signals INTA# through INTD# are routed to the I/O APIC inputs PIRQA through PIRQD? The ideal scenario is pictured in Fig. 2.57, where each individual PCI interrupt line is routed to an interrupt controller as a separate input. But such a solution is possible only if only up to four PCI devices exist because the I/O APIC has only four available interrupt inputs.

PCI interrupt signals can be routed to I/O APIC interrupt pins (PIRQs) in several ways. A straightforward method of connecting (hardwiring) these lines from PCI devices to the PIRQs would be to connect all INTA# interrupts to PIRQA, all INTB# interrupts to PIRQB, all INTC# interrupts to PIRQC, and all INTD# interrupts to PIRQD. Figure 2.58 illustrates this method of PCI interrupt routing. As mentioned

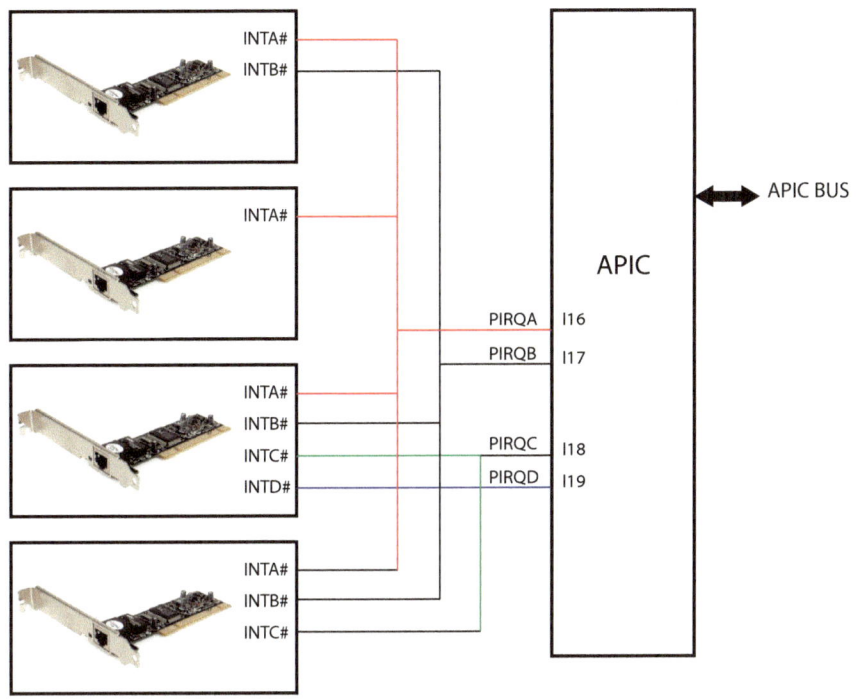

Fig. 2.58 Unbalanced routing of PCI interrupt lanes

above, the most common case is when a PCI device has only one function, and its interrupt must be connected to the INTA# pin. Therefore, if we decided to route all PCI interrupt lanes as we've written, almost all system devices would share interrupt input PIRQA. As shown in Fig. 2.58, the PIRQA request line is heavily weighted (four PCI devices). Suppose this lane is connected to the IRQ16 of the APIC. This way, every time a processor has a signal that there is an interrupt on the IRQ16 input, it has to poll all of the device drivers of the PCI devices connected to that IRQ16 line (PIRQA) if they have asserted an interrupt. If there are many of those devices, it will surely decrease system response to the interrupt. In this case, lanes PIRQB-PIRQD would stand idle most of the time.

The optimal way of PCI interrupt routing is that each PIRQ should have the same number of connected PCI functions. We should also consider that some functions trigger interrupts very rarely and some almost constantly (e.g., Ethernet controller). Hence, we may connect the PIRQs more randomly so that each will share about the same number of actual PCI legacy interrupts.

Figure 2.59 illustrates one method of doing this. This illustration shows how legacy interrupt traces are physically routed across the PCI slots. Although the physical interrupt lines wire each PIRQ to every slot, each PIRQ connects differently to the pins in each slot. Here, wire PIRQA connects interrupts INTA# in the PCI slot 1, INTB# in the PCI slot 2, INTC# in the PCI slot 3, and INTD# in the PCI slot 4.

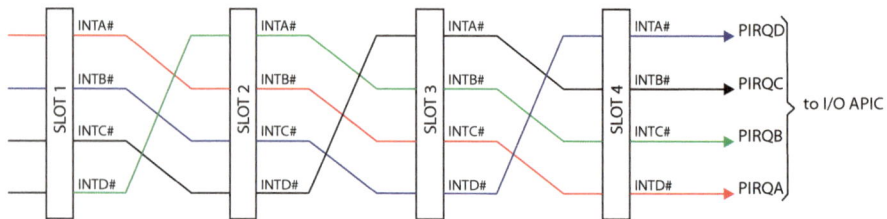

Fig. 2.59 Round-robin routing of PCI legacy interrupt traces (lanes)

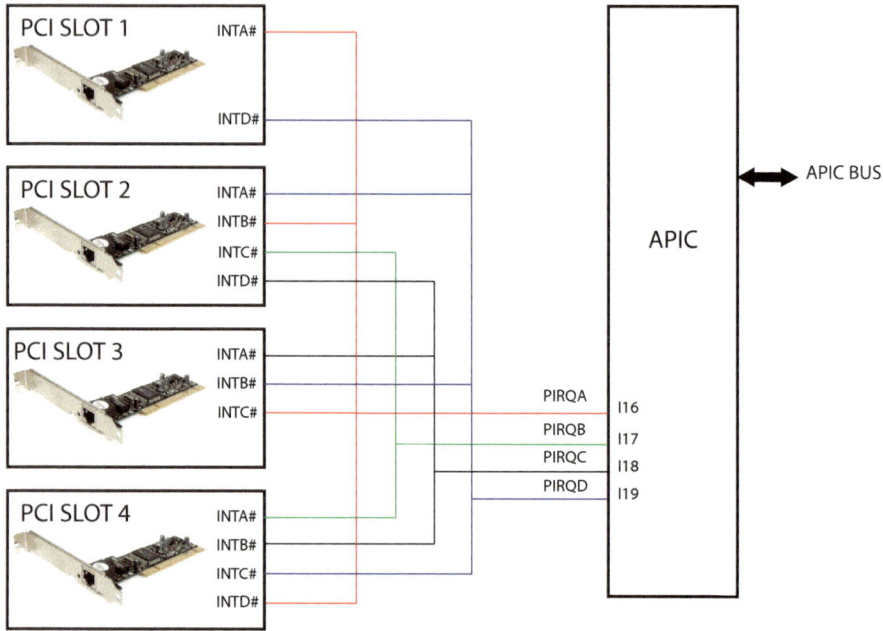

Fig. 2.60 Common round-robin routing of PCI interrupt lanes

Likewise, wire PIRQB connects interrupts INTB# in the PCI slot 1, INTC# in the PCI slot 2, INTD# in the PCI slot 3, and INTA# in the PCI slot 4, etc.

Figure 2.60 shows a 4-slot PCI system with the following cards installed:

1. Card 1 installed in Slot 1. This card includes two PCI functions, of which one generates interrupts on IRQA#, and the other generates interrupts on IRQD#. The IRQA# pin of this card connects to PIRQA.
2. Card 2 installed in Slot 2. This card includes four PCI functions, thus generating interrupts on IRQA# through IRQD#. The IRQA# pin of this card connects to PIRQB.

3. Card 3 installed in Slot 3. This card includes three PCI functions, which generate interrupts on IRQA# through IRQC#. The IRQA# pin of this card connects to PIRQC.

4. Card 4 installed in Slot 4. This card includes four PCI functions, thus generating interrupts on IRQA# through IRQD#. The IRQA# pin of this card connects to PIRQD.

A practical use for this is that one may change the interrupt routing of a PCI card by inserting it in a different slot. In the above example, INTA# of a PCI card will be connected to wire PIRQA if the card is inserted into slot 1, but INTA# will be connected to wire PIRQB when inserted into slot 4.

2.9.3 Message Signaled Interrupts

As we described in the previous section, a PCI device can use up to four dedicated interrupt pins to signal an interrupt request to the I/O APIC. This method is referred to as Legacy INTx interrupts. Each PCI device can have up to four PCI legacy interrupts. After the I/O APIC has received an interrupt request, it forwards it to LAPICs by means of the APIC messages. But why not implement this "interrupt message" functionality into a PCI device itself? This way, we could eliminate the need for interrupt traces and interrupt sharing. This method is referred to as **Message Signaled Interrupts (MSI)** and is described in this section. **Message Signaled Interrupts** are special memory writes sent over the system bus. In essence, the MSI interrupts do not differ from ordinary PCI memory write transactions. But they are recognized by LAPICs from the address they write to. The data sent in these transactions contain an interrupt vector.

Using a separate signal for PCI INTx interrupts raises several issues. First, many x86 systems require separate physical traces on the motherboard to connect the signals to interrupt controller input pins. Second, the interrupt signals should be routed cleverly to distribute interrupt requests evenly across the PIRQ lines. However, the largest issue can arise when a device writes data to memory and raises a pin-based interrupt to signal the CPU that the data has been written. The interrupt may arrive before all the data has arrived in memory. In order to ensure that all the data has arrived in memory, the interrupt handler must read a special register on the PCI device, which raises the interrupt. This read will not be completed until pending transactions between the CPU and the PCI device are completed. PCI transaction ordering rules require that all the data arrive in memory before the value may be returned from this special register. Thus, this dummy read guarantees that all the effects of the event that triggered the interrupt will be visible to the CPU. But this dummy read adds extra latency and work to the interrupt handler. Using message signaled interrupts avoids this problem as the interrupt message is a memory write transaction. Hence, it cannot pass the data writes, so when the interrupt is raised, the interrupt handler knows that all the data has been successfully written into memory.

To summarize, the advantages of MSI interrupts versus the legacy interrupts are:

- the MSI interrupts eliminate the need for interrupt traces between PCI devices and the I/O APIC,
- the MSI interrupts eliminate multiple PCI functions sharing the same PIRQ,
- the MSI interrupts eliminate the need for device polling in the interrupt handlers,
- the MSI interrupts eliminate the need to perform a read from a device's register to force all posted memory writes to be flushed to memory.

When a PCI function supports MSI, it generates an interrupt request to the processor's LAPIC by writing a predefined data item to a predefined memory address in the LAPIC. This PCI write transaction containing predefined data and a predefined address is referred to as an interrupt message. MSI was introduced as an optional component in revision 2.2 of the PCI spec in 1999. However, with the introduction of the PCIe specification in 2004, implementation of MSI became mandatory from a hardware standpoint. It is worth noting that MSI interrupts can't work without LAPIC, but MSI eliminates the devices' need to use the IO-APIC, allowing every device to write directly to the CPU's LAPIC.

Each PCI function that generates MSI must contain two addressable registers (Message Data register and Message Address register) where BIOS or OS stores this predefined data and addresses during the initialization. When a PCI function asserts an interrupt using MSI, it performs a PCI write operation that writes the content of the Message Data register to the address specified in the Message Address register.

The following is the sequence for MSI delivery and servicing:

1. A device needing servicing from the CPU generates an MSI, writing the interrupt vector number directly into the Local-APIC of the CPU servicing it.
2. The interrupted CPU begins running the handler associated with the interrupt vector number it received. The device is serviced without any need to check and clear an IRQ pending bit

The format of the Message Address register is presented in Fig. 2.61. Fields in the Message Address Register are as follows:

- Bits 31-20 contain a fixed value for interrupt messages (0xFEE). This value locates interrupts at the 1-MByte area with a base address of 0xFEE00000. All accesses to this region are directed as interrupt messages.
- Destination ID—This field contains an 8-bit destination ID. It identifies the target LAPIC.
- Redirection hint (RH)—When this bit is set, the message is directed to the processor with the lowest interrupt priority among processors that can receive the interrupt. When this bit is reset, the interrupt is directed to the processor listed in the Destination ID field.
- Destination mode (DM)—This bit indicates whether the Destination ID field should be interpreted as logical or physical LAPIC for delivery of the lowest priority interrupt. If RH is 0, then the DM bit is ignored. If RH is 1 and DM is 0, only the processor listed in the Destination ID field is considered for delivery of

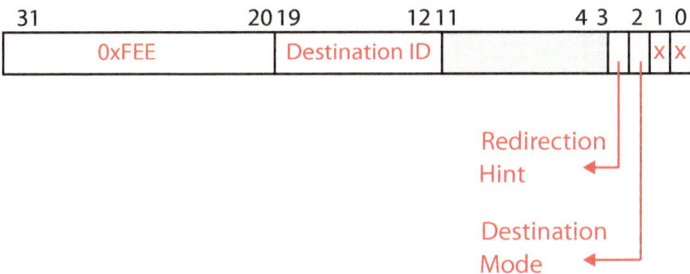

Fig. 2.61 Layout of the MSI memory address register

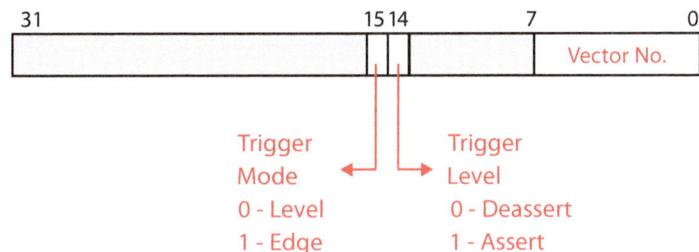

Fig. 2.62 Layout of the MSI memory data register

that interrupt (this means no redirection). If RH is 1 and DM is 1, the redirection is limited to only those processors that are part of the logical group of processors based on the Destination ID field in the message.

Figure 2.62 illustrates the format of the Message Data register. The fields in the Message Data Register are:

- Vector number. This 8-bit field contains the interrupt vector number associated with the message. Values range from 0x10 to 0xFE. The software must guarantee that the field is not programmed with vectors 0x00 to 0x0F, as these are reserved for non-external interrupts and exceptions.
- Trigger Mode. If this bit is 0, the interrupt is edge-triggered. If this bit is 1, the interrupt is level-triggered.
- Level. For edge-triggered interrupts, this field is ignored. For level-triggered interrupts, this bit reflects the active state of the interrupt input.

Direct Memory Access

3

CHAPTER GOALS

Have you ever wondered how information travels between input-output devices and main memory in a computer system? This chapter explains the Direct Memory access (DMA) I/O technique used in modern computer systems, including those using the Intel and ARM microprocessors. This chapter also aims to demystify the DMA controller internals and its programming with various peripherals. From this chapter, you should gain a basic understanding of DMAs, including:

- Understand the basic concept of Direct Memory Access (DMA) and its role in computer systems.
- Explore the need for DMA and its advantages over traditional Programmed I/O (PIO) methods.
- Distinguish between programmed IO, interrupt-driven IO, and DMA transfers.
- Learn about the integration of DMA into modern computer system architectures.
- Understand how DMA controllers interact with the CPU, memory, and peripheral devices.
- Explain the operation of the signals used in direct memory access controllers.
- Examine the operation of DMA controllers and their role in managing data transfers between peripherals and memory.
- Understand the different DMA transfer modes, including single transfer, block transfer, and scatter-gather DMA.
- Explore the mechanisms for initiating, controlling, and completing DMA transfers in various computing environments.

© The Author(s), under exclusive license to Springer Nature Switzerland AG 2024
P. Bulić, *Understanding Computer Organization*, Undergraduate Topics in Computer Science, https://doi.org/10.1007/978-3-031-58075-8_3

- Explain the function of the Intel 8237 DMA controller when used for DMA transfers.
- Explain the function of the DMA controller used in STM Cortex-M-based systems.
- Explain the function of bus-mastering (also referred to as first-party DMA).
- Explore real-world examples and case studies illustrating the use of DMA in diverse computing scenarios.
- Learn from practical examples of DMA-enabled applications.

3.1 Introduction

Direct Memory Access (DMA) emerges as a pivotal mechanism in computer systems and data transfer, revolutionizing the efficiency and speed of data movement within a computing environment. DMA catalyzes system performance by alleviating the burden on the CPU and enabling seamless, high-speed data transfers between peripherals and memory without CPU intervention.

Traditionally, data transfer between peripherals (such as network interfaces, storage devices, and I/O controllers) and system memory necessitated CPU involvement at every process step. This process, called programmed input/output, resulted in significant CPU overhead, latency, and inefficiency, limiting the overall system throughput and responsiveness.

DMA addresses these limitations by introducing a dedicated data transfer engine, separate from the CPU, capable of autonomously managing data transfers between peripherals and memory. By offloading data transfer tasks from the CPU to specialized DMA controllers, DMA significantly reduces CPU overhead and improves system performance, efficiency, and responsiveness.

At its core, DMA operates on the principle of direct access to system memory, enabling peripherals to read from or write to specific memory locations without CPU intervention. This direct access eliminates the need for the CPU to orchestrate each data transfer, allowing it to focus on executing critical tasks and enhancing overall system throughput. Throughout this exploration of Direct Memory Access (DMA), we will delve into its underlying principles, mechanisms, and applications across various computing domains. From its fundamental operation to advanced features and optimizations, we will uncover the transformative impact of DMA on system performance, efficiency, and scalability.

3.2 Programmed Input/Output

The programmed I/O was the most straightforward type of I/O technique for exchanging data between I/O devices and memory. This data transfer method requires the least amount of hardware. With programmed I/O, data transfers between I/O devices and memory are accomplished by the central processing unit (CPU). In the case of programmed I/O, *the I/O device does not have direct access to the main memory*. The I/O devices have memory-mapped registers. This means that the CPU accesses the I/O device's registers using LOAD/STORE instructions.

A transfer from an I/O device to the main memory (or vice versa) requires the execution of several instructions by the CPU. This includes a LOAD instruction to transfer the data from the I/O device's data register(s) to the CPU and a STORE instruction to transfer the data from the CPU to the main memory. Besides, the CPU must continuously sense the I/O device's status. When the CPU issues a command to the I/O device, it must wait until the I/O operation is complete or new data is available. For example, before reading the data from the I/O device with the LOAD instruction, the CPU must first read the status register (also with a LOAD instruction) of the I/O device to check if the I/O device has new data. Similarly, before writing the data to the I/O device, the CPU must first read the status register (also with a LOAD instruction) of the I/O device to check if the I/O device is prepared to accept new data. As the CPU is faster than the I/O module, the problem with programmed I/O is that the CPU has to wait a long time for the I/O device to be ready for either reception or transmission of data. The CPU stays in the program loop until the I/O unit indicates that it is ready for data transfer. This process of waiting and checking the status of the I/O device is known as **polling or busy waiting**. As a result, polling severely degrades the performance of the entire system. This situation can be avoided by using an interrupt-driven I/O, which we discuss in the next section.

Let's look at how programmed I/O would work if you were copying information from an I/O device to the main memory. Figure 3.1 illustrates a simplified block diagram of a computer system using programmed I/O. The I/O device shares the data, address, and control bus with the main memory. Although modern computer systems

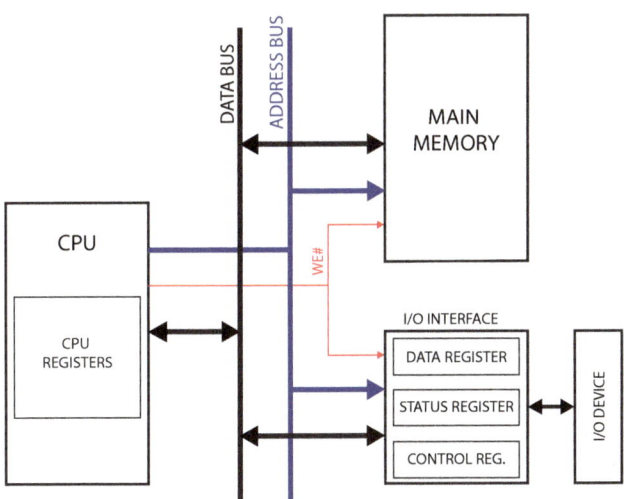

Fig. 3.1 A simplified block diagram of a computer system with programmed I/O

have more buses organized hierarchically, we can still simplify this discussion by assuming that there is only one bus in the system. The I/O devices in modern computer systems are memory-mapped, meaning that the CPU accesses these devices through a well-defined I/O interface. The I/O interface of an I/O device contains a set of registers, each with its unique address from the global address space. The CPU reads and writes to these I/O registers in the same way as it reads or writes to the main memory: using the LOAD and STORE instructions.

The I/O device in Fig. 3.1 has three memory-mapped registers: a control register, a status register, and a data register. The control register is used to program the I/O device, e.g., set the data rate, parity check, etc. The status register reflects the status of the I/O device, e.g., the I/O device is ready to accept new data, or the I/O device has new data, etc. The data register is used to transfer data to/from the I/O device. In programmed I/O mode, the CPU constantly checks the status register to see if new data is available. Thus, the CPU would read the status register with the LOAD instruction and check a particular bit, which flags that the I/O device has new data. The CPU would perform the polling operation inside a program loop. In the case new data is available, the CPU first transfers data from the data register into an internal register with the LOAD instruction. Then, the CPU would transfer data from the internal register into the memory with a STORE instruction. Listing 3.1 illustrates the programmed I/O transfer from the I/O device to the main memory:

```
                        ; wait for new data
    busy_wait:          lw r1, status_reg;
                        beq r1, r0, busy_wait;

                        ; CPU transfers data
    transfer:           lw r2, data_reg
                        sw rw, mem_addr
```

Listing 3.1 Programmed I/O data transfer

While not in use anymore, programmed I/O mode transfers were used in older hard drives a few decades ago when so-called DMA transfers didn't exist. For example, programmed I/O was used by the Western Digital WD1003, the hard disk controller used by the first PCs. Programmed I/O is still used now in some low-end and embedded computer systems. Also, the Intel 80286, 80386, and 80486 microprocessors used in personal computers were well suited to programmed I/O since they can move data blocks with a single *String Move* instruction. This data move instruction allowed programmed I/O transfers to reach speeds of about 2.5 Mbytes/s. In an embedded system where the CPU has nothing else to do, busy waiting is reasonable. However, polling is inefficient in a more sophisticated computer system where the CPU has to do other things. Hence, a better I/O transfer method is needed.

3.3   Interrupt-Driven I/O

A disadvantage of polling is that the CPU must continuously sense the I/O device's status in a loop. The waiting may significantly slow down the system's capability of executing other instructions and processing other data. The so-called interrupt-driven I/O could be more efficient. In interrupt-driven I/O, the I/O device, when ready for a new transfer, initiates the data transfer by interrupting the CPU. The CPU then executes the interrupt service program that transfers the data. Similarly, as in programmed I/O, the transfer from an I/O device to the main memory (or vice versa) requires the execution of a LOAD instruction to transfer the data from the I/O device's data register(s) to the CPU and a STORE instruction to transfer the data from CPU to the main memory. But now, there is no need to wait in a loop and read the I/O device's status. The interrupt-driven I/O technique requires more complex hardware but makes far more efficient use of CPU time and capacities.

For the transfer from an I/O device to memory, the device interrupts the CPU when new data has arrived and is ready to be retrieved by the system processor. As most I/O devices have memory-mapped registers, the interrupt service program will then read the device's data register into a CPU register and store the data from the CPU register to the memory location.

For the transfer from a memory location to an I/O device, the device delivers an interrupt either when it is ready to accept new data or to acknowledge a successful previous data transfer. The interrupt service program will then read the memory location into a CPU register and store the data from the CPU register to the device's data register.

Hence, in the interrupt-driven I/O, the CPU continuously works on given tasks. When the I/O device is ready for the data transfer, such as when someone types a key on the keyboard or a serial communication interface is prepared to transmit a new byte, it interrupts the CPU from its work to take care of the data transfer. The CPU can work continuously on a task without checking the input devices, allowing the devices themselves to interrupt it as necessary.

The interrupt-driven I/O is adequate for simple computer systems, but some situations in modern computing complicate the picture. For example, what if the CPU executes some critical task that should not be interrupted? What if the CPU executes an interrupt service program corresponding to an interrupt with a priority higher than the current interrupt request? In such a case, the handling of the current interrupt request from an I/O device should be deferred.

3.4   Direct Memory Access

We have seen two different methods used to transfer data between I/O devices and the main memory: polling and interrupt-driven I/O. Both techniques work well with low-bandwidth devices and some low-end computer systems, and both methods use the CPU to move data. While transferring data, the CPU can not perform other operations, making both methods inappropriate for modern, high-speed computer systems.

An alternative mechanism is to offload the CPU and have **another device transfer data directly to or from the main memory**—without involving the CPU. This mechanism is called **direct memory access (DMA)**. DMA is a feature that allows systems to access the main memory without any help from the processor. The special device that performs the DMA transfer is a **DMA controller**. A DMA controller offloads the CPU tremendously as it fulfills a memory transfer without intervention from the processor. When the transfer is finished, it signals the CPU with an interrupt.

The DMA controller transfers data between the main memory and an I/O device independent of the CPU. The CPU only initializes the DMA controller. A DMA transfer is fulfilled in the following steps:

1. The CPU initializes the DMA controller: it provides the source and destination addresses of the data to be transferred, the number of bytes to be transferred, and the type of transfer to perform (we will discuss these types later).
2. When data is available, **the I/O device requests the DMA transfer** from the DMA controller. The DMA controller then requests the bus from the CPU. When the CPU grants access to the bus, the DMA controller becomes the bus master and starts the transfer. During the transfer, the DMA controller supplies the memory addresses and the control signals needed to complete the transfer. If the request from the I/O device requires more than one transfer, the DMA controller will automatically generate the next memory address(es) and complete the entire DMA transfer of hundreds of thousands of bytes without involving the CPU. The modern DMA controllers usually contain FIFO buffers that help them deal with different timings and delays during a transfer.
3. When the DMA transfer is complete, the DMA controller interrupts the CPU. The CPU can then decide if more transfers are required and reinitialize the DMA controller for new DMA transfers.

I'm sure you are now wondering how the CPU accesses the main memory during a DMA transfer. Well, (usually) it does not. But wait, how does the CPU fetch instructions and data from the main memory if the DMA controller occupies the memory bus? The CPU in modern computer systems never directly accesses the main memory—it always accesses the L1, L2, and L3 caches first, and only if there is a miss in the L3 cache the memory controller transfers the cache line to/from the main memory. Thus, we rely on the temporal and spatial data locality and assume that there is a very high probability that instructions and operands needed by the CPU are already in the cache(s). So, the DMA transfer usually does not prevent the CPU from fetching instructions and data. By using caches, the CPU leaves most of the memory bandwidth free for use by a DMA controller. In the case of the cache miss, modern systems rely on multitasking: in that case, the OS would perform a switch to another (ready) task.

Summary: Direct Memory Access

Direct memory access (DMA) is a mechanism that allows us to offload the CPU and to have a DMA controller transfer data directly between a peripheral device and the main memory.

A DMA transfer starts with a peripheral device placing a DMA request to the DMA controller. The DMA controller then requests the bus from the CPU and starts the transfer. When the DMA transfer is complete, the DMA controller interrupts the CPU.

Because of the use of cache and memory hierarchy in modern computer systems, a DMA transfer does not prevent the CPU from fetching instructions and data.

Let's look at how a DMA controller transfers data between the main memory and an I/O device. Figure 3.2 illustrates a simplified block diagram of a system with a DMA controller. The DMA controller is connected to the data, address, and control buses. It also has six control signals: DREQ, DACK, HOLD, HLDA, END, and WE#.

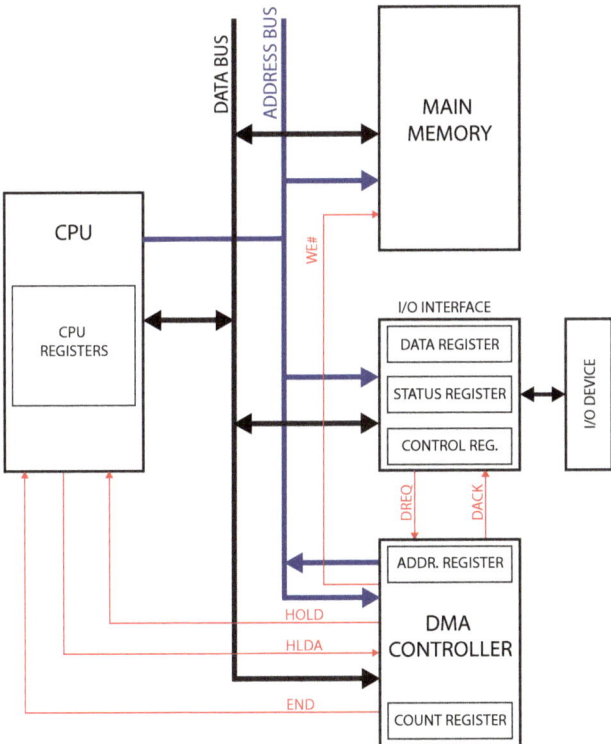

Fig. 3.2 A simplified block diagram of a computer system with a DMA controller

Two control signals, DREQ (DMA request) and DACK (DMA acknowledge), are used between the DMA controller and an I/O device to request and acknowledge a DMA transfer. Two other control signals, HOLD (Hold Request) and HLDA (Hold Acknowledge), are used between the DMA controller and the CPU to request and acknowledge a DMA transfer. The WE# signal selects between the memory read or memory write operations, and the END signals the CPE that the DMA transfer is finished and the data is ready for further processing. The DMA controller has two registers: the **address register** and the **count register**. The address register holds the memory location address from/to which the data is to be transferred. The count register holds the number of data words to be transferred. Both registers are memory-mapped, and the CPU is responsible for their initialization.

Before any data transfer occurs, the CPU should initialize the DMA controller's address and count registers. The CPU writes the memory location address to/from which the data is to be transferred into the address register and the number of data words to be transferred into the count register. During the transfer, the DMA controller will decrement the count register after each data word is transferred. It will also increment or decrement the address register, depending on the mode of operation, and automatically store/read the data to/from consecutive memory locations. When the count register signals that there is no more data to be transferred, the DMA controller will activate the END signal. This signal is usually connected to a CPU interrupt input and raises an interrupt when a transfer is completed.

Let's suppose the data transfer of one data word from the main memory to the I/O device. Firstly, the CPU writes the memory location address that holds the data into the address register and the value of 1 into the count register, indicating that only one data word is to be transferred. The following steps are then required to accomplish the DMA transfer (Fig. 3.3):

1. The I/O device is ready to receive data, so it asserts the DREQ signal.
2. The DMA controller requests the bus (requests the DMA transfer) from the CPU by asserting the HOLD signal.
3. The CPU relinquishes the control of the main memory. It voluntarily places all its bus signals at a high-impedance state and asserts the HLDA signal to indicate the bus is granted.
4. The DMA controller places the memory address from the address register on the address bus and puts the WE# signal into the high state to indicate the read access.
5. The main memory places the requested data onto the data bus.
6. The DMA controller asserts the DACK signal. This indicates that the I/O device can fetch data from the data bus. The DMA controller also decrements the count register.
7. The I/O device latches data from the data bus into its data register.
8. As the count register now indicates that there is no data left to transfer, the DMA controller de-asserts the HOLD signal to return the control over the bus to the CPU and activates the END signal to raise a CPU interrupt.
9. The CPU de-asserts the HLDA signal and eventually starts to service the interrupt request.

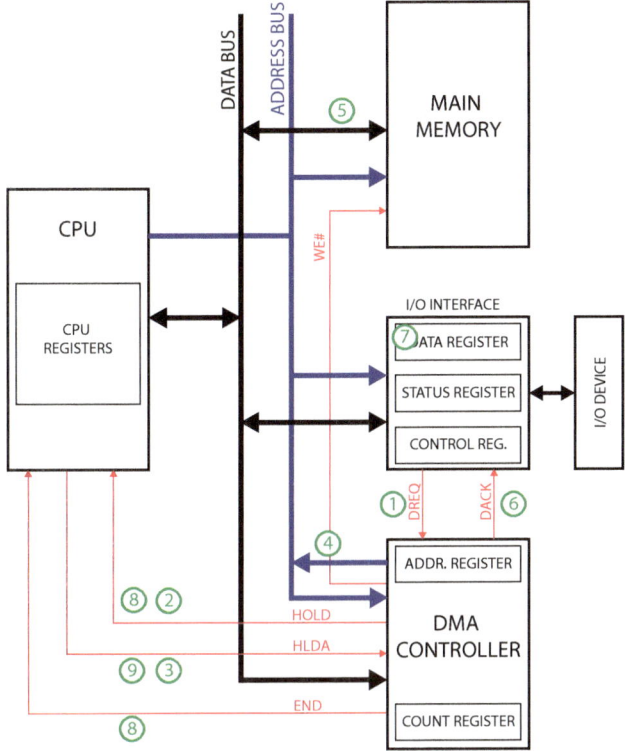

Fig. 3.3 A DMA transfer

Usually, only one DMA controller in the computer system is used for DMA transfers to/from several I/O devices. In that case, **the DMA controller has a separate pair of DREQ and DACK signals for each I/O device**. This separate pair (DREQ, DACK) is called **DMA channel**.

The DMA transfer described above is referred to as **"Fly-by" DMA**. This means that the data, which is transferred between an I/O device and memory, does not pass through the DMA controller. "Fly-by" DMA refers to the DMA transfer between an I/O device and memory in which the data flows into/out of the memory using address lines for only one side of the transfer (the memory side). The other side (the I/O device) is "addressed" by the DACK signal, i.e., the DACK signal selects the I/O device involved in the DMA transfer, which should then latch data from the bus or place data onto the bus.

The way that the DMA function is implemented varies between computer architectures, and there is also another type of DMA transfer referred to as **"Fly-through" DMA**. In "Fly-through" DMA, both source and destination addresses must be specified. The data flows through the DMA controller, which now has a FIFO buffer to store the data temporarily. The "Fly-through" DMA controller first places the source address onto the address bus, reads the data from the source into its internal FIFO, then places the destination address onto the address bus and writes the data from its FIFO into the destination.

"Fly-by" DMA is much faster because "Fly-through" DMA results in two bus transfers: one from the source to the internal FIFO and the other from the internal FIFO to the destination. On the other hand, "Fly-through" DMA enables memory-to-memory DMA transfers, which are impossible with "Fly-by" DMA controllers.

Summary: DMA controllers

Each DMA transfer is driven by at least the DMA controller's two internal registers: the address register and the count register.

A DMA channel is a pair of two control signals between a peripheral device and the DMA controller: DMA request (DREQ) and DMA acknowledge (DACK).

In "Fly-by" DMA transfers, the data, which is transferred between an I/O device and memory, does not pass through the DMA controller. Only the memory address must be specified, while the DACK signal selects the peripheral device. Only one memory transaction is needed to accomplish a DMA transfer.

In "Fly-through" DMA, both source and destination addresses must be specified. The data flows through the DMA controller, which has a FIFO buffer to store the data temporarily. The "Fly-through" DMA controller first places the source address onto the address bus, reads the data from the source into its internal FIFO, then places the destination address onto the address bus and writes the data from its FIFO into the destination. Two memory transactions are required to accomplish one DMA transfer.

3.5 Real-World DMA Controllers

So far, we have learned that a DMA controller is a special device that transfers data between an I/O device and the main memory without involving the CPU. Modern DMA controllers also support memory-to-memory DMA transfers, thus allowing efficient data transfers between two memory regions. For example, on most modern computer systems, the C library function memcpy() is implemented using the DMA transfer. This section will describe two real-world DMA controllers and their functionality: the Intel 8237A DMA "fly-by" controller used in older Intel PCs and the "fly-through" DMA controller used in modern ARM Cortex-M-based systems.

3.5.1 Intel 8237A DMA Controller

The Intel PC DMA subsystem is based on the Intel 8237A DMA controller. The Intel 8237A contains four DMA channels that can be programmed independently, and any of the channels may be active at any moment. These channels are numbered as 0, 1, 2, and 3. The Intel 8237A DMA controller moves one byte in each transfer and is very similar to the DMA controller described in Fig. 3.2.

The Intel 8237A is depicted in Fig. 3.4. It has two signals for each channel, named DRQ (DMA request) and DACK (DMA acknowledge). There are additional signals with the names HRQ (Hold Request), HLDA (Hold Acknowledge), EOP# (End of Process), and the bus control signals MEMR# (Memory Read) and MEMW# (Memory Write). Table 3.1 provides a full 8237A signals description.

The Intel 8237A DMA controller is a "fly-by" DMA controller. Subsequently, the DMA can only transfer data between an I/O device and a memory, but not between two I/O devices or two memory locations. Actually, the Intel 8237A controller does allow two channels to be connected to allow memory-to-memory DMA operations, but nobody in the PC industry used this DMA controller this way since it is faster to move data between memory locations using the CPU. Each DMA channel is activated only when an I/O device connected to that DMA channel requests a transfer by asserting the DRQ line.

Each channel in the 8237A DMA controller has two internal registers that control the transfer: the **count register** and the **address register**. Both registers are programmable by the CPU. The count register holds the number of bytes to be transferred, while the address register holds the (initial) memory address. When a data byte is transferred, the address register is decremented or incremented, depending on how it is programmed. The count register is decremented after each transfer. When the value in the count register goes from zero to 0xFFFF, the EOP# output signal is activated.

The 8237A is designed to operate in two major cycles. These are called Idle and Active cycles. When no channel is requesting DMA transfer, the 8237A controller enters the Idle cycle. In this cycle, the 8237A samples the DREQ lines every clock

Fig. 3.4 Intel 8237A DMA controller. The signals used to initialize the DMA controller are not shown

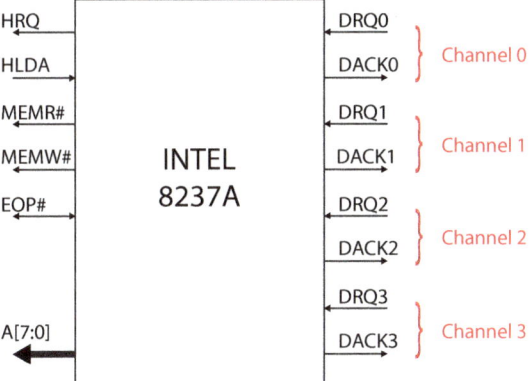

Table 3.1 Intel 8237A signals description

Signal name	Signal description
HRQ	Hold request is an output used to request the bus
HLDA	Hold acknowledge is an input that signals that the CPU has granted the bus
DREQ[3:0]	DMA request inputs are used to request a DMA transfer for each of the four DMA channels
DACK[3:0]	DMA acknowledge outputs acknowledge the a channel DMA request and select the I/O device during the DMA transfer
A[7:0]	These pins are outputs and are used to provide the DMA transfer memory address
EOP#	End-of-process is a bidirectional active-low signal used as an input to terminate a DMA transfer or as an output to signal the end of a DMA transfer
MEMR#	Memory read is an active-low output used to read data from the selected memory location during a DMA transfer
MEMW#	Memory write is an active-low output used to write data to the selected memory location during a DMA transfer

cycle to determine if any channel is requesting a DMA transfer. When a channel requests a DMA service by asserting its DREQ signal, the 8237A asserts the HRQ signal to the microprocessor requesting the bus and enters the Active cycle. It is in the Active cycle that the DMA transfer will take place.

There are three modes of operation: single mode, block mode, and demand mode. In single mode, the device is programmed to make one transfer only. Single-mode transfer releases the HRQ signal after each byte is transferred. In this mode, DRQ must be held active until DACK becomes active. If the DRQ is held active, the 8237A again requests the bus with the HRQ signal. Upon receipt of a new HLDA, another single transfer will be performed.

Block mode automatically transfers the number of bytes indicated by the count register. The count register will be decremented, and the address register will be decremented or incremented following each transfer. In block mode, the DMA controller is activated by DREQ to continue making transfers until the count register goes from 0x0000 to 0xFFFF, or an external EOP# is activated. DREQ needs to be held active only until DACK becomes active.

In demand mode, the DMA controller transfers data until an external EOP# is asserted or until DREQ goes inactive. This mode is used when there is a block of data to be transferred, but the I/O device does not have a high data capacity, and the transfer should be paused until the I/O device is ready again. During the time when the transfer is paused, the CPU is allowed to use the bus, and the intermediate values of address and word count are preserved. When the I/O is ready to continue, the DMA transfer is re-established by activating the DREQ signal.

Intel 8237A was also used for a DRAM memory refresh. The content of the DRAM memory must be regularly refreshed in order to preserve the data. We will detail DRAM operations in the next chapter. For now, it suffice to say that a row of data in DRAM is refreshed by reading it into an internal row register and writing it back. To refresh one data row in DRAM, the 8237A DMA controller reads data from memory onto the data bus. During this dummy read, data from the DRAM array is read into the row register. This automatically leads to the refresh of one memory cell row. However, no I/O devices fetch the data, as no device issued a DRQ. Hence, the DMA controller does not assert a DACK.

Summary: Intel 8237A DMA Controller

The Intel 8237A controller is a "fly-by" DMA controller. Subsequently, the DMA can only transfer data between an I/O device and a memory. Each DMA transfer requires only one memory transaction. It was used in Intel-based PC systems.

It contains four DMA channels, and any of the channels may be active at any moment. Each channel in The 8237A DMA controller has two internal registers that control the transfer: the count register and the address register. Both registers are programmable by the CPU.

3.5.2 STM32H7 Series DMA Controller

This subsection describes the direct memory access (DMA) controller available in the STM32H7 Arm Cortex-M7 core-based series of systems-on-chips. The STM32H7 series DMA controller allows data transfers to occur in the background without the intervention of the Cortex-M7 processor. During this operation, the processor can execute other tasks, and it is only interrupted when a whole data block is transferred and available for processing. The STM32H7 series DMA controller is a fly-through DMA controller and supports the transfer of large amounts of data with no significant impact on system performance. The DMA controller can do automated memory-to-memory, peripheral-to-memory, and peripheral-to-peripheral transfers.

Figure 3.5 illustrates the simplified block diagram of the STM32H7 series DMA controller. The STM32H7 series DMA controller features three ports: a programming port for DMA programming and two ports (peripheral and memory ports) that allow the DMA to initiate data transfers between different I/O devices and memory. A **port** connects the data, address, and control bus. Thus, one port comprises data lines, address lines, and control signals, which are not depicted in Fig. 3.5 due to simplicity.

Each STM32H7 series DMA controller supports up to eight streams. A **stream** is an active DMA transfer between a peripheral device and memory, two peripheral devices, or between two memory blocks. Each stream is associated with a peripheral device that can trigger a data transfer request when ready. More than one enabled DMA stream must not serve the same peripheral request.

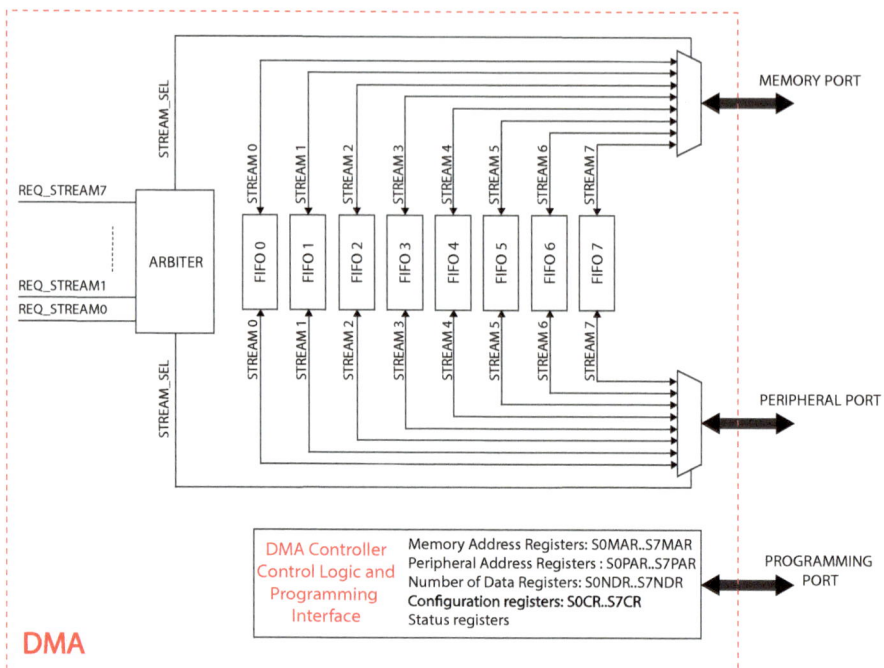

Fig. 3.5 Simplified block diagram of a STM32H7 series DMA controller

The DMA controller contains an arbiter for handling the priority between DMA streams. Stream priority is software-configurable (four levels: very high, high, medium, low). The arbiter selects the stream with the highest priority. The lowest stream number gets priority if two or more DMA streams have the same software priority level.

The programmable 8-channel DMAMUX multiplexer block (Fig. 3.6) enables routing DMA request lines between the peripherals and the DMA controller. Each channel contains one 128-to-1 multiplexer and selects a unique DMA request line from peripherals. Each DMA stream is driven by one DMAMUX output channel (request). Any DMAMUX output request can be individually programmed to select the DMA request source signal from up to 128 available request input signals. The assignment of DMAMUX request multiplexer inputs to the DMA request lines from peripherals are detailed in the product reference manual.

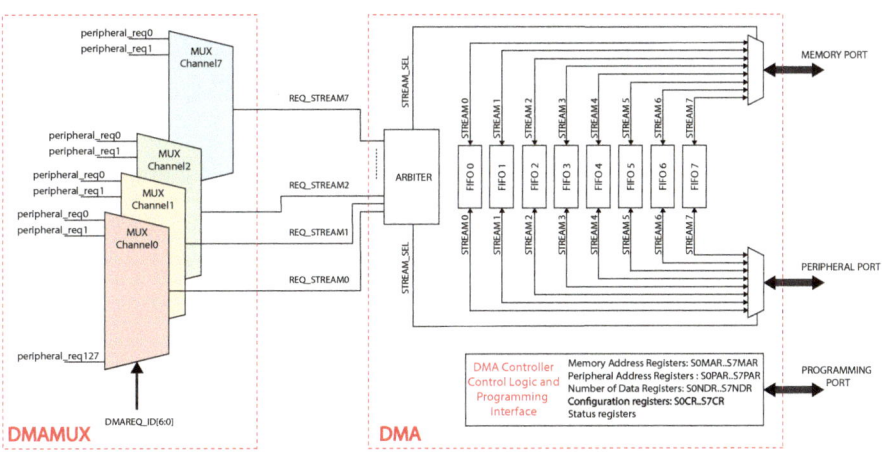

Fig. 3.6 Simplified block diagram of DMAMUX block and STM32H7 DMA controller

When a peripheral is ready, it sends a DMA request to the DMAMUX, which routes the request to a selected DMA stream request input of the DMA controller. Depending on the stream priority, the DMA controller will serve the DMA request. As this is a "fly-through" DMA controller, the data flows through the DMA controller, which has a FIFO buffer associated with each stream. The FIFO buffer is used to temporarily store the data and amortize the difference in transmission speeds of two peripheral devices. Standard block transfer is accomplished by the DMA controller performing a sequence of memory transfers. Each transfer involves a load operation from a source address into the FIFO, followed by a store operation from the FIFO to a destination address.

The DMA controller's control logic and programming interface are accessed through the programming port. The programming interface comprises a set of memory-mapped registers per stream. Each stream is characterized by four registers: Memory Address Register (SxMAR), Peripheral Address Register (SxPAR), Number of Data Register (SxNDR), and Configuration Register (SxCR). All these registers are memory-mapped and will be discussed in the following subsections.

STM32H7 devices embed two DMA controllers, offering up to 16 streams in total (eight per controller), each dedicated to managing memory access requests from one or more peripherals.

Summary: STM32H7 series DMA controller

The STM32H7 series DMA controller is a "fly-through" DMA controller used in the STM32H7 Arm Cortex-M-based systems.

The STM32H7 series DMA controller features three ports: a programming port for DMA programming and initialization and two ports (peripheral and memory ports) that allow the DMA to initiate data transfers between different I/O devices and memory.

Each STM32H7 series DMA controller supports up to eight *streams*. A stream is an active DMA transfer between a peripheral device and memory, two peripheral devices, or between two memory blocks.

Each DMA stream is driven by one DMAMUX output channel (request).

As this is a "fly-through" DMA controller, the data flows through the DMA controller, which has a FIFO buffer associated with each stream.

STM32H7 devices embed two DMA controllers.

3.5.2.1 Peripheral and Memory Addresses

Each DMA transfer is defined by a source address and a destination address. Both addresses should be aligned to transfer size. The transfer size value defines the volume of data to be transferred from source to destination. Each stream has a pair of registers to store these addresses: Peripheral Address Register (SxPAR—*Stream x Peripheral Address Register*) and Memory Address Register (SxMAR—*Stream x Memory Address Register*). Before each transfer, the CPU should initialize both registers with valid addresses. It is possible to configure the DMA to automatically increment the source and/or destination address after each data transfer.

3.5.2.2 Transfer Size, Type and Mode

The transfer size and the transfer mode also define each DMA transfer. The transfer size is a value that defines the volume of data to be transferred from source to destination. This value is stored in the so-called Number of Data Register (NDR). Each stream has its Number of Data Register, labeled as SxNDTR. Each SxNDTR is a 16-bit register, and the number of data items to be transferred is software-programmable from 0 to 65535. After each transfer, the value in SxNDTR is decreased by the amount of the transferred data; thus, SxNDTR contains the number of data transfers still to be performed.

The STM32H7 series DMA controller can perform two transfer types: normal type and circular type. In normal type, once the SxNDTR register reaches zero (the transfer has been completed), the stream is disabled. This means that the CPU should reinitialize the DMA controller in order to activate the stream again. In the circular type, the DMA controller can handle circular buffers and continuous data flow. In this type, the SxNDTR register is reloaded automatically with the previously programmed value when a transfer has been completed.

Each STM32H7 series DMA controller is capable of performing three different transfer modes:

1. peripheral to memory,
2. memory to peripheral,
3. memory to memory.

3.5.2.3   FIFOs and Burst Transfers

Each stream has a 4x32-bit FIFO that temporarily stores data coming from the source before transmitting them to the destination. The DMA FIFOs help reduce memory access and do burst transactions, which optimize the transfer bandwidth. They also allow independent source and destination transfer width (byte, half-word, word). When the data widths of the source and destination are not equal, the DMA automatically packs/unpacks the necessary transfers to optimize the bandwidth. For example, the data from the source can be transferred into FIFO as bytes or 16-bit half-words and then transferred to the destination from FIFO as bytes, 16-bit half-words, or 32-bit words.

Because of the internal FIFOs, the DMA controller is capable of burst transfers of length 4x, 8x, or 16x data units. A data unit can be a byte, a 16-bit half-word, or a 32-bit word. The burst size on the DMA peripheral port must be set according to the peripheral needs/capabilities. The size of the burst is software-configurable, usually equal to half the FIFO size of the peripheral.

3.5.2.4   DMA Transactions

A DMA transaction consists of a sequence of a given number of data transfers. The number of data items to be transferred and their width (8-bit, 16-bit, or 32-bit) are software-programmable. Each DMA transfer consists of three operations:

1. loading from the peripheral data register or a location in memory addressed through the DMA_SxPAR or DMA_SxM0AR register
2. storage of the data loaded to the peripheral data register or a location in memory addressed through the DMA_SxPAR or DMA_SxM0AR register
3. post-decrement of the DMA_SxNDTR register containing the number of transactions that still have to be performed.

The peripheral sends a request signal to the DMA controller through the DMA-MUX block. The DMA controller serves the request depending on the channel priorities. As soon as the DMA controller accesses the peripheral, an acknowledgment signal is sent to the peripheral by the DMA controller, which in turn releases its request. Once the peripheral de-asserts the request, the DMA controller releases the acknowledge signal.

Figure 3.7 illustrates a peripheral-to-memory DMA transaction. Each time a peripheral request occurs, the stream initiates a transfer from the source (address is in SxPAR) to fill the FIFO. Then, the contents of the FIFO are drained and stored

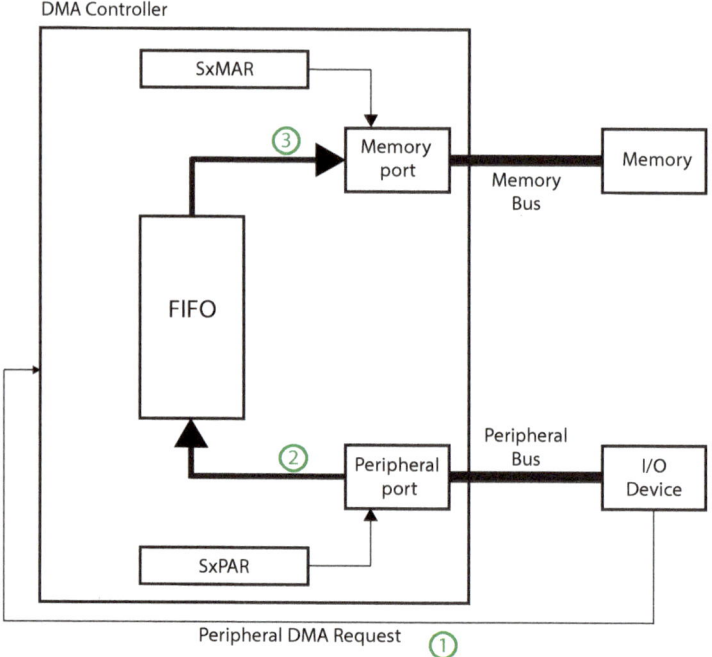

Fig. 3.7 A peripheral-to-memory DMA transaction

in the destination (address is in the SxMAR). The transfer stops once the SxNDTR register reaches zero or when the enable bit in the SxCR register is cleared by software (stream disabled).

Figure 3.8 illustrates a memory-to-peripheral DMA transaction. In this mode, the stream immediately initiates transfers from the source (address is in SxMAR) to fill the FIFO entirely, and the SxMAR register is incremented/decremented. The DMA controller does not wait for a DMA request from a peripheral device to read from memory. When a peripheral request occurs, the contents of the FIFO are drained and stored in the destination (address is in the SxPAR). The DMA controller reloads the empty internal FIFO again with the next data to be transferred from memory (address is in SxMAR). The transfer stops once the SxNDTR register reaches zero or when the enable bit in the SxCR register is cleared by software (stream disabled).

3.5.2.5   Programming and Using the STM32H7 Series DMA Controller

Programming and using the STM32H7 series DMA controller is relatively easy. Each stream is controlled using four memory-mapped registers: memory address register (SxMAR), peripheral address register (SxPAR), number-of-data register (SxNDTR), and configuration register (SxCR). Once set, the DMA controller handles data transfers and increments memory addresses without disturbing the CPU. To configure the DMA controller and a DMA stream, the following procedure should be applied:

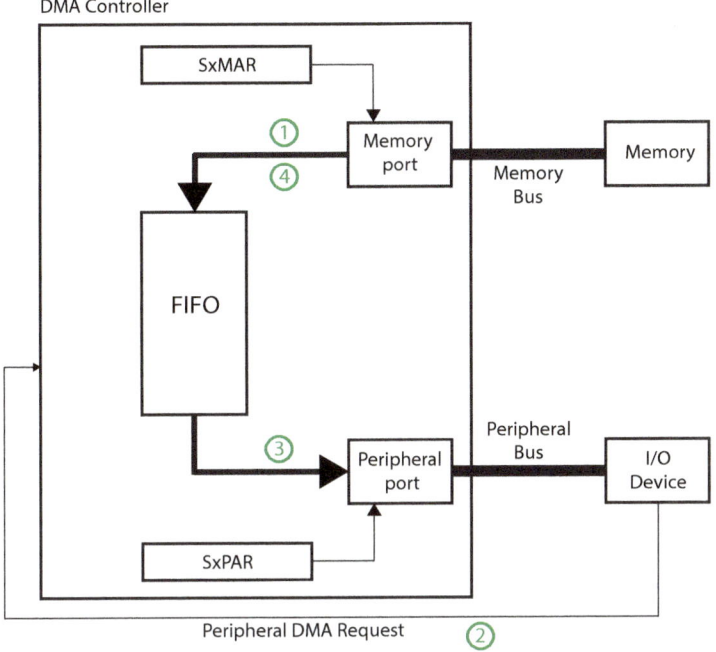

Fig. 3.8 A memory-to-peripheral DMA transaction. In this mode, the stream immediately initiates transfers from the memory to entirely fill the FIFO. When a peripheral request occurs, the contents of the FIFO are stored in the peripheral device

1. If the stream is enabled, disable it by resetting the stream enable bit in the SxCR register, then read this bit to confirm that there is no ongoing stream operation. When the EN bit is 0, the stream is ready to be configured.
2. Set the peripheral port register address in the SxPAR register. After the peripheral DMA request, the data will be moved from/to this address to/from the peripheral port.
3. Set the memory address in the SxMAR register. After the peripheral DMA request, the data will be written to or read from this memory address.
4. Configure the total number of data items to be transferred in the SxNDTR register. After each (burst) transfer, this value is decremented accordingly.
5. Use DMAMUX to route a peripheral DMA request line to the DMA stream request signal.
6. Configure the stream priority, the data transfer direction, single or burst transactions, peripheral and memory data widths, circular/normal transfer type, and interrupts in the SxCR register.
7. Activate the stream by setting the stream enable bit in the SxCR register.

As soon as the stream is enabled, it can serve any DMA request from the peripheral connected to the stream, and DMA transactions using the stream can be performed.

Summary: STM32H7 series DMA transfers

Source and destination addresses define each DMA transfer. Each stream has a pair of registers to store these addresses: Peripheral Address Register (SxPAR -Streamx Peripheral Address Register) and Memory Address Register (SxMAR -Stream x Memory Address Register)

Each DMA transfer is also defined by the transfer size and the transfer mode. Each stream has its Number of Data Register (SxNDR), which stores the transfer size.

The STM32H7 series DMA controller can perform two transfer types: normal type and circular type.

FIFOs allow independent source and destination transfer width and burst transfers.

Each DMA transfer consists of two transactions on the bus: loading from the peripheral data register or a location in memory and storage of the data to the peripheral data register or a location in memory.

3.6   Bus Mastering DMA

So far, we have learned that we can use a special piece of hardware, a DMA controller, to transfer large amounts of data between a peripheral device and memory. This approach is sometimes referred to as **third-party DMA**. Third-party DMA requires an independent DMA controller, which is built into motherboard chipsets, to move data between a peripheral device (referred to as the *first party*) and system RAM (referred to as the *second party*). Here, the DMA controller is shared by multiple peripheral devices, which is why it is viewed as a third-party DMA. As we have learned previously, each "fly-through" DMA transfer (the type of DMA used in most of today's computer systems) requires two memory transactions: one to load the data from the source and one to store the data to the destination.

   The better approach to DMA transfers would be having only one memory transaction per DMA transfer while avoiding third-party "fly-by" DMA controllers. This is possible with the latest I/O devices built into modern computer systems, where each I/O device can act as a *bus master*, i.e., each device can directly access any other I/O device or memory on the bus. Indeed, each modern I/O device now contains its own integrated DMA controller, which is not shared by other I/O devices. This highest performing DMA type is called first-party DMA or **Bus Mastering DMA**. Peripheral devices, which support the Bus Mastering technology, have the ability to move data to and from system memory without the intervention of the CPU or a third-party DMA controller.

   Bus Mastering allows data to be transferred much faster than third-party DMA. This is because half as many bus cycles are needed. The third-party DMA requires

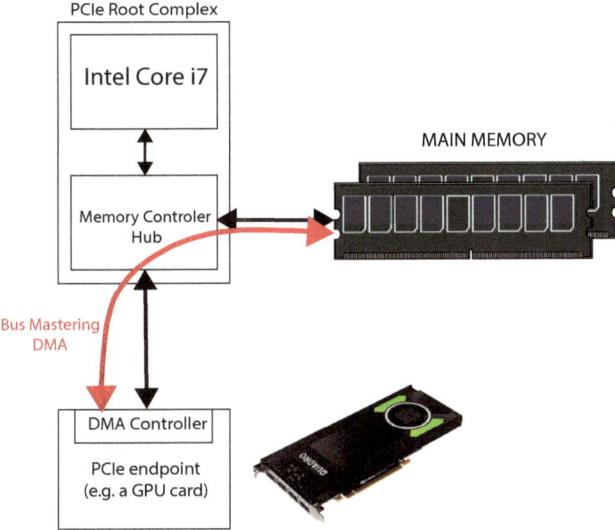

Fig. 3.9 Bus Mastering in an Intel-based system. Bus Mastering is the feature integrated into PCIe endpoint devices. A DMA transfer either transfers data from an endpoint device into system memory or from system memory into the endpoint device on the PCI Express bus. The DMA request is always initiated by the integrated DMA controller in the endpoint device after being initialized from the application driver (i.e., receiving parameters that define DMA transaction and memory buffer address)

the DMA controller to alternately read a segment of data from one device (this can be a peripheral device or system memory) and write it to the other device. Each data segment requires at least one bus cycle to be read and one bus cycle to be written. Bus mastering devices only require bus cycles when accessing system memory, so half as many bus cycles are needed. Because of this, devices that support Bus Mastering can move data many times faster than third-party DMA. While bus mastering theoretically allows one peripheral device to communicate directly with another, in practice, almost all peripherals master the bus exclusively to perform peripheral-to-memory and memory-to-peripheral transfers.

Bus Mastering is used in the computer systems with a PCI Express (PCIe) bus. A Bus Mastering DMA implementation is by far the most common type of DMA found in systems based on PCI Express and resides within the peripheral device, which is called Bus Master because it initiates the movement of data to and from system memory. Figure 3.9 shows a typical Intel system architecture. The system includes the CPU core(s) and a memory controller hub, forming the so-called PCIe root complex. The system in Fig. 3.9 also contains the main memory and one PCIe peripheral device (e.g., a GPU card). The peripheral device is connected to the PCIe bus. PCIe peripheral devices are called PCI **endpoint devices**. The memory controller hub also acts as a bridge between the PCIe bus, the CPU bus, and the main memory bus. A DMA transfer either transfers data from an endpoint device into system memory or from system memory into the endpoint device on the PCI Express bus. The DMA request is always initiated by the integrated DMA controller in the endpoint device after being initialized from the application driver (i.e., receiving parameters that define DMA transaction and memory buffer address).

Summary: Bus Mastering

Bus Mastering DMA is the highest-performing DMA type. Peripheral devices, which support the Bus Mastering technology, have the ability to move data to and from system memory without the intervention of the CPU or a third-party DMA controller.

Bus Mastering is used in the computer systems with a PCI Express (PCIe) bus.

PCIe peripheral devices are called PCI **endpoint devices**. Endpoint devices have their own integrated DMA controller, which is not shared by other endpoint devices.

Main Memory

4

CHAPTER GOALS

In this chapter, we will cover the modern memory design and operations in memory chips and modules that enable efficient data transfer between the memory controller and the so-called DIMMs, i.e., memory modules used in modern computer systems. To fully understand the organization and operation of modern memory chips, we need to start with some fundamental digital building blocks. Then, we gradually build memory components, arrays, operations inside the memory chips, timings, and techniques to boost the performance of memory chips. At the end of the chapter, you should fully understand modern DDR SDRAM chips, DDR memory technology, memory timings, DIMM modules, and multi-channel architecture.

From this chapter, you should gain a basic understanding of the design and operation of computer memory and storage circuits, including:

- Static memory circuits using the six-transistor cell.
- Dynamic memory circuits, including the one-transistor cell.
- Understand the fundamental principles and operation of Dynamic Random Access Memory (DRAM).
- Examine the organization of DRAM chips and modules, including row and column addressing schemes.
- Understand the different access modes of DRAM, including page mode, fast page mode, and burst mode.
- Explore the historical evolution and key advancements in DRAM technology.

© The Author(s), under exclusive license to Springer Nature Switzerland AG 2024
P. Bulić, *Understanding Computer Organization*, Undergraduate Topics in Computer Science, https://doi.org/10.1007/978-3-031-58075-8_4

CHAPTER GOALS

- Explore the implications of DRAM organization and access modes on memory access latency and bandwidth.
- Learn about the various timing parameters and specifications that characterize DRAM performance.
- Explore the impact of DRAM timing parameters on memory access speed and system performance.
- Understand how architectural enhancements improve DRAM performance, reliability, and scalability.
- Explore the evolution of Double Data Rate (DDR) memory standards and their impact on DRAM technology and learn about different DDR generations, including DDR2, DDR3, DDR4, and DDR5.
- Learn from practical examples of SDRAM usage.
- Understand the basic concept and purpose of Dual In-Line Memory Modules (DIMMs) in computer systems.
- Learn about DIMM slots, memory channels, and population rules on motherboards.

4.1   Introduction

In the intricate ecosystem of computer systems, Random Access Memory (RAM) emerges as a cornerstone, playing a pivotal role in storing and accessing data for rapid retrieval and manipulation. As the primary volatile memory component in modern computing devices, RAM is a dynamic workspace where programs, applications, and data reside during active use, facilitating seamless multitasking, efficient data processing, and responsive system performance.

At its essence, RAM embodies the essence of immediacy and accessibility, offering fast, byte-addressable storage that enables rapid read and write operations. Unlike non-volatile storage mediums such as hard disk drives (HDDs) or solid-state drives (SSDs), which store data persistently but incur latency penalties for access, RAM provides near-instantaneous access to stored data, making it indispensable for tasks that demand high-speed data manipulation and real-time responsiveness.

In this chapter, we focus on the main memory used in modern computer systems as one in Fig. 4.1. Figure 4.1 illustrates the memory hierarchy in the Intel i7-860 based system. Intel i7-860 is an out-of-order execution processor that includes four cores. The L1 and L2 caches are separate for each core, while the L3 cache is shared among the cores on a chip. The L1 cache is the 32 KB, four-way set-associative cache. There are two L1 caches per core: instruction (I) and data (D). The L2 cache is the 256 KB, eight-way set-associative cache. Finally, the L3 cache is the 8 MB, 16-way set-associative cache.

A CPU core directly accesses only its L1 cache. If a hit in L1 occurs, the data is returned after an initial latency of 4 cycles. If the L1 cache misses, the L2 cache

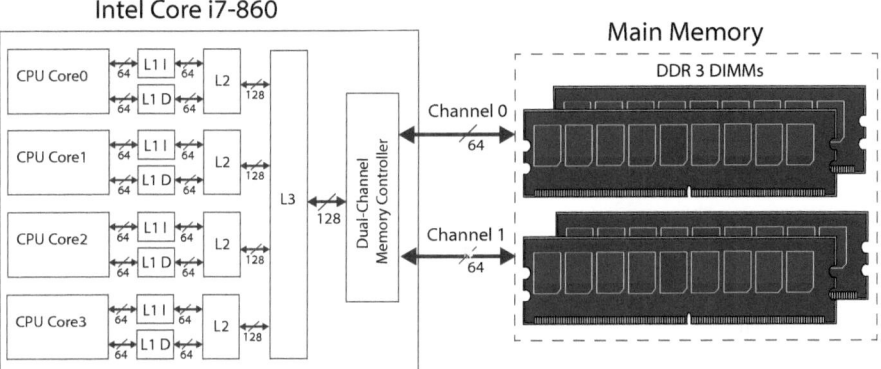

Fig. 4.1 Intel i7-860 memory hierarchy

is accessed. If a hit in L2 occurs, the block of size 64B is returned after an initial latency of 10 cycles at a rate of 8 bytes per clock cycle. If the L2 cache misses, the L3 cache is accessed. If a hit occurs in L3, the 64-byte block is returned after an initial latency of 35 cycles at a rate of 16 bytes per clock. If L3 misses, memory access is initiated—the on-chip memory controller must get the block of size 64B from the main memory.

The main memory is implemented using DDR SDRAM memory chips placed on the printed circuit boards called Dual In-Line Memory Module (DIMM). The memory controller on i7-800 supports two 64-bit memory channels. Each channel is used to access eight 8-bit memory chips placed on one side of DIMM (64 bits per access). Two 64-bit memory channels are used simultaneously as one 128-bit channel (since there is only one memory controller, and the same address of the missing block in L3 is sent on both channels) to fill the missing block in L3. Thus, the memory controller fills the 64-byte cache block at a rate of 16 bytes (124 bits) per memory clock cycle.

Have you struggled to read the description of the memory hierarchy in the Intel i7-860-based system? Don't worry; at the end of this chapter, you should be able to understand it. Let us now begin our journey into the world of modern memory. Throughout this exploration of the main memory, we will delve into its fundamental characteristics, architecture, and functionality, unraveling the intricate mechanisms that govern its operation within the broader context of computer systems.

4.2 Basics of Digital Circuits: A Quick Review

Before looking under the hood of modern memory chips used in computer systems, we should apprehend some basic concepts from digital electronics like MOS transistors used as logical switches and MOS inverters. The aim is to understand the operations in modern memory chips and not fall into the physical equations of elec-

Fig. 4.2 nMOS and pMOS
transistor symbols

tronic circuits. Therefore, the description of the basic concepts of digital circuits will
be significantly simplified.

The basic building block of all digital circuits is the MOS transistor. MOS is an
acronym for Metal-Oxid-Semiconductor and indicates the manufacturing process
used to make transistors. The MOS transistor has three terminals: gate (G), drain
(D), and source (S). The gate terminal is a control input: it controls the flow of
electrical current between the source and drain terminals. There are two types of
MOS transistors: nMOS and pMOS. Figure 4.2 shows the symbols of both MOS
transistors. We will consider only the type of operation where MOS transistors act
as logical switches.

4.2.1   MOS Transistor as a Switch

Consider first an nMOS transistor. No current flows between the drain and the source
if the gate terminal is grounded (logical 0). Hence, we say the transistor is OFF. If
the gate voltage is high and corresponds to logic 1, a conducting path of electrons is
formed from the source to drain, and current can flow. We say the transistor is ON.

The reverse holds for a pMOS transistor. When the gate is at a positive voltage that
corresponds to logic 1, there is no current flow, so the transistor is OFF. A sufficiently
low gate voltage that corresponds to logic 0 forms a conducting path from source to
drain, so the transistor is ON.

In summary, the gate of a MOS transistor controls the flow of current between
the source and drain. Simplifying this to the extreme allows us to **view the MOS
transistors as ON/OFF switches**. When the gate of an nMOS transistor is 1, the
transistor is ON, and the current flows between the source and the drain. When the
gate is 0, the nMOS transistor is OFF, and no current flows between the source and
the drain. A pMOS transistor is just the opposite, being ON when the gate is low and
OFF when the gate is high. Figure 4.3 illustrates this switch model.

4.2.2   CMOS Inverter

The most straightforward logic gates that can be built using MOS transistors are
inverters. An inverter is built from two **complementary** MOS transistors, one nMOS,
and one pMOS, hence the name *complementary MOS (CMOS) inverter*. Figure 4.4
shows the schematic and the switch-level model for a CMOS inverter or NOT gate
using one nMOS transistor and one pMOS transistor. The bar at the top of the
schematic indicates a supply voltage (Vdd), and the triangle at the bottom indicates

Fig. 4.3 Switch-level models of nMOS and pMOS transistors

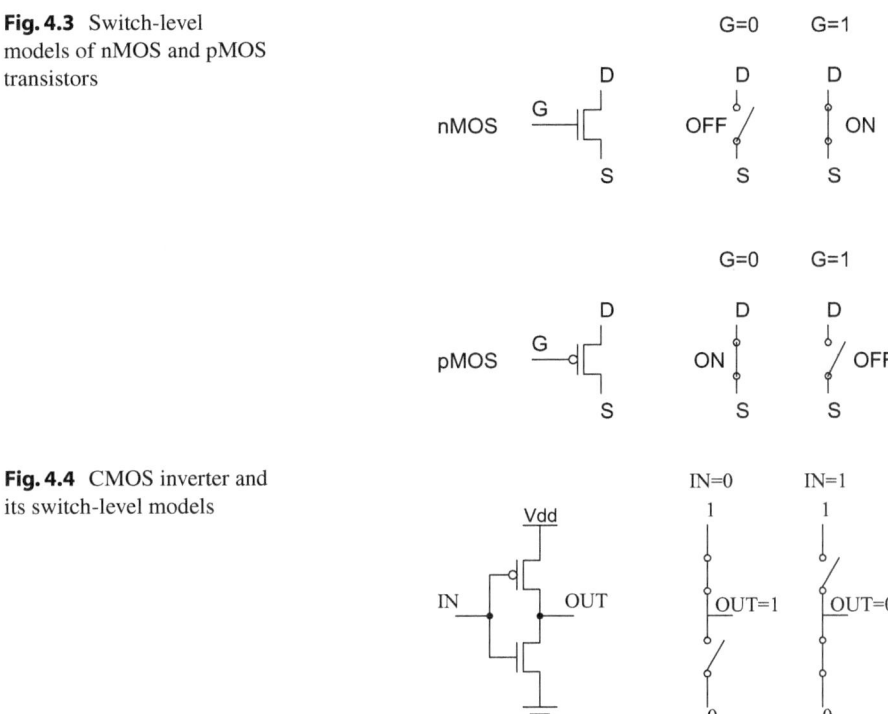

Fig. 4.4 CMOS inverter and its switch-level models

the ground terminal (GND). The input IN connects both transistors' gates. When the input IN is 0, the nMOS transistor is OFF, and the pMOS transistor is ON. Thus, the output OUT is pulled to logic 1 because it is connected to Vdd through the pMOS transistor. Conversely, when IN is 1, the nMOS is ON, the pMOS is OFF, and OUT is pulled down to '0', because it is connected to GND through the nMOS transistor.

4.2.3 Bistable Element

Now that we are familiar with MOS transistors and CMOS inverters, it is time to learn how we can store one bit of information in a MOS digital circuit, i.e., how to form a 1-bit storage (memory) cell using MOS transistors and inverters. The fundamental building block of memory is a **bistable element**—a logic element with two stable states. Figure 4.5 shows the bistable element composed of two inverters, I1 and I2. The inverters are cross-coupled, meaning that the input of I1 is the output of I2 and vice versa.

If $Q = 0$, I2 receives a FALSE input, producing a TRUE output on $\overline{Q}$. I1 receives a TRUE input, producing a FALSE output on Q. This is consistent with the original assumption that Q = 0, so the circuit is in a stable state. If $Q = 1$, I2 receives a TRUE input, producing a FALSE output on $\overline{Q}$. I1 receives a FALSE input, producing a TRUE output on Q. This is consistent with the original assumption that Q = 1, so

Fig. 4,5 A bistable element

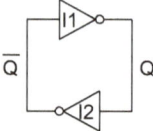

the circuit is again in a stable state. Because the cross-coupled inverters have two stable states, 0 and 1, the circuit is said to be **bistable**. The state of the cross-coupled inverters is contained in one binary state variable, Q. Specifically if $Q = 0$, it will remain 0 forever, and if $Q = 1$, it will remain 1 forever. Although the cross-coupled inverters can store a bit of information, they are not practical because the user has no inputs to control the state. So, we have to expand the bistable element with circuitry, which provides inputs to control the value of the state variable. One such element that can accept the inputs to control the value stored in the bistable is a *static RAM cell.*

> **Summary: Transistors, Inverter and Bistable**
>
> The gate of a MOS transistor controls the flow of current between the source and drain. Simplifying this to the extreme allows us to *view the MOS transistors as ON/OFF switches.*
>
> An inverter is built from two *complementary* MOS transistors, one nMOS and one pMOS.
>
> The fundamental building block of memory is a *bistable element*. It is composed of two cross-coupled inverters. It stores one bit of information.

4.3 SRAM Cell

Static random-access memory (static RAM or SRAM) is a type of random-access memory (RAM) that uses a bistable element to store one bit of information. This is the type of memory used as the building block of most caches because of its superior performance over other memory structures, specifically DRAM, which we will cover later. SRAM is faster and more expensive than DRAM; it is typically used for CPU cache and registers, while DRAM is used for a computer's main memory.

A typical SRAM cell is made up of six MOS transistors—two complementary pairs that form two cross-coupled inverters (bistable), and two *access* nMOS transistors that serve as a switch used to control the state of the bistable element during the read and write operations. Figure 4.6 shows an SRAM cell. Each bit in an SRAM cell is stored in the bistable element composed of four transistors that form two cross-coupled inverters. As we have already learned, this cross-coupled connection creates regenerative feedback that allows it to *store a single bit of data indefinitely*

Fig. 4.6 SRAM cell

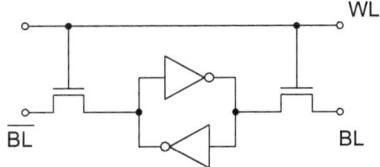

(a) A basic structure of an SRAM cell.

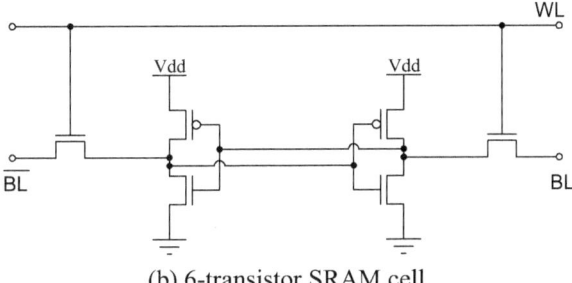

(b) 6-transistor SRAM cell.

provided that power is supplied to the SRAM cell. The SRAM cell also has two bit lines that control both the input and output of the data from the cell. The first bit line (BL), holds the same value that is stored in the cell. The second bit line ($\overline{\text{BL}}$) holds the inverse of the value that is stored in the cell.

When the word line (WL) is not selected (WL = 0), the cell is in standby mode. Setting the word line to a logic high enables access to nMOS transistors. This connects the cell with both bit lines and allows the cells to be read or written. The SRAM cell is read by asserting a WL and detecting the voltage difference at the bit lines BL and $\overline{\text{BL}}$. The SRAM cell is written by setting the content on the bit lines BL and $\overline{\text{BL}}$ and asserting the word line.

Due to the ability to store the information indefinitely and the high speed of SRAM cells, they are used to implement caches and registers in microprocessors. Furthermore, the main advantage of SRAM is that it uses the same fabrication process as the microprocessor core, simplifying the integration of cache and CPU registers onto the processor die. On the other hand, the main **disadvantages of SRAM cells are price, low density, and high operational power consumption**. These disadvantages prevent the usage of SRAM cells in the main computer memory.

Since SRAM cells are not used to build the main memory, we will end up dealing with and learning about SRAM cells at this point, and we will dive deep into DRAM cells. By contrast, DRAM typically uses a different process that is not optimal for logic circuits, making the integration of CPU logic and DRAM harder than the integration of CPU logic and SRAM. But DRAMs are smaller, cheaper, and consume less power, which makes them the better candidate for implementing the main memory.

Summary: SRAM cell

A *SRAM cell* uses a bistable element to store one bit of information. It is made up of a bistable and two access nMOS transistors that serve as a switch used to control the state of the bistable element during the read and write operations.

Due to the ability to store the information indefinitely and the high speed of SRAM cells, they are used to implement caches and registers in microprocessors.

4.4 DRAM Cell

Dynamic Random Access Memory (DRAM) is the main memory used for all computers. To pack more bits per chip, a DRAM cell consists only of a single MOS transistor (T) and a storage capacitor (C), as shown in Fig. 4.7. The data in the cell can be read or written through the bit line (BL) terminal. In contrast to SRAMs, DRAMs store their contents as a charge on a capacitor C. This way, the DRAM cell is substantially smaller than the SRAM cell. The transistor T acts as a switch between the storage capacitor and the bit line. The word line (WL) terminal is used to switch the transistor T on/off. Reading the bit from the DRAM cell discharges the capacitor and thus destroys the information. Even if we do not read the DRAM cell, the charge leaks from the capacitor because the cell transistor does not entirely disconnect the storage capacitor from the bit line. Even though the transistor is switched off, a tiny current flows from the capacitor to the bit line and discharges the capacitor. Therefore, the charge (information) must be refreshed several times each second. Hence the name *dynamic*.

Fig. 4.7 A DRAM cell

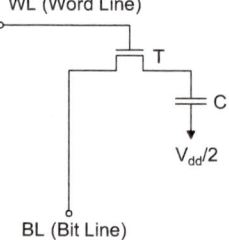

Summary: SRAM cell

A *SRAM cell* uses a bistable element to store one bit of information. It is made up of a bistable and two access nMOS transistors that serve as a switch used to control the state of the bistable element during the read and write operations.

Due to the ability to store the information indefinitely and the high speed of SRAM cells, they are used to implement caches and registers in microprocessors.

4.4.1   Basic Operation of DRAM

The transistor T acts as a switch between the storage capacitor C and the bit line BL. One capacitor node is connected to $V_{dd}/2$. The voltage across the capacitor is either $+V_{dd}/2$, if the capacitor stores "1", or $-V_{dd}/2$, if the capacitor stores "0". The charge stored in a capacitor is equal to capacitance times voltage across the capacitor:

$$Q = C \times V_{dd}/2. \tag{4.1}$$

In a 90 nm DRAM process technology, the capacitance of a DRAM storage cell is 30 fF. If we assume $V_{dd} = 3.3V$, then

$$Q = 30fF \times 3.3V/2 = 34.5fC.$$

As you may recall from physics class, one electron equals a charge of $1.6 \cdot 10^{-19}C$; thus, the storage capacitor stores only 210000 electrons! Even though the transistor has a very high resistance when switched off, the charge on the capacitor leaks away through the switched-off transistor in tens to hundreds of milliseconds. Therefore, DRAM storage cells should be regularly refreshed to avoid data loss.

The data is written into a memory cell by placing the "1" or "0" charge into the storage capacitor. To write data into a cell, we first set the bit line to Vdd ("1") or to GND ("0") and assert the word line to connect the capacitor to the bit line. The storage capacitor then retains the stored charge after the word line is de-asserted and the transistor is turned off. The electric charge on the storage capacitor slowly leaks off, so without intervention, the data on the chip would soon be lost. This capacitor will be accessed for either a new write, a read, or a refresh (Fig. 4.8).

To read data from the DRAM cell, the bit line is first precharged to $V_{dd}/2$. The word line is then driven high to connect a cell's storage capacitor to its bit line. This causes the transistor to conduct, transferring charge from the storage cell to the connected bit line (if the stored value is "1") or from the connected bit line to the storage cell (if the stored value is "0"). This process is depicted in Fig. 4.8. In both cases, information stored in the DRAM cell is lost. Thus, reading from DRAM is a **destructive operation**. The bit lines are relatively long because they connect storage cells in all memory words, and they act as a capacitor with relatively high capacitance (the capacitance of the bit lines is ten times the capacitance of the storage capacitor).

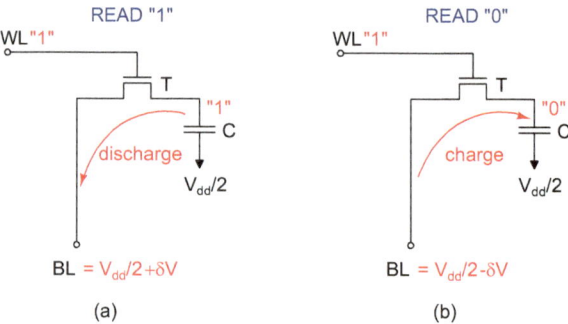

Fig. 4.8 Reading from a DRAM cell. **a** Reading "1" from a DRAM cell discharges the storage capacitor and slightly increases the voltage of the bit line. **b** Reading "0" from a DRAM cell charges the storage capacitor and slightly decreases the voltage of the bit line. In both cases, information is lost

Fig. 4.9 A DRAM cell with a sense amplifier

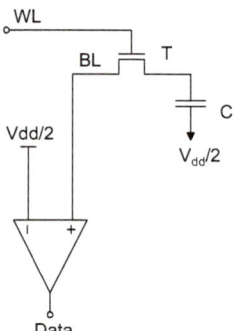

According to the charge-sharing equation (capacitive voltage divider), the voltage swing (the magnitude of a voltage difference) δV on the bit line during readout is

$$\delta V = \frac{V_{dd}}{2} \frac{C}{C + C_{BL}}, \tag{4.2}$$

where C is the capacitance of the storage capacitor and C_{BL} is the capacitance of the bit line. If the capacitance of the bit line is ten times the capacitance of the storage capacitor and $V_{dd} = 3.3V$, the voltage difference δV on the bit line during the read operation is only 150 mV! When dealing with such a tiny voltage swing, **correctly detecting the bit value is quite a challenge**. Thus, we need a special circuit to sense this small voltage swing. Sensing is necessary to read the cell data properly. A special circuit used to detect the voltage swing and read the data is a **sense amplifier**.

To sense the voltage swing on the bit line, a sense amplifier is used, as presented in Fig. 4.9. A sense amplifier has two inputs. One input is connected to the bit line, and the other input is tied to $V_{dd}/2$. The sense amplifier detects the voltage difference at its inputs and outputs 0 at the Data terminal if the voltage on the bit line is less than $V_{dd}/2$, or 1 otherwise.

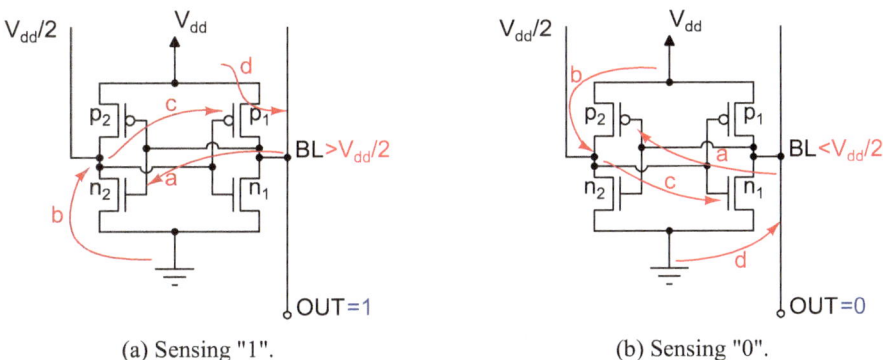

(a) Sensing "1".   (b) Sensing "0".

Fig. 4.10   A simplified structure and operation of a sense amplifier. **a** Sensing "1". **b** Sensing "0"

4.4.2   Basic Operation of Sense Amplifiers

A sense amplifier is a simple circuit made up of two cross-coupled CMOS inverters—so it is an SRAM cell. Figure 4.10 shows a sense amplifier built from cross-coupled CMOS inverters. The bit line (BL) is initially precharged to $V_{dd}/2$. During a read, the bit line changes its voltage by a small amount, δV. If the voltage of the bit line is higher than $V_{dd}/2$ (Fig. 4.10a), the n2 nMOS transistor begins to conduct and pulls the precharged line down to "0". This, in turn, causes the p1 pMOS transistor to conduct. After a small delay, BL is pulled high, and OUT = 1. On the other hand, if the voltage of the bit line is lower than $V_{dd}/2$ (Fig. 4.10b), the (p2) pMOS transistor begins to conduct and pulls the precharged line up to "1". This, in turn, causes the n1 nMOS transistor to conduct. After a small delay, BL is pulled down to "0", and OUT = 0. The feedback that occurs from the cross-connected inverters thereby amplifies the small voltage difference between the BL and the precharged input reference until the bit line is entirely at the lowest voltage or the highest voltage.

We have just learned that the main function of sense amplifiers is to sense the tiny voltage swing on the bit lines that occurs when an access transistor is turned on and a storage capacitor places its charge on the bit line. The second function of sense amplifiers is to restore the value of cells after the voltage on the bit lines is sensed. Recall that turning on the access transistor allows a storage capacitor to share its stored charge with the bit line. However, the process of sharing the charge from a storage cell discharges that storage cell. Thus, the information in the cell is lost and cannot be read again. But this information is stored in the sense amplifier, as the sense amplifier is a bistable circuit made up of two cross-coupled inverters. As such, it can store information as long as the supply voltage is present. Consequently, after sensing, the sense amplifier is used to write back the bit value to the storage cell. This operation is referred to as **precharge**.

Summary: DRAM cell

Dynamic Random Access Memory (DRAM) is the main memory used for all computers. DRAMs store their contents as a charge on a capacitor. A DRAM cell consists only of a storage capacitor and a single nMOS transistor that acts as a switch between the storage capacitor and the bit line.

Reading from a DRAM cell is a *destructive operation*. Besides, the charge on the capacitor leaks away through switched-off transistors in tens to hundreds of milliseconds. Thus, DRAM cells should be regularly *refreshed*.

A sense amplifier is a special circuit used to detect the tiny voltage swing on the bit line and read the data. The sense amplifier is also used to write back the bit value to the storage cell. This operation is referred to as *precharge*.

4.5   DRAM Arrays and DRAM Banks

DRAM is usually arranged in a rectangular **memory array** of storage cells organized into rows and columns. Figure 4.11 shows a simplified basic structure of a DRAM cell array containing R-by-C cells. **DRAM arrays usually contain many hundreds or thousands of cells in height and width**. The cells of a DRAM are accessed by a **row address** and a **column address**. The row address lines (i.e., the word lines) are connected to the gates of the nMOS transistors, and the column lines are connected to the sense amplifiers.

The array size represents a trade-off between density and performance. Larger arrays contain more bits of information, but they also require longer word lines and bit lines. Longer word and bit lines have a higher capacitance. An array that contains thousands of cells in height and width has an order of magnitude higher capacitance on a bit line than in a cell, so the bit line voltage swing δV during a read is tiny, which is hard to detect. Besides, due to a higher capacitance, larger arrays are slow. A typical array size in a recent DRAM is 8K words (rows) by 1024 bits (columns).

A DRAM memory chip usually has 4-16 DRAM arrays that are accessed simultaneously. Hence, a DRAM chip transmits or receives a number of bits equal to the number of arrays each time the memory controller accesses the DRAM. Each array provides a single bit to the output pin. DRAM chips are described as xN, where N refers to the number of memory arrays and output pins. For example, in a simple organization, a x8 DRAM (pronounced 'by eight') indicates that the DRAM has at least eight memory arrays and that a column width is 8 bits (each column read or write access transmits 8 bits of data). This means that the DRAM transmits or receives eight bits each time the memory controller accesses the DRAM. A set of memory arrays accessed simultaneously is referred to as a **bank**.

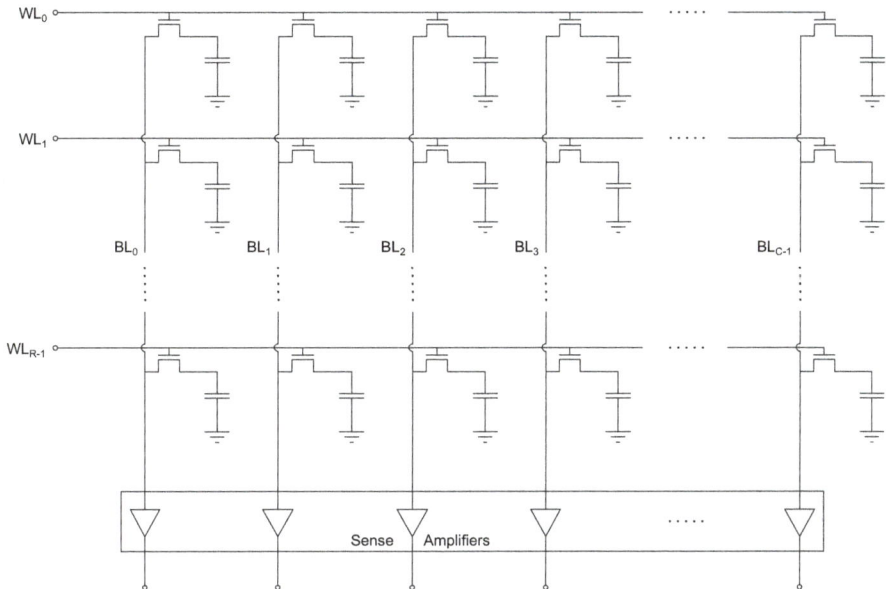

Fig. 4.11 A simplified structure of a DRAM array

Summary: DRAM Arrays and DRAM Banks

DRAM is arranged in a rectangular *memory array* of storage cells organized into rows and columns.

The cells of a DRAM are accessed by a *row address* and a *column address*.

A **bank** is a set of N memory arrays accessed simultaneously, forming an N-bit width column. Usually, there are 4, 8, or 16 DRAM arrays in a bank.

4.6 DRAM Chips

Figure 4.12 presents the basic structure of a DRAM chip. As we have learned, the DRAM memory is organized as a rectangular matrix of rows and columns. The DRAM chip in Fig. 4.12 contains a bank of 8 arrays. Each array has 1024-by-256 storage cells. All arrays in a bank are accessed at the same time, so the DRAM chip in Fig. 4.12 reads or transmits eight bits in a single access (D0 to D7). The components identifying the row and column are referred to as the **row address decoder** and the **column selector**. The row address decoder is used to activate the appropriate word line from the given row address. The column selector is used to select the appropriate column from the given column address.

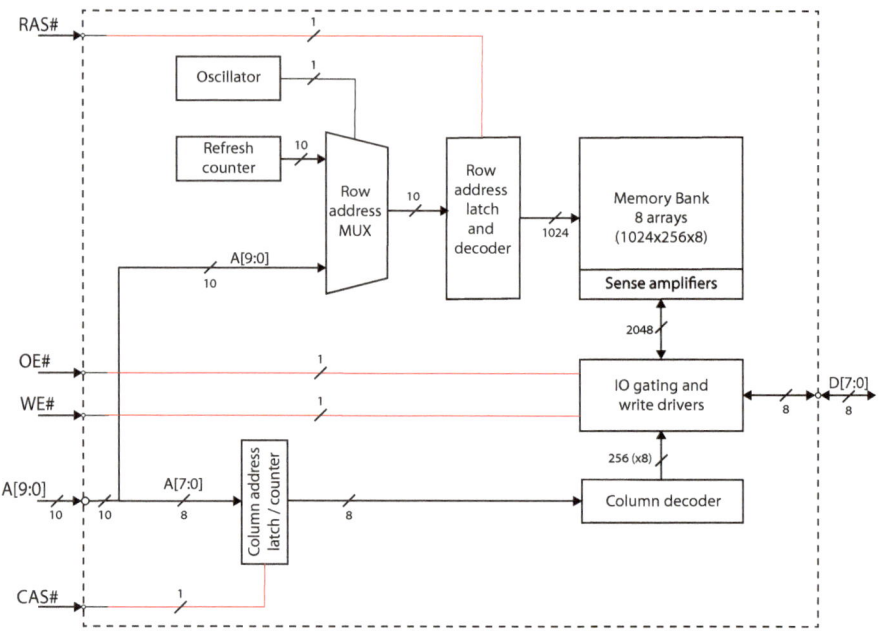

Fig. 4.12 Simplified structure of a 256K × 8-bit DRAM chip

As the capacity of DRAMs is large, the DRAM chips would require a large number of address lines to address a row and a column. For example, to address a cell in a 32256-by-1024 array, we need 15 bits to select a word and 10 bits to select a column. Such a large number of address bits could be an issue. The solution is to **multiplex the address lines**. Firstly, the row address is applied to the address lines, then the column address follows. In such a way, the number of address pins is cut almost in half. The same holds for DRAM in Fig. 4.12. Instead of having 18 address bits (10 for the row and 8 for the column), only 10 address bits are used. To indicate which of two addresses is currently on the bus, we need **two additional control signals**: the **row access strobe (RAS)** and the **column access strobe (CAS)**. When the RAS signal is activated, the address bits A0 to A9 are latched into the **row address latch**. Similarly, when the CAS signal is activated, the address bits A0 to A7 are latched into the **column address latch**.

Two more control signals are required to transfer data into and from a DRAM chip appropriately. The **write enable (WE)** signal is used to choose a read or write operation. A low voltage level signifies that a write operation is desired; a high voltage level is used to choose a read operation. During a read operation, the **output enable**

Fig. 4.13 256K × 8-bit
DRAM chip pinout

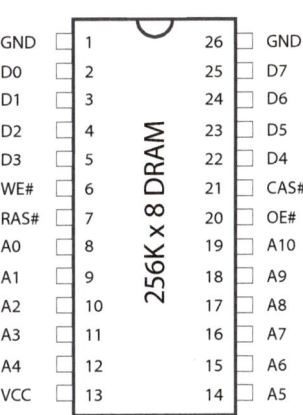

(**OE**) signal is used to prevent data from appearing at the output until needed. When OE is low, data appears on the data outputs as soon as it is available. OE is kept high during a write operation. Figure 4.13 illustrates a pinout diagram of a 256K × 8-bit DRAM from Fig. 4.12.

Summary: DRAM Chips

DRAM chips contain at least one memory bank. The *row address decoder* is used to activate the appropriate word line from the given row address. The *column selector* is used to select the proper column from the given column address.

As the number of address bits required to select rows and columns can be quite large, the *address lines are multiplexed*. To indicate which of the two addresses is currently on the bus, we need two additional control signals: the *row access strobe (RAS)* and the *column access strobe (CAS)*.

The *write enable (WE)* signal is used to choose a read or a write operation. During a read operation, the *output enable (OE)* signal is used to prevent data from appearing at the output until needed.

> **FALLACY: Memories (DRAMs) are physically organized as a liner vector of memory words.**
>
> It is a common and erroneous belief that memory is physically organized as a vector of memory words (and not as a rectangular array of rows and columns). Such an organization of memory would otherwise be ideal. A memory array would be just one long vector of memory cells, and there would be only one memory cell in a word. All memory cells would then be connected to the same bit-line. In that case, a DRAM array would contain R-by-1 memory cells. The memory with 8-bit words would then be composed of eight parallel R-by-1 memory arrays. In this case, the row address would already be the column address because there would be only one column in a row. The memory addresses would not be multiplexed, and we would not need the RAS and CAS signals. Wouldn't that be great? However, it is physically impossible to make such a memory because, in such a memory, the bit lines would be extremely long and would have huge capacitance. The capacitance of such long bit lines is several thousand times greater than the capacitance of the memory cells, and it is impossible to detect a tiny voltage swing.

4.7 Basic DRAM Operations and Timings

The most challenging aspect when working with DRAMs is resolving the timing requirements. **DRAMs are generally asynchronous, responding to input signals whenever they occur**. The DRAM works appropriately as long as the signals are applied in the proper sequence, with signal durations and delays between signals that meet the specified limits. The following signals control the DRAM operations:

1. Row Address Strobe (RAS). RAS is active low. To enable RAS, a transition from a high voltage to a low voltage is required. The voltage must remain low until RAS is no longer needed. During a complete memory cycle, there is a minimum amount of time that RAS must be active (t_{RAS}). There is a minimum amount of time that RAS must be inactive before activating it again, called the RAS precharge time (t_{RP}). t_{RP} tells us how fast the row can be precharged before we can engage another RAS.
2. Column Address Strobe (CAS). CAS is used to latch the column address and to initiate the read or write operation. It is active low. The memory specification lists the minimum amount of time CAS must remain active (t_{CAS}). For most memory operations, there is also a minimum amount of time that CAS must be inactive before activating it again, called the CAS precharge time (t_{CP}).
3. Write Enable (WE). The write enable signal is used to choose a read operation or a write operation. It is active low.

4. Output Enable (OE). It is active low. When OE is low during a read operation, data appears on the data outputs as soon as it is available. During a write operation, OE should be high.
5. Address. The addresses are used to select a memory location on the chip. The address pins on a memory device are used for both row and column selection (multiplexing).
6. Data In or Out. The DRAM memory device's data pins are used for input and output. Data at data pins are stored in the selected memory cells during a write operation. During a read operation, data from the selected memory cells appear on the data pins once access is completed, and OE is low.

4.7.1   Reading Data from DRAM Memory

To read the data from a DRAM memory cell, we must select the DRAM memory cell by applying its row and column addresses to the address input pins. The charge on the selected DRAM cell must then be sensed by the sense amplifier and sent to the data output (pins). In terms of timing, the following steps must occur (Fig. 4.14):

1. The row address must be applied to the address input pins on the memory device before RAS goes low.
2. RAS must go from high to low and remain low for the prescribed amount of time (t_{RAS}). When RAS goes low, the **memory row** addressed by the row address **is open**, and the charge from the cells in the selected row starts to flow to the bit lines.
3. The column address must be applied to the address input pins on the memory device before CAS goes low.
4. WE must be set high for a read operation to occur before the transition of CAS and remain high after the transition of CAS.
5. Only after the prescribed amount of time (t_{RCD}), CAS must go from high to low and remain low for the prescribed amount of time (t_{CAS}). RAS-to-CAS delay (t_{RCD}) time ensures that the charge from the selected cells is on the bit lines and properly sensed by the sense amplifiers.
6. Data appears on the data output pins of the memory device. The time the data appears on the output pins is called CAS latency (t_{CL}).
7. Before the read cycle can be considered complete, CAS and RAS must return to their inactive states. A new read or write access can start only after the prescribed time (t_{RP}—Row Precharge).

The read access lasts for a **row cycle time** (t_{RC}):

$$t_{RC} = t_{RAS} + t_{RP}. \tag{4.3}$$

The row cycle time, t_{RC}, determines the minimum time a memory row takes to complete a full cycle, from row activation up to the precharging of the active row. This is an interval between accesses to different rows in a given set of DRAM arrays.

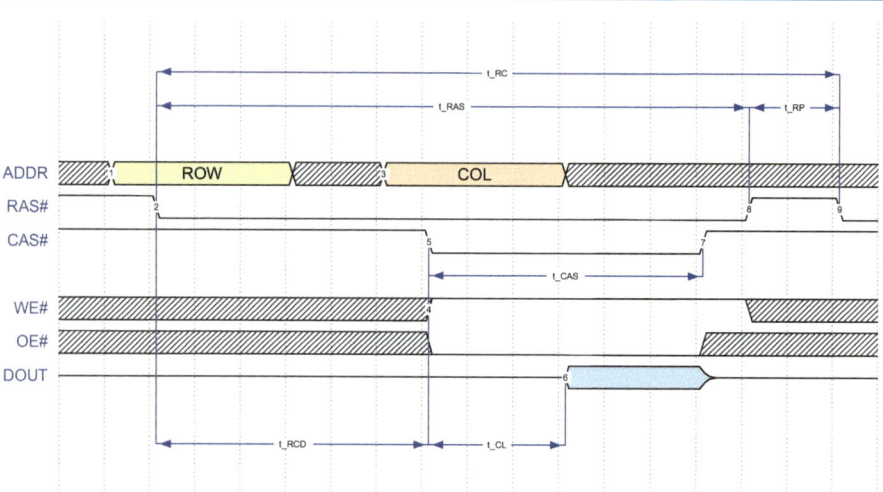

Fig. 4.14 Simplified DRAM read cycle

4.7.2 Writing Data to DRAM Memory

To write to a DRAM memory cell, the row and column address for the DRAM cell must be selected, and data must be presented at the data input pins. The sense amplifier either charges the memory cell's capacitor or discharges it, depending on whether a 1 or 0 is to be stored. In terms of timing, the following steps must occur (Fig. 4.15):

1. The row address must be applied to the address input pins on the memory device before RAS goes low.
2. RAS must go from high to low and remain low for the prescribed amount of time (t_{RAS}). When RAS goes low, the **memory row** addressed by the row address is open.
3. Data must be applied to the data input pins before CAS goes low.
4. The column address must be applied to the address input pins on the memory device after RAS goes low and before CAS goes low.
5. WE must be set low for a write operation to occur.
6. Only after the prescribed amount of time (t_{RCD}), CAS must switch from high to low and remain low for a prescribed amount of time (t_{CAS}).
7. Before the write cycle can be considered complete, CAS and RAS must return to their inactive states. A new read or write access can start only after the prescribed amount of time (t_{RP}).

The write access also lasts for a **row cycle time** (t_{RC}) (Fig. 4.15).

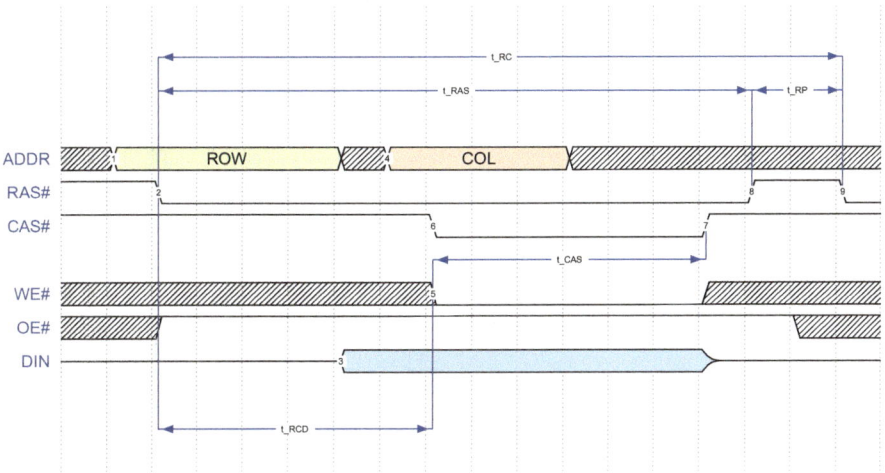

Fig. 4.15 Simplified DRAM write cycle

4.7.3 Refreshing the DRAM Memory

Since DRAM memory cells are capacitors, their charge can leak away over time. If the charge is lost, the data is lost! To prevent the loss of data, DRAMs must be refreshed, i.e., the charge on the individual memory cells must be restored. **DRAMs are refreshed one row at a time**. The frequency of refresh depends on the silicon technology used to manufacture the memory chip and the design of the memory cells. **Most of today's DRAMs require a refresh to occur every 64 ms**.

Reading or writing a memory cell has the effect of refreshing the selected cell because after reading/writing, the entire row is precharged. Unfortunately, not all cells are read or written within a 64 ms time frame. Hence, each row in the array must be accessed and restored during the refresh interval. The refresh cycles are distributed across the entire refresh interval of 64 ms in such a way that all rows are refreshed within the required interval. If, for example, a DRAM array has 4096 rows, every 15.6 microseconds, a new row must be refreshed. At the end of the 64 ms interval, the process begins again.

DRAMs use an **internal oscillator** to determine the refresh frequency and a **counter** to keep track of which row is to be refreshed and initiate the refresh periodically. Such an auto-initiated refresh is referred to as **self refresh**. To refresh one row of the memory array, the so-called **CAS-before-RAS refresh** is used. The following steps form the CAS-before-RAS refresh:

1. CAS must switch from high to low while the WE signal remains in a high state (equivalent to read).
2. After the prescribed delay, RAS must switch from high to low.
3. The internal counter determines which row is to be refreshed and applies the row address at the address pins

4. After the required delay, CAS returns to a high level.
5. After the necessary delay, RAS returns to a high level.

Summary: DRAM Operations and Timings

DRAMs are asynchronous systems, responding to input signals whenever they occur. The DRAM will work properly as long as the input signals are applied in the proper sequence, with signal durations and delays between signals that meet the specified limits.

Typical operations in DRAMs are read, write, and refresh. All these operations are initiated and controlled by the input signal sequence.

The read and write accesses last for a *row cycle time* (t_{RC}):

$$t_{RC} = t_{RAS} + t_{RP}.$$

DRAMs must be refreshed in order to prevent the loss of data. DRAMs are refreshed one row at a time. DRAMs use an *internal oscillator* to determine the refresh frequency and initiate a refresh and a *counter* to keep track of the row to be refreshed. Such an auto-initiated refresh is referred to as *self refresh*. Self-refresh uses the so-called CAS-before-RAS sequence.

Summary: Important timings in DRAMs.

Name	Symbol	Description
Row Active Time	t_{RAS}	The minimum amount of time RAS is required to be active (low) to read or write to a memory location.
CAS latency	t_{CL}	This is the time interval it takes to read the first bit of memory from a DRAM with the correct row already open.
Row Address to Column Address Delay	t_{RCD}	The minimum time required between activating RAS and activating CAS. It is the time interval between row access and data ready at sense amplifiers.
Random Access Time	t_{RAC}	This is the time required to read any random memory cell. It is the time to read the first bit of memory from an DRAM without an active row. $t_{RAC} = t_{RCD} + t_{CL}$.
Row Precharge Time	t_{RP}	After a successful data retrieval from the memory, the row that was used to access the data needs to be closed. This is the minimum amount of time that RAS must be inactive.
Row Cycle Time	t_{RC}	This is the time associated with single rad or write cycle. $t_{RC} = t_{RAS} + t_{RP}$

4.8   Improving the Performance of DRAMs

As mentioned earlier, one DRAM access is divided into row access and column access. Let's first look at how we read two consecutive columns from the same row in classic DRAMs. A timing diagram for reading two consecutive columns, A and B, in the same row X is shown in Fig. 4.16. Although both columns are in the same row X, we have to repeat the entire reading cycle from Fig. 4.14 to read each column. Wouldn't it be better to keep the entire row 'open' once the amplifiers sensed all bits in that row? Actually, the sense amplifiers can act like a row buffer to keep the row data. That way, we don't have to access the row every time and then close it after reading each column. Exactly this solution was used for one of the first performance enhancements in DRAM memories. But wait, how often do we access two or more consecutive columns from the same row? Very often, indeed, due to **temporal and spatial locality**. All methods used to improve the performance of a DRAM chip and to decrease the access time rely on the ability to access all of the data stored in a row without having to initiate a completely new memory cycle.

4.8.1   Fast Page Mode DRAM

Fast Page Mode DRAM is a minor modification to the first-generation DRAMs that allows faster access to data in the same row. The performance of read and write accesses to a row was improved by avoiding the inefficiency of opening and precharging the same row repeatedly to access different columns in the same row. Fast Page Mode DRAM eliminates the need for a row address if data is in the previously accessed row. In the Fast Page Mode DRAM, after a row has been opened by holding RAS low, the row bits are kept by the sense amplifiers, and multiple reads or writes could be performed to any of the columns in the open row. Each column access is initiated by asserting CAS and presenting a column address.

We start a regular read operation using Fast Page Mode by addressing the row (same steps 1 through 6 as in Fig. 4.14). Once the row data is valid, we switch CAS high but leave RAS low. There is a minimum amount of time that CAS must

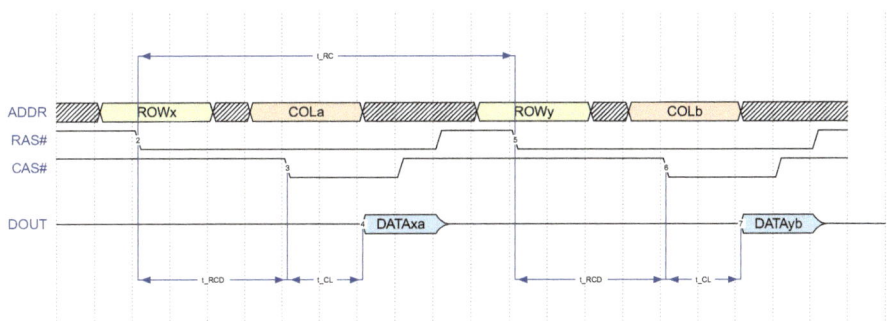

Fig. 4.16 Simplified read timing for two columns in the same row for conventional DRAM

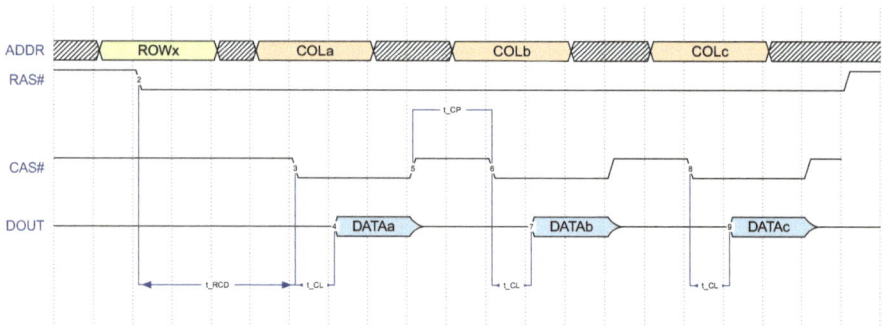

Fig. 4.17 Simplified timing diagram illustrating a read cycle in Fast Page Mode DRAM

be inactive, called the CAS precharge time (t_{CP}). When CAS has been inactive (high) for the required amount of time (t_{CP}), we repeat steps 3 through 6 of the read operation from Fig. 4.14. We can continue in this way until a new row address is required or the chip needs to be refreshed. Figure 4.17 is a simplified timing diagram that illustrates a Fast Page Mode read cycle.

Let's use an example to illustrate how fast page mode impacts the system's performance. In this example, we compare two scenarios: 4 memory accesses in the same row without fast page mode and four memory accesses in the same row with fast page mode. We assume that t_{RC} is 70 ns, t_{RCD} is 20 ns, and t_{CL} is 15 ns. In the first scenario, the data from the fourth column will be available after $3 \cdot t_{RC} + t_{RCD} + t_{CL} = 245$ ns. In the second scenario, we also assume that CAS should remain high for 5 ns before going down again (t_{CP} is 5 ns) and that data is kept valid for 20 ns. Now, the data from the fourth column will be available after $t_{RCD} + 3 \cdot (t_{CL} + 20 + t_{CP}) = 140$ ns.

4.8.2   Extended Data Output DRAM

The second change to improve the performance is Extended Data Out (EDO) DRAM. EDO is very similar to FPM. The primary advantage of EDO DRAMs over FPM DRAMs is that the data outputs are not disabled when CAS goes high on the EDO DRAM, allowing the data from the current read cycle to be present at the outputs while the next read cycle begins, i.e., data is still present on the output pins, while CAS is changing and a new column address is latched. This enables the shorter time required for CAS to be active (low), allowing a certain amount of overlap in operation (pipelining), resulting in faster access (cycle) time. Figure 4.18 is a timing diagram that illustrates an EDO mode read cycle. Let's now illustrate how EDO impacts the system's performance using the same example as before, i.e., four memory accesses in the same row with EDO. Assuming that we keep the data valid for 20 ns, the data from the fourth column becomes available after $t_{RCD} + t_{CL} + 3 \cdot 20 = 95$ ns.

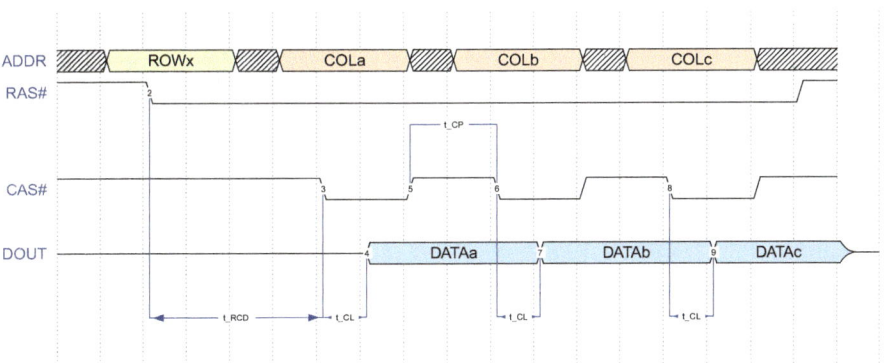

Fig. 4.18 Simplified read timing for two columns in the same row for conventional DRAM

Summary: FPM and EDO DRAMs

Due to *temporal and spatial locality*, we often access two or more consecutive columns from the same row.

All methods used to improve the performance of a DRAM chip and decrease the access time rely on the ability to access all of the data stored in a row without initiating a completely new memory cycle.

Fast Page Mode DRAM eliminates the need for a row address if data is located in the row previously accessed.

In *EDO DRAMs*, data remains on the output pins while CAS is changing, and a new column address is latched. This allows a certain amount of overlap in operation (pipelining), resulting in faster access time.

4.9 Synchronous DRAM

Originally, DRAMs we have just covered and were produced from the early 1970s to early 1990s had an asynchronous interface in which input control signals directly affect internal functions. The **synchronous DRAM (SDRAM)** device represents a significant improvement over the DRAM devices. In particular, SDRAM devices differ from previous generations of DRAM devices in two significant ways:

1. The **clock signal** was added to the SDRAM device; hence, the SDRAM device has a synchronous device interface, where commands instead of signals are used to control internal latches and
2. SDRAM devices contain **multiple independent banks**.

Besides, SDRAMs typically   also have a programmable **mode register** to hold the number of bytes requested and hence can send many bytes over several cycles per request without sending any new addresses. This type of transfer is referred to as **burst mode**.

SDRAMs have a clock signal, and all internal actions occur on its negative edge. As we saw, in DRAM devices, the RAS, CAS, and WE signals from the memory controller directly control internal latches and input/output buffers, and these signals can arrive at the DRAM device's pins at any time. The DRAM devices then respond to the RAS, CAS, and WE signals immediately. On the contrary, in SDRAM devices, the RAS, CAS, and WE signals do not directly control internal latches and buffers. In SDRAM devices, these signals form a **command bus** used to transmit **commands** to the internal state machine, which **executes the commands at the falling edge of the clock signal**. In this way, the control of internal latches and input/output buffers moved from the external memory controller into the state machine in the SDRAM device's control logic. The RAS, CAS, and WE names were retained for signals on the command bus, which transmits commands, although these specific signals no longer control latches and buffers internal to the SDRAM device.

The second feature that significantly differentiates the SDRAM device from the DRAM devices is that the SDRAM devices contain multiple banks. The presence of multiple independent banks in each SDRAM device means that while one bank is busy with a row activation command or a precharge command, the memory controller can send a new command to a different bank. Multiple banks now enable the interleaving of memory requests to different banks in a single SDRAM device. SDRAM devices contain either 2, 4, or 8 independent banks. One to three bank address inputs (BA0, BA1, and BA2) determine which bank the command refers to.

4.9.1 Functional Description

Figure 4.19 shows the simplified block diagram of an SDRAM device with two independent banks. The hash (#) beside a signal name denotes the signal is active low. Each bank has its row address latch and decoder, column decoder, and sense amplifiers. Each bank in the SDRAM device in Fig. 4.19 consists of eight DRAM arrays of size 4096-by-1024 bits. The address now consists of a bank number (BA0), a row address (A[11:0]), and a column address (A[9:0]).

In an SDRAM device, commands are decoded on the rising edge of the clock signal (CLK) and executed on the falling edge of CLK if the chip-select signal (CS) is active. The command is asserted on the command bus by the external memory controller. The command bus consists of WE, CAS, and RAS signals. All these signals are active low. Although the signal lines retain the function-specific names from DRAMs, they only form a command bus. Table 4.1 shows the command set of the SDRAM device and the input signal combinations on the command bus that designate the commands. The table also shows that as long as CS is not active, the SDRAM device ignores the signals on the command bus.

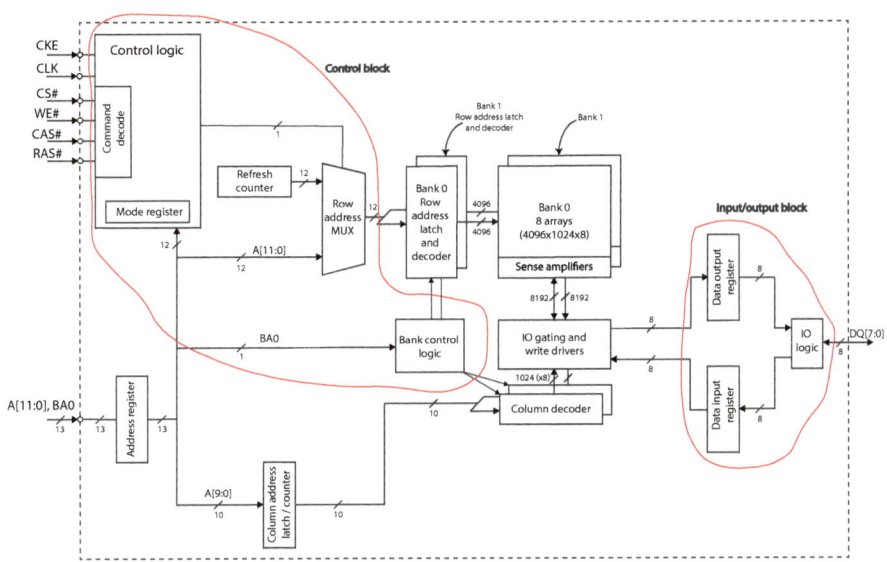

Fig. 4.19 Simplified block diagram of a SDRAM device with two banks

Table 4.1 SDRAM commands

Command	CS#	RAS#	CAS#	WE#	Address
COMMAND INHIBIT	H	X	X	X	X
NO OPERATION (NOP)	L	H	H	H	X
ACTIVE (select bank and activate row)	L	L	H	H	Bank/row
READ (select bank and column, and start READ burst)	L	H	L	H	Bank/col
WRITE (select bank and column, and start WRITE burst)	L	H	L	L	Bank/col
PRECHARGE (deactivate row in bank)	L	L	H	L	Bank/row
AUTO REFRESH	L	L	L	H	X
LOAD MODE REGISTER	L	L	L	L	Code

The control block in Fig. 4.19 consists of control logic, a multiplexor to select a row address, a refresh counter, and bank control logic. The refresh counter keeps track of the row to be refreshed. The multiplexor is used to select a row address to be transferred into the row address latch and decoder. The address is either an address coming from the refresh counter (in case the control logic performs a refresh cycle) or an address on the external address bus coming from the DRAM controller. Control logic contains a command decoder, a finite state machine that executes commands, and the mode register. The mode register is a programmable 10-bit register whose individual bits determine:

Fig. 4.20 Simplified state diagram of the internal state machine

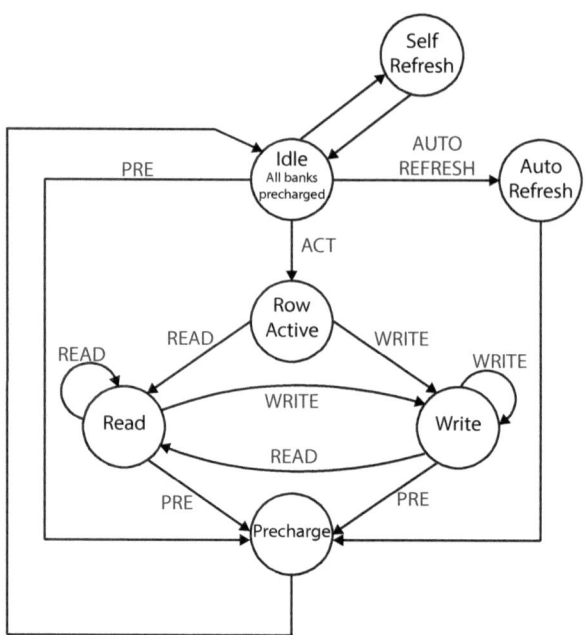

- CAS latency (CL). CL is t_{CL} rounded up to the nearest number of clock cycles,
- the length of the burst transfer,
- and the order of memory words in the burst transfer.

The control logic receives a command from the command bus. Then, depending on the command type and values contained in the respective fields of the mode register, the control logic performs specific sequences of operations to execute the command. These operations are performed by the internal state machine on successive clock cycles without requiring clock-by-clock control from the memory controller. Figure 4.20 illustrates a simplified state diagram of the internal state machine. After the initialization of the mode register, the internal state machine is in the Idle state with all banks and rows precharged. The SDRAM chip will regularly perform the self-refresh if no command is issued to SDRAM. The internal counter drives the self-refresh operation. To start memory access, the memory controller should first issue the ACTIVE command. This will eventually open a row/bank, and the internal state machine waits in the Active state for additional commands. To read data, the memory controller should issue the READ command, and to write data into memory, the memory controller should issue the WRITE command. Then, the internal state machine enters the Read or Write state and uses the column address to generate the appropriate internal signals to access the column. The READ or WRITE commands can be followed by any number of READ or WRITE commands, or the PRECHARGE command can be issued to restore the data and close the open bank/row. After the precharge operation has been executed, the internal state machine will wait in the IDLE state.

For example, in the case of the ACTIVE command, the state machine passes the row address to the row address latch and the decoder through the multiplexor. The address bit BA0 determines the bank which will be accessed. The bank control block, which acts as a decoder, selects the appropriate row address latch, the decoder, and the appropriate column decoder based on the BA0 bit. The selected row is then opened, and its content is transferred into the sense amplifiers. When the memory controller asserts a READ command, the internal state machine drives the bank control logic, which selects the appropriate column decoder based on the BA0 bit. The column decoder then selects the word from the sense amplifiers of the chosen bank. Each bank has its own column decoder—this feature is especially useful when interleaving transfers from two (or more) active banks. The SDRAM device in Fig. 4.19 allows two rows of the DRAM to be opened simultaneously. Memory accesses between two opened banks can be interleaved to hide RAS-to-CAS delay and row precharge time. When an address that designates a new bank is sent for the first time, the row in that bank must be opened. But when subsequent access specifies the same row in an already open bank, the access can happen quickly, sending only the column address. This feature requires that each bank has its own row address latch, sense amplifiers, and a column decoder. For example, while one row is accessed, the memory controller can send an ACTIVE command to a different bank and, in such a way, transfer a new row into the sense amplifiers. This row can then be read or written to without waiting for t_{RCD}. Later, we will learn how data is transferred to/from the SDRAM chip and how the burst transfers and bank interleaving can speed up memory transactions.

Summary: SDRAMs

SDRAM devices have a synchronous device interface, where commands, instead of signals, control internal latches.

In SDRAM devices, signals CAS, RAS, WE, and CS form a *command bus* used to transmit *commands* to the *internal state machine*.

SDRAM devices contain multiple independent banks.

SDRAMs can transfer many columns over several cycles per request without sending any new addresses. This type of transfer is referred to as *burst mode*.

4.9.2 Basic Operations and Timings

Now that we are familiar with the basic functionality of SDRAMs, we will present four basic operations in SDRAMs: ACTIVE, READ, WRITE, and PRECHARGE.

4.9.2.1 Activate (Open) Row

A row in a bank must be opened before any READ or WRITE commands can be issued to that bank within the SDRAM. This is accomplished via the ACTIVE command. The purpose of the ACTIVE command is to open (activate) a row in a

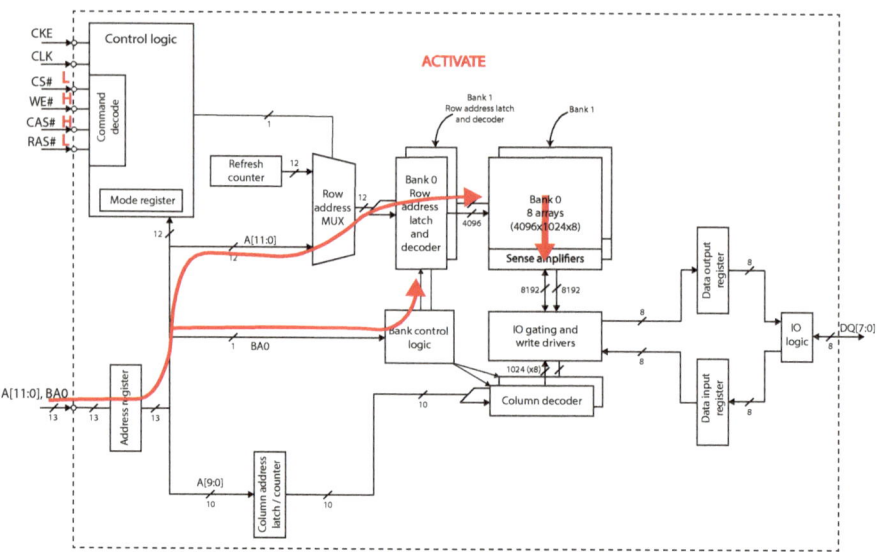

Fig. 4.21   The progression of the ACTIVE command

selected bank and move data from the DRAM arrays to the sense amplifiers of the
open bank. Figure 4.21 illustrates the progression of the ACTIVE command. The
address A11-A0 from the address bus is stored in the row address latch and decoder
of the selected bank. The address bit BA0 selects the bank and its row address latch
and decoder. Then, the entire row of data is read into the sense amplifiers. Similarly
to DRAMs, two timings are associated with the ACTIVE command: *Row Address
to Column Address Delay* (t_{RCD}) and *Row Active Time* t_{RAS}. t_{RCD} is the time it
takes for the ACTIVE command to move data from the DRAM cell arrays to the
sense amplifiers that hold the entire row of data. After t_{RCD}, a column read or write
access commands can be issued to move data between the sense amplifiers and the
memory controller through the input/output block and data bus (Fig. 4.22). Row
address to column address delay, t_{RCD}, should be divided by the clock period and
rounded up to the nearest whole number to determine the earliest clock edge after
the ACTIVE command on which a READ or WRITE command can be issued. For
example, a t_{RCD} of 20 ns with a 125 MHz clock (8ns period) results in 2.5 clock
periods, rounded to 3. A subsequent ACTIVE command to a different row in the
same bank can only be issued after the previous active row has been precharged.

Row active time, t_{RAS}, is the minimum amount of time that must elapse before
the PRECHARGE command can be issued to the open row. t_{RAS} is also referred to
as ACTIVE-to-PRECHARGE time.

4.9.2.2   Read

Figure 4.23 illustrates the progression of a column read command. A column read
command moves data from the sense amplifiers of a selected bank to the memory
controller through IO gating and write drivers and data output register. The address

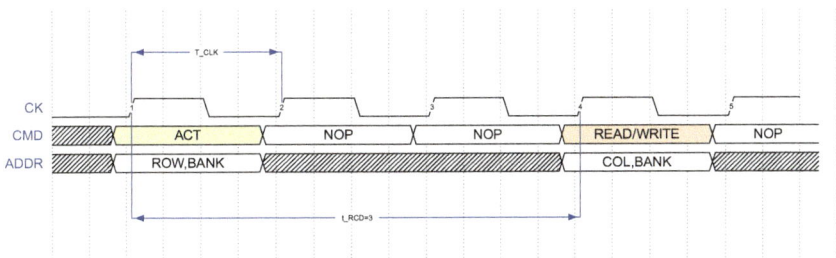

Fig. 4.22 Meeting t_{RCD}

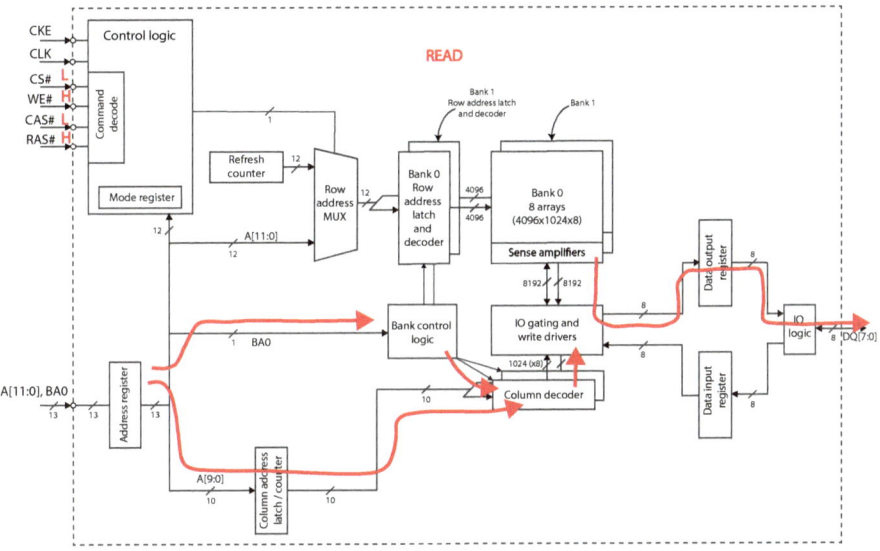

Fig. 4.23 The progression of the READ command

A[9:0] from the address bus is stored in the column address latch and column decoder of the selected bank. The address bit BA0 selects the bank and its column decoder and sense amplifiers. Then, the selected 8-bit data is read from the sense amplifiers and output to DQ pins. There are two (timing) parameters associated with a column read command: CAS latency (CL) and burst length (BL).

CL is the time it takes for the SDRAM device to move the requested data from the sense amplifiers through IO gating and output register onto the data DQ bus. For SDRAMs, the **CAS latency (CL) is the delay, in clock cycles**, between the registration of a READ command and the availability of the output data. In modern SDRAMs, the CAS latency can be set to two or three clocks. If a READ command is registered at clock edge n, and the CL is m clocks, the data will be available by

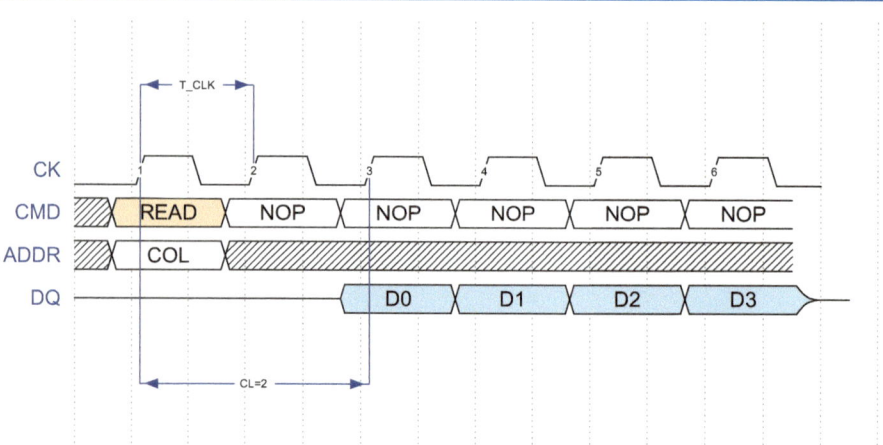

Fig. 4.24 The READ burst with CL = 2 and BL = 4

clock edge $n + m$. Now, we can combine the timing parameters, t_{RCD} and CL, to form a **random access time** (t_{RAC}).

$$t_{RAC} = t_{RCD} + CL \tag{4.4}$$

Random access time, t_{RAC}, denotes the speed at which the SDRAM device can move data from the DRAM arrays into the memory controller.

Modern memory systems move data in relatively short bursts, and the burst length (BL) is programmable. The burst length determines the maximum number of column locations that can be accessed for a given READ or WRITE command. Typically, BL is 2, 4, or 8. Read bursts are initiated with a READ command, as shown in Fig. 4.24. The starting column and bank addresses are provided with the READ command. During READ bursts, the valid data from the starting column address is available following the CAS latency after the READ command. Each subsequent data will be valid by the next positive clock edge. Upon completion of a burst, assuming no other commands have been initiated, the DQ signals will go to High-Z.

Data from a fixed-length READ burst can be followed immediately by data from a new READ or WRITE command. In such a way, a continuous flow of data can be maintained. SDRAM devices use a pipelined architecture, and therefore, a READ command can be initiated on any clock cycle following a READ command. The new READ command should be issued x cycles before the clock edge at which the last desired data element is valid, where $x = CL - 1$. This is shown in Fig. 4.25 for CL = 2 and BL = 4. Full-speed random read accesses can be performed to the same bank, or each subsequent READ can be performed to a different open bank (bank interleaving).

4.9.2.3 Write

Figure 4.26 illustrates the progression of the WRITE command. The WRITE command moves data from the DQs pins through IO gating and write drivers and data

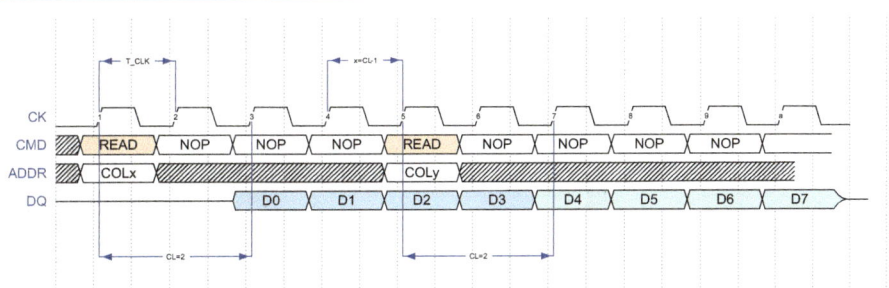

Fig. 4.25 Two consecutive READ bursts with CL = 2 and BL = 4

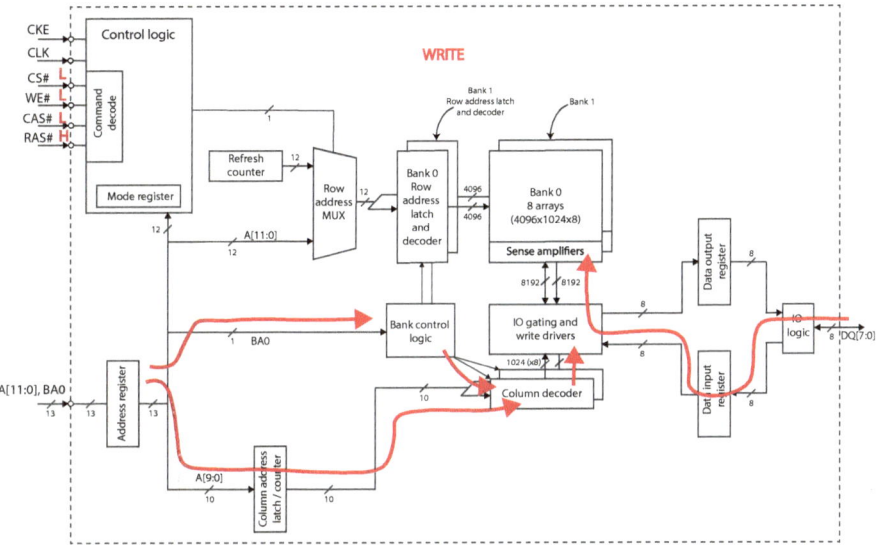

Fig. 4.26 The progression of the WRITE command

input register to the sense amplifiers of a selected bank. The column address A[9:0] from the address bus is stored in the column address latch and column decoder of the selected bank. The address bit BA0 selects the bank and its column decoder and sense amplifiers.

Figure 4.27 shows a write burst with BL = 4. The starting column and bank addresses are provided with the WRITE command, which initiates write bursts. During write bursts, the first valid data is registered with the WRITE command. Subsequent data are registered on each successive positive clock edge. Upon completion of a fixed-length burst, assuming no other commands have been initiated, the DQ pins remain at High-Z, and any additional input data is ignored.

Data from a fixed-length WRITE burst can be followed immediately by data from a new READ or WRITE command. In such a way, a continuous flow of data can be maintained. Figure 4.28 shows two consecutive write bursts with BL = 2.

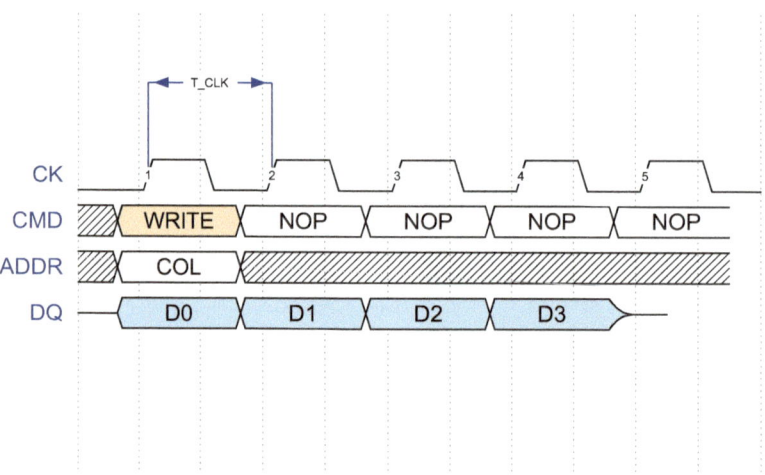

Fig. 4.27 The WRITE burst with BL = 4

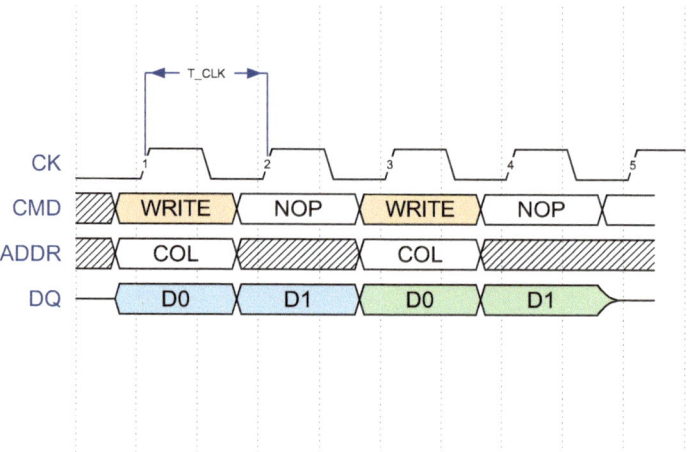

Fig. 4.28 Two consecutive WRITE bursts with BL = 2

4.9.2.4 Precharge

So far, we have seen that accessing data in an SDRAM device is a two-step process. First, the ACTIVE command opens a row in a selected bank and moves data from the DRAM cells in that row to the sense amplifiers. The data then remains in the sense amplifiers and can be transferred to or from SDRAM using the READ and WRITE commands. The PRECHARGE command is used to deactivate the open row in a particular bank or the open row in all banks—it restores data in the row, resets the sense amplifiers and the bit lines, and prepares the sense amplifiers for another row access. Figure 4.29 illustrates the progression of the PRECHARGE command. The address A[11:0] from the address bus is stored in the row address latch and decoder

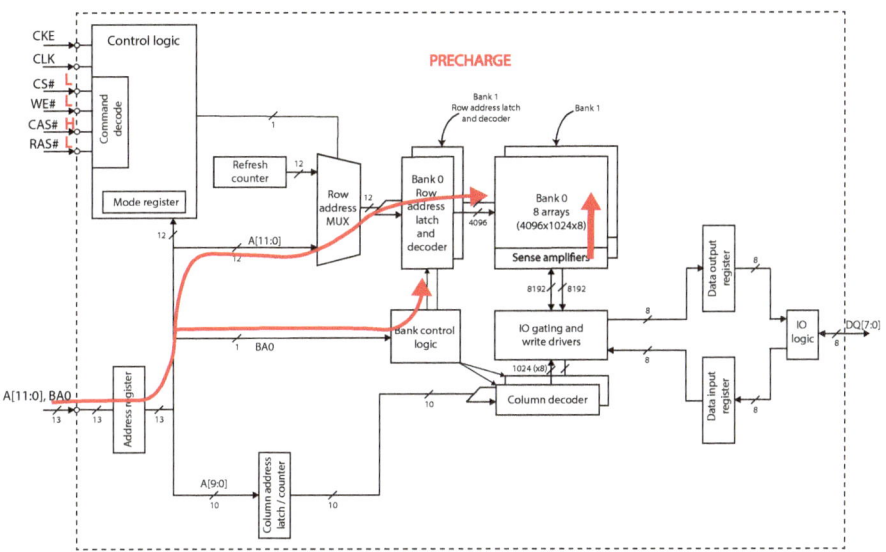

Fig. 4.29 The progression of the PRECHARGE command

of the selected bank. The address bit BA0 selects the bank and its row address latch and decoder. Then, the selected bank is precharged.

The timing parameter associated with the (row) PRECHARGE command is row precharge time, t_{RP}. The bank(s) will be available for a subsequent access row precharge time (t_{RP}) after the PRECHARGE command is issued. Recall that t_{RAS} is the minimum amount of time that the row should remain open before issuing the PRECHARGE command (i.e., ACTIVE-to-PRECHARGE time). Now, we can combine the timing parameters, t_{RP} and t_{RAS}, to form a **row cycle time** (t_{RC}):

$$t_{RC} = t_{RAS} + t_{RP} \tag{4.5}$$

Row cycle time, t_{RC}, denotes the speed at which the SDRAM device can bring data from the DRAM arrays into the sense amplifiers, restore the data to the DRAM cells, and be ready for another ACTIVE command. t_{RC} is the fundamental limitation to the speed at which data may be retrieved from different rows within the same SDRAM bank.

A PRECHARGE command may follow a READ or WRITE burst to the same bank. In the case of PRECHARGE after READ, the PRECHARGE command should be issued $x = CL - 1$ cycles before the clock edge at which the last data element in a burst is valid. This is shown in Fig. 4.30 for CL = 2. In the case of PRECHARGE after WRITE, the PRECHARGE command should be issued at least one clock period after the positive clock edge at which the last input data is registered, regardless of frequency (Fig. 4.31).

Following the PRECHARGE command, a subsequent command to the same bank cannot be issued until t_{RP} is met. The disadvantage of the PRECHARGE command

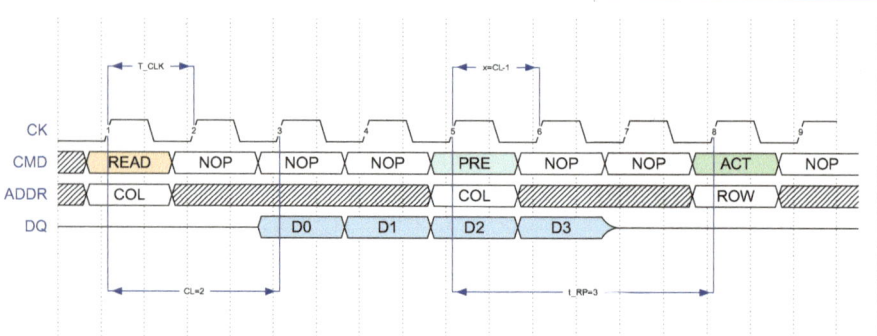

Fig. 4.30 READ to PRECHARGE

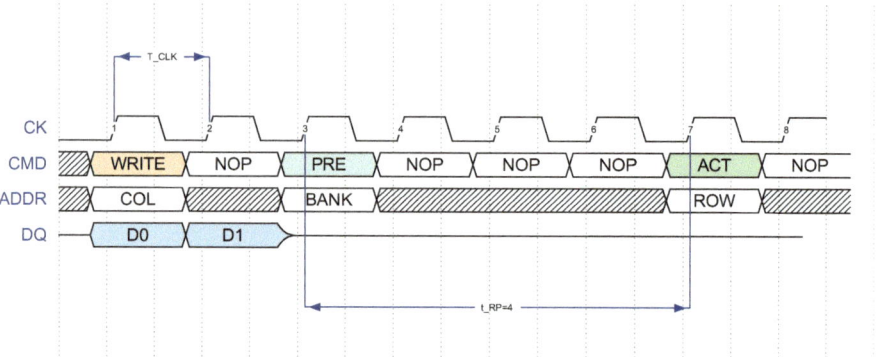

Fig. 4.31 WRITE to PRECHARGE

is that it requires that the command and address buses be available at the appropriate time to issue the command.

4.9.3 Case Study: Using the STM32F Flexible Memory Controller to Access SDRAM

The Flexible Memory Controller (FMC) found in STM32 microcontrollers consists of the following main blocks:

1. The interface to the CPU's Advanced High-performance Bus (AHB),
2. The NOR Flash/SRAM memory controller,
3. The SDRAM memory controller, and
4. NAND Flash controller.

The block diagram of the FMC is shown in Fig. 4.32. The AHB interface allows the CPU (and other bus master peripherals) to access the external memories through the FMC controller. Two primary purposes of FMC are to translate transactions on the

Fig. 4.32 FMC block diagram

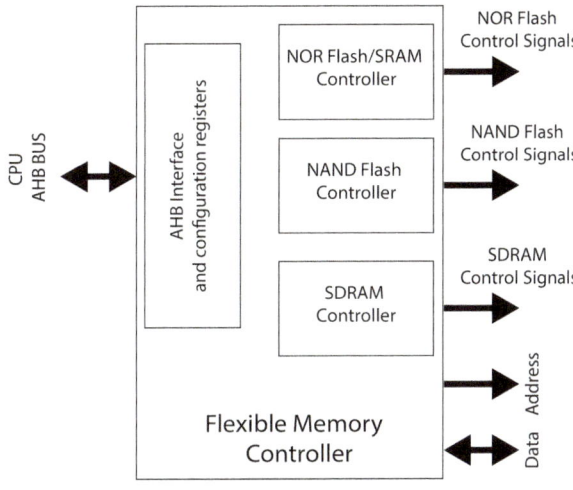

Fig. 4.33 Memory regions accessible from the FMC controller

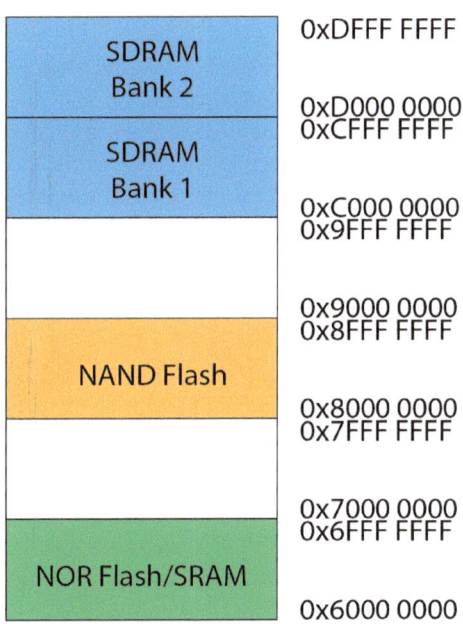

high-speed CPU bus (namely AHB bus) into the appropriate external protocol and to meet the access time requirements of the external memory devices.

From the FMC (or microprocessor) point of view, the external memory is divided into six fixed-size regions of 256 Mbytes each, called banks (Fig. 4.33). The first bank is used to address NOR Flash memory devices. The third bank is used to address NAND Flash memory devices. The last two banks are used to address two SDRAM devices (one device per bank). The address bit 28 on the AHB bus (internal AHB

Fig. 4.34 FMC SDRAM
Controller block diagram
and signals

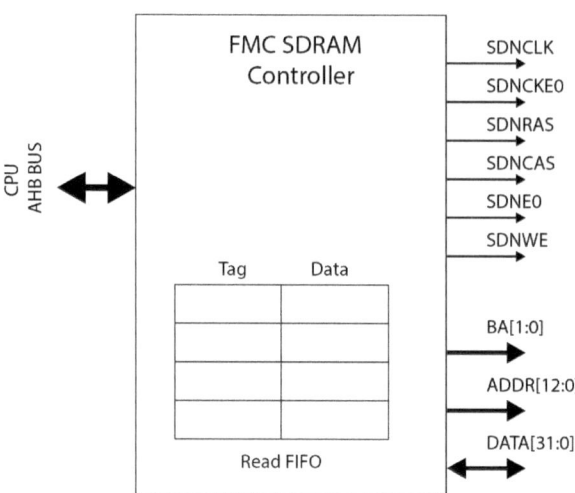

address line 28) selects one of the memory devices (banks). Let us focus only on the FMC SDRAM controller and an SDRAM device in the fifth bank.

All external memories share the addresses, data, and control signals with the controller, and each external device is accessed utilizing a unique chip-select signal. The FMC performs only one access at a time to an external device. Here, we will describe only the SDRAM controller and its use to interface a 128 Mbit SDRAM memory chip. All AHB transactions, in this case, translate into the SDRAM device protocol.

The FMC SDRAM controller supports SDRAM devices of up to 256 Mbytes. It can issue a 13-bit row address, an 11-bit column address, and a 2-bit bank address. The memory accesses can be 8-bit, 16-bit, and 32-bit. We will use Micron's 1 Meg x 32 x 4 banks MT48LC4M32B2 SDRAM chip, organized as 4096 rows x 256 columns x 32 bits per bank. Hence, the memory controller would issue a 12-bit row address, an 8-bit column address, and a 2-bit bank address.

The SDRAM controller in Fig. 4.34 accepts single and burst read and write requests and translates them into single memory accesses. In both cases, the SDRAM controller keeps track of the active row in each bank to be able to perform consecutive read and write accesses. The FMC SDRAM controller comprises a read FIFO (6 lines x 32 bits). It is used to read data in advance—the memory controller anticipates READ commands to the open row if the RBURST bit is set in the FMC_SDCRx register and stores data in the FIFO. Two bits RPIPE[1:0] in the FMC_SDCRx register defines how much data will be anticipated and stored into the FIFO during the read access. If we set both RPIPE[1:0] bits to zero, four data will be anticipated during a single read access. The first read data will be transmitted to the AHB bus, and the other three will be stored in the read FIFO buffer. The read FIFO buffer stores a 14-bit address tag for each line to identify its content: 11 bits for the column address, 2 bits for the internal bank in the active row, and 1 bit for the SDRAM device. Each

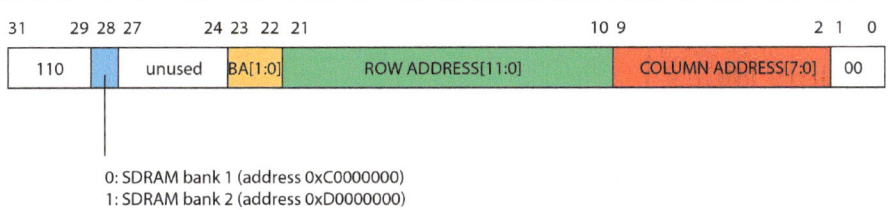

0: SDRAM bank 1 (address 0xC0000000)
1: SDRAM bank 2 (address 0xD0000000)

Fig. 4.35 Address maping for a 128-bit SDRAM (4096 rows x 256 columns x 4 banks x 32 bit)

time a read request occurs, the SDRAM controller checks if the address matches one of the address tags in the read FIFO buffer. In such a case, data are directly read from the FIFO buffer. Otherwise, a new read command is issued to the SDRAM device, and new data is read to the FIFO buffer.

The FMC SDRAM controller periodically issues auto-refresh commands to refresh the SDRAM. The programmer should initialize the internal counter value in the FMC_SDRTR. This value defines the number of memory clock cycles between two refresh cycles (refresh rate). When this counter reaches zero, the FMC SDRAM controller issues the auto-refresh command. If there is ongoing memory access, the auto-refresh request is delayed until the memory access finishes; otherwise, the auto-refresh request takes precedence. If the memory access request occurs during an auto-refresh operation, the request is buffered and processed when the auto-refresh is complete.

For our particular case, where the FMC SDRAM controller is used to access the MT48LC4M32B2 SDRAM chip, the 32-bit memory address from the AHB bus is mapped into the SDRAM address as presented in Fig. 4.35. This Fig. illustrates how the 32-bit addresses issued by the CPU on the AHB bus map to the 26-bit addresses issued by the SDRAM controller to the SDRAM device.

In order to use the FMC SDRAM controller with an external SDRAM device residing in the SDRAM Bank 1, we should:

1. first, initialize the FMC SDRAM controller according to the used SDRAM device, and
2. secondly, initialize the SDRAM device.

The first step involves programming two FMC SDRAM controller configuration registers, SDRAM Control Register 1 (FMC_SDCR1) and SDRAM Timing Register 1 (FMC_SDTR1). The bits in FMC_SDCR1 (Fig. 4.36) define the SDRAM clock period, CAS Latency, whether the FMC anticipates READ commands (burst read), data bus width and the internal organization of the SDRAM chip (rows, columns and banks).

The bits in FMC_SDTR1 define SDRAM timing parameters, e.g. RAS-to-CAS delay, row-precharge delay, etc. In order to correctly set the bits in these two registers, we should consult the datasheet for a particular SDRAM chip (Fig. 4.37).

31	30	29	28	27	26	25	24	23	22	21	20	19	18	17	16
15	14	13	12	11	10	9	8	7	6	5	4	3	2	1	0
	RPIPE[1:0]		RBURST	SDCLK		WP	CAS		NB	MWID		NR		NC	
rw	rw	rw	rw	rw	rw	rw	rw	rw	rw	rw	rw	rw	rw	rw	rw

Fig. 4.36 Control register (FMC_SDCR)

31	30	29	28	27	26	25	24	23	22	21	20	19	18	17	16
				TRCD				TRP				TWR			
				rw	rw	rw	rw	rw	rw	rw	rw	rw	rw	rw	rw
15	14	13	12	11	10	9	8	7	6	5	4	3	2	1	0
TRC				TRAS				TXSR				TMRD			
rw	rw	rw	rw	rw	rw	rw	rw	rw	rw	rw	rw	rw	rw	rw	rw

Fig. 4.37 Timing register (FMC_SDTR)

The second step initializes the SDRAM chip. During the SDRAM chip initialization, the FMC controller sends several predefined commands to the SDRAM chip. We should write these commands into the FMC SDRAM Command Mode Register (FMC_SDCMR) to send them. The required initialization steps are described in the datasheet for a particular SDRAM chip and involve the following:

1. providing stable CLOCK signal,
2. performing a PRECHARGE ALL command, which puts all rows in all banks into an idle state,
3. issuing several AUTO REFRESH commands
4. issuing several NOP commands before SDRAM is ready for access.

Instead of directly setting bits in the FMC SDRAM configuration registers, we will rather use the HAL library. The HAL library abstracts most of the FMC SDRAM controller hardware details. The FMC SDRAM controller is abstracted in HAL with the `SDRAM_HandleTypeDef` C structure. The two most important members of this structure are the C reference to `FMC_SDRAM_TypeDef Instance` structure and `FMC_SDRAM_InitTypeDef Init`. The `Instance` is a reference to the SDRAM registers (it holds the base address of the FMC SDRAM registers), while the `Init` structure allows for FMC SDRAM controller configuration. The `FMC_SDRAM_InitTypeDef Init` structure is defined as follows:

```
typedef struct
{
  uint32_t SDBank;
  uint32_t ColumnBitsNumber;
  uint32_t RowBitsNumber;
  uint32_t MemoryDataWidth;
  uint32_t InternalBankNumber;
  uint32_t CASLatency;
  uint32_t WriteProtection;
  uint32_t SDClockPeriod;
  uint32_t ReadBurst;
} FMC_SDRAM_InitTypeDef;
```

Listing 4.1 FMC SDRAM `FMC_SDRAM_InitTypeDef` C structure

Let us briefly describe the elements of the `FMC_SDRAM_InitTypeDef` Init structure:

- `SDBank`: Specifies the SDRAM memory device that will be used (bank 1 or bank 2 according to Fig. 4.33).
- `ColumnBitsNumber`: Defines the number of bits of the column address.
- `RowBitsNumber`: Defines the number of bits of the row address.
- `MemoryDataWidth`: Defines the memory device width.
- `InternalBankNumber`: Defines the number of the device's internal banks.
- `CASLatency`: Defines the SDRAM CAS latency in the number of memory clock cycles.
- `WriteProtection`: Enables/Disables the SDRAM device to be accessed in write mode.
- `SDClockPeriod`: Defines the SDRAM Clock Period for SDRAM devices. The SDRAM clock period can be HCLK/2 or HCLK/3, where HCLK is the clock period on the CPU's AHB bus.
- `ReadBurst`: Enables the SDRAM controller to anticipate the next read commands during the CAS latency and stores data in the Read FIFO.

Besides the `FMC_SDRAM_InitTypeDef` C structure, which abstracts the content of FMC_SDCR1 register, the `FMC_SDRAM_TimingTypeDef` C structure is used to abstract the content of FMC_SDTR1 register. It is defined as follows:

```
typedef struct
{
  uint32_t LoadToActiveDelay;
  uint32_t ExitSelfRefreshDelay;
  uint32_t SelfRefreshTime;
  uint32_t RowCycleDelay;
  uint32_t WriteRecoveryTime;
  uint32_t RPDelay;
  uint32_t RCDDelay;
} FMC_SDRAM_TimingTypeDef;
```

Listing 4.2 FMC SDRAM `FMC_SDRAM_TimingTypeDef` C structure

The elements of the FMC_SDRAM_TimingTypeDef C structure are self-explanatory (they represent the particular timings for SDRAM chips), and there is no need to describe them.

The code Listing 4.3 shows the FMC SDRAM controller initialization.

```
uint8_t Init_SDRAM(void)
{
  static uint8_t sdramstatus = SDRAM_ERROR;
  /* SDRAM device configuration */
  sdramHand.Instance = FMC_SDRAM_DEVICE;

  /* Timing configuration for 100Mhz as SDRAM clock frequency
     (System clock is up to 200Mhz) */
  /* These parameters are from the MT48LC4M32B2 Data Sheet,
     Table 18 and Table 19 */
  sdramTiming.LoadToActiveDelay    = 2;     // t_MRD
  sdramTiming.ExitSelfRefreshDelay = 7;     // t_XSR
  sdramTiming.SelfRefreshTime      = 5;     // t_RAS
  sdramTiming.RowCycleDelay        = 7;     // t_RC
  sdramTiming.WriteRecoveryTime    = 2;     // t_WR
  sdramTiming.RPDelay              = 2;     // t_RP
  sdramTiming.RCDDelay             = 2;     // t_RCD


  sdramHand.Init.SDBank            = FMC_SDRAM_BANK1;
  sdramHand.Init.ColumnBitsNumber  = FMC_SDRAM_COLUMN_BITS_NUM_8;
  sdramHand.Init.RowBitsNumber     = FMC_SDRAM_ROW_BITS_NUM_12;
  sdramHand.Init.MemoryDataWidth   = FMC_SDRAM_MEM_BUS_WIDTH_32;
  sdramHand.Init.InternalBankNumber = FMC_SDRAM_INTERN_BANKS_NUM_4;
  sdramHand.Init.CASLatency        = FMC_SDRAM_CAS_LATENCY_3;
  sdramHand.Init.WriteProtection  = ↵
    FMC_SDRAM_WRITE_PROTECTION_DISABLE;
  sdramHand.Init.SDClockPeriod     = FMC_SDRAM_CLOCK_PERIOD_2;
  sdramHand.Init.ReadBurst         = FMC_SDRAM_RBURST_ENABLE;
  sdramHand.Init.ReadPipeDelay     = FMC_SDRAM_RPIPE_DELAY_0;

  /* SDRAM controller initialization */

  if(HAL_SDRAM_Init(&sdramHand, &sdramTiming) != HAL_OK)
  {
    sdramstatus = SDRAM_ERROR;
  }
  else
  {
    sdramstatus = SDRAM_OK;
  }

  /* Once the FMC SDRAM Ctrl is initialized, we can access
     and initialize the SDRAM chip */
  /* SDRAM initialization sequence */
  SDRAM_Initialization_sequence(REFRESH_COUNT);

  return sdramstatus;
}
```

Listing 4.3 FMC SDRAM controller initialization

Firstly, we set the SDRAM timing parameters (in the FMC_SDTR1 register) considering the 100MHz SDRAM clock, and then we set the SDRAM configuration (in the FMC_SDCR1 register). To initialize the FMC SDRAM controller (that is to copy the elements of both C structures into the appropriate fields of the FMC_SDCR1 and FMC_SDTR1 registers), we call the HAL function HAL_SDRAM_Init(SDRAM_HandleTypeDef *hsdram, FMC_SDRAM_ TimingTypeDef *Timing).

After the FMC SDRAM initialization, we should initialize the SDRAM chip. SDRAMs must be powered up and initialized in a predefined manner. This is a necessary step required to put all SDRAM rows in the idle state (precharge all rows) and prepare the SDRAM chip for accepting and executing the commands. The SDRAM initialization sequence is described in the SDRAM datasheet in detail. The code Listing 4.4 shows the FMC SDRAM chip initialization. Briefly, the initialization procedure contains four steps:

1. Enable the stable SDRAM clock.
2. Wait for at least 100us prior to issuing any command.
3. Perform a PRECHARGE ALL command.
4. Issue at least two AUTO REFRESH commands.
5. The SDRAM is now ready for mode register programming. Because the mode register will power up in an unknown state, it should be loaded with desired bit values prior to applying any operational command.

```
/**
  * @brief  Init the SDRAM device.
  * SDRAMs must be initialized in a predefined manner. Operational ↵
    procedures
  * other than those specified in the SDRAM Data Sheet may result in ↵
    undefined operation.
  * @param  RefreshCount: SDRAM refresh counter value
  * @retval None
  */
void SDRAM_Initialization_sequence(uint32_t RefreshCount)
{
  __IO uint32_t tmpmrd = 0;

  /* Step 1: Configure a clock configuration enable command */
  sdramCmd.CommandMode            = FMC_SDRAM_CMD_CLK_ENABLE;
  sdramCmd.CommandTarget          = FMC_SDRAM_CMD_TARGET_BANK1;
  sdramCmd.AutoRefreshNumber      = 1;
  sdramCmd.ModeRegisterDefinition = 0;


  /* Send the Clock Configuration Enable command to the target bank*/
  /* The command is sent as soon as the Command MODE field in the
     CMR is written */
  HAL_SDRAM_SendCommand(&sdramHand, &sdramCmd, SDRAM_TIMEOUT);

  /*
   * Once the clock is stable, the SDRAM requires a 100us delay
   * prior to issuing any command
   */

  /* Step 2: Insert 100 us minimum delay */
  /* Inserted delay is equal to 1 ms due to systick time base unit */
  HAL_Delay(1);
```

```
/*
 * Once the 100us delay has been satisfied, a PRECHARGE command
 * should be applied. All banks must then be precharged,
 * thereby placing the device in the all banks idle state. .
 */
/* Step 3: Configure a PALL (precharge all) command */
sdramCmd.CommandMode            = FMC_SDRAM_CMD_PALL;
sdramCmd.CommandTarget          = FMC_SDRAM_CMD_TARGET_BANK1;
sdramCmd.AutoRefreshNumber      = 1;
sdramCmd.ModeRegisterDefinition = 0;

/* Send the Precharge All command to the target bank */
/* The command is sent as soon as the Command MODE field
   in the CMR is written */
HAL_SDRAM_SendCommand(&sdramHand, &sdramCmd, SDRAM_TIMEOUT);

/*
 * Once in the idle state, at least two AUTO REFRESH cycles must
 * be performed. If desired, more than two AUTO REFRESH
 * commands can be issued in the sequence.
 */
/* Step 4: Configure an Auto Refresh command */
sdramCmd.CommandMode            = FMC_SDRAM_CMD_AUTOREFRESH_MODE;
sdramCmd.CommandTarget          = FMC_SDRAM_CMD_TARGET_BANK1;
sdramCmd.AutoRefreshNumber      = 8;
sdramCmd.ModeRegisterDefinition = 0;

/* Send the Auto-refresh commands to the target bank */
/* The command is sent as soon as the Command MODE
   field in the CMR is written */
HAL_SDRAM_SendCommand(&sdramHand, &sdramCmd, SDRAM_TIMEOUT);


/*
 * The SDRAM is now ready for mode register programming.
 * Because the mode register will power up in an unknown state,
 * it should be loaded with desired bit values prior to
 * applying any operational command. Using the LMR command,
 * program the mode register.
 */
/* Step 5: Program the external memory mode register */
tmpmrd = (uint32_t)SDRAM_MODEREG_BURST_LENGTH_1          |\
                   SDRAM_MODEREG_BURST_TYPE_SEQUENTIAL   |\
                   SDRAM_MODEREG_CAS_LATENCY_3           |\
                   SDRAM_MODEREG_OPERATING_MODE_STANDARD |\
                   SDRAM_MODEREG_WRITEBURST_MODE_SINGLE;

sdramCmd.CommandMode            = FMC_SDRAM_CMD_LOAD_MODE;
sdramCmd.CommandTarget          = FMC_SDRAM_CMD_TARGET_BANK1;
sdramCmd.AutoRefreshNumber      = 1;
sdramCmd.ModeRegisterDefinition = tmpmrd;

/* Send the Load Mode Register command to the target bank */
/* The command is sent as soon as the Command MODE field in
   the CMR is written */
HAL_SDRAM_SendCommand(&sdramHand, &sdramCmd, SDRAM_TIMEOUT);

/*
 * Wait for at least tMRD time. This is automatically performed by
 * the FMC SDRAM controller. At this point the DRAM is ready for
 * any valid command.
 */
```

```
/* Step 6: Set the refresh rate counter in Refresh Timer register ←
   */
/* This 13-bit field defines the refresh rate of the SDRAM device.
   It is expressed in number of memory clock cycles. */
HAL_SDRAM_ProgramRefreshRate(&sdramHand, RefreshCount);
}
```

Listing 4.4 SDRAM initialization sequence

To enable the above procedure, the FMC SDRAM controller provides a special register called Command Mode register (FMC_SDCMR), illustrated in Fig. 4.38. It contains four fields. The MODE field defines the command issued to the SDRAM chip. The possible commands are, for example, "CLK ENABLE", "PRECHARGE ALL", "AUTO REFRESH", and "LOAD MODE REGISTER". The CTB1 and CTB2 fields select the SDRAM chip to which the command is sent. As soon as the MODE field is written, the FMC SDRAM controller will issue the corresponding command to SDRAM chips selected by CTB1 and CTB2 command bits. The NRFS field defines how many consecutive Auto-refresh commands are issued in the fourth step of the initialization sequence, the MRD field contains the content that should be written to the SDRAM Mode Register. The mode register is a 12-bit special register inside the SDRAM chip and is used to define the specific mode of operation of the SDRAM. This definition includes the selection of a burst length (BL), a burst type, a CAS latency (CL), an operating mode and a write burst mode, as shown in Fig. 4.39. The mode register is programmed from the FMC SDRAM controller via the "LOAD MODE REGISTER" command and retains the stored information until it is programmed again or the SDRAM device loses power.

The initialization of the SDRAM device is performed by sending a series of commands from the FMC_SDCMR register to the SDRAM device. Each command contains the actual instruction and its parameters. To facilitate the SDRAM chip initialization, HAL provides the FMC_SDRAM_CommandTypeDef C structure and HAL_SDRAM_SendCommand function. The FMC_SDRAM_CommandTypeDef C structure abstracts the content of FMC_SDCMR register and is defined as follows:

31	30	29	28	27	26	25	24	23	22	21	20	19	18	17	16
												MRD			
										rw	rw	rw	rw	rw	rw
15	14	13	12	11	10	9	8	7	6	5	4	3	2	1	0
MRD							NRFS				CTB1	CTB2	MODE		
rw	rw	rw	rw	rw	rw	rw	rw	rw	rw	rw	rw	rw	rw	rw	rw

Fig. 4.38 Command Mode register (FMC_SDCMR)

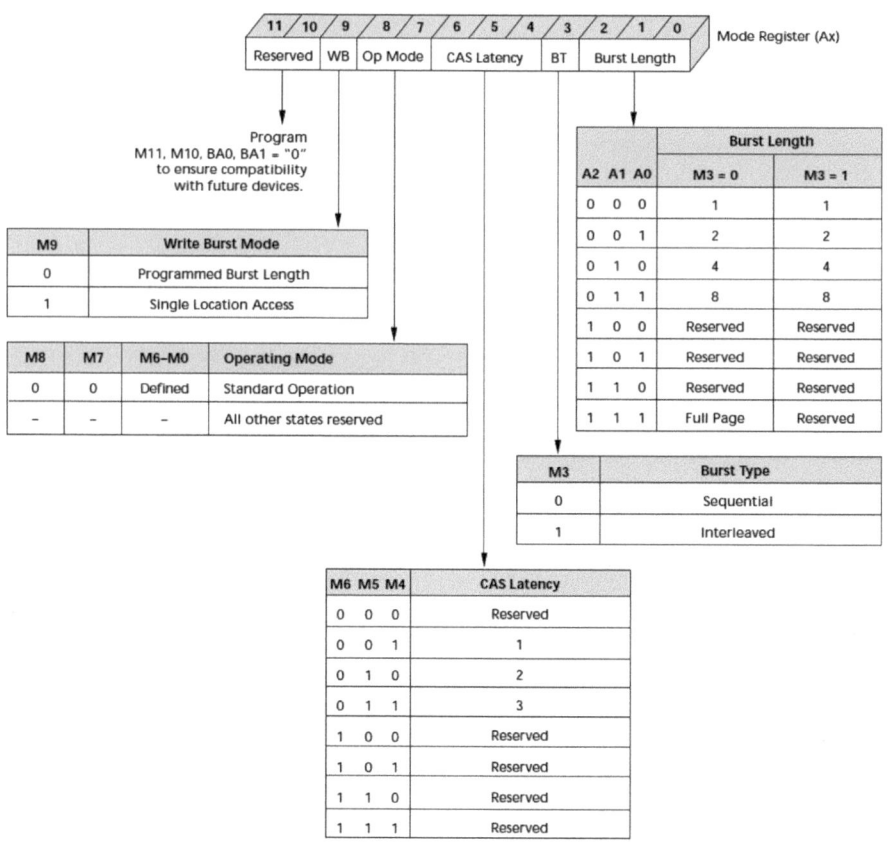

Fig. 4.39 SDRAM mode register

```
typedef struct
{
  uint32_t CommandMode;
  uint32_t CommandTarget;
  uint32_t AutoRefreshNumber;
  uint32_t ModeRegisterDefinition;
} FMC_SDRAM_CommandTypeDef;
```

Listing 4.5 FMC SDRAM `FMC_SDRAM_CommandTypeDef` C structure

Let us briefly describe the elements of the `FMC_SDRAM_CommandTypeDef` `Init` structure:

- `CommandMode`: Defines the command issued to the SDRAM device.
- `CommandTarget`: Defines which SDRAM device (1 or 2) the command will be issued to.

- AutoRefreshNumber: Defines the number of consecutive auto-refresh commands issued in auto-refresh mode.
- ModeRegisterDefinition: Defines the SDRAM Mode register content.

In order to send a command to the SDRAM device, we first fill the fields in the FMC_SDRAM_CommandTypeDef Init structure and then call the HAL_SDRAM_SendCommand function.

At the end of the SDRAM chip initialization, we set the auto-refresh period in the FMC SDRAM controller. The AUTO REFRESH command is used during the regular operation of the SDRAM to refresh its content. This command is nonpersistent, so it must be issued each time a refresh is required. If memory access is in progress, the auto-refresh request is delayed. The refresh controller inside the SDRAM chip generates the address of the row that should be refreshed. For example, the 128Mb SDRAM requires 4096 AUTO REFRESH commands every 64ms. To ensure that each row is refreshed according to this requirement, the SDRAM controller must issue an AUTO REFRESH command every 15.625us. The FMC SDRAM controller provides the Refresh Timer register (FMC_SDRTR). This register holds the 13-bit refresh rate in number of SDRAM clock cycles. This 13-bit field should be set immediately after the initialization of SDRAM. The 13-bit refresh rate is calculated as follows. As the SDRAM clock runs at 100 Mhz (10 ns period), 15.625 us equals 1562 SDRAM clock periods. We should subtract at least 20 SDRAM clock periods from this value to obtain a safe margin if an auto-refresh request occurs when a read request has been accepted. Hence, the 13-bit refresh rate in the FMC_SDRTR register corresponds to 1542.

To demonstrate the different scenarios when using the FMC SDRAM controller, we copy a matrix of size 64 rows times 256 columns from the external SDRAM to the internal SRAM. The elements of the matrix are 32-bit unsigned integers. In the first scenario (Listing 4.6), the matrix is accessed in row-major order, while in the second scenario (Listing 4.7), the matrix is accessed in column-major order. The constants PA3_SDRAM_DEVICE_ADDR and SDRAM_COLS in Listings 4.6 and 4.7 equal 0xC0008000 and 256, respectively. Hence, the matrix is read from the SDRAM startin at address 0xC000800.

```
void SDRAM_mat_row_access_test(void){
  volatile uint32_t address;

  for (int i = 0; i<MAT_ROWS; i++) {
    for(int j=0; j<SDRAM_COLS; j++) {
      address = PA3_SDRAM_DEVICE_ADDR + ((i*SDRAM_COLS + j)<<2);
      matrixB[i][j] = *(uint32_t*)address;
    }
  }
}
```

Listing 4.6 Read matrix from SDRAM in row-major order

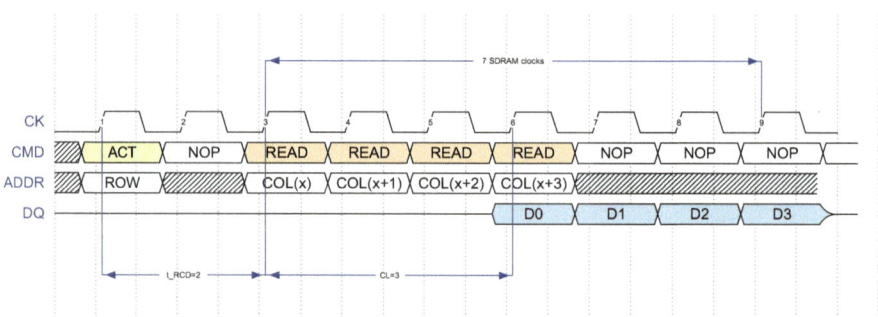

Fig. 4.40 Using row-major order to read a matrix, the SDRAM controller anticipates four consecutive READ command to the active SDRAM row for each read initiated from the CPU

```
void SDRAM_mat_col_access_test(void){
  volatile uint32_t address;

    for (int i = 0; i<SDRAM_COLS; i++) {
    for(int j=0; j<MAT_ROWS; j++) {
      address = PA3_SDRAM_DEVICE_ADDR + ((j*SDRAM_COLS + i)<<2);
      matrixB[j][i] = *(uint32_t*)address;
    }
  }
}
```

Listing 4.7 Read matrix from SDRAM in column-major order

Figure 4.40 illustrates one read issued from the CPU for the first scenario (row-major order access). The FMC SDRAM controller does not support SDRAM burst reads or writes (the only allowable burst length is 1). Instead, it supports burst reads on the CPUs AHB bus by utilizing the internal FIFO. Hence, it anticipates four READ commands to fill in the internal FIFO. The FIFO content is then transferred to the CPU using the AHB burst read of length 4.

In the second scenario, the matrix is accessed using column-major order. Figure 4.41 illustrates two consecutive reads issued from the CPU. As the CPU reads data from consecutive rows in each iteration, the CPU controller first reads four consecutive words from the active SRAM row and fills the internal FIFO, but it only returns one word to the CPU over the AHB bus. As the CPU starts another read from the next row, the SDRAM controller first precharges the active row. It then waits for two SDRAM clock periods (Row Precharge time) before activating the next row.

It is obvious that row-major order access is considerably faster than column-major order access. A rough estimate of the access time for row-major order access considering an already open row is seven (7) SDRAM clock periods per four words. On the other side, a rough estimate of the access time for column-major order access is eight (8) SDRAM clock periods per word. Recall that only one word is transferred to the CPU, although the SDRAM controller anticipates four consecutive reads from the active row.

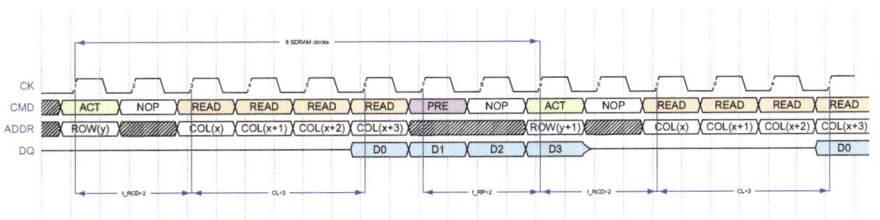

Fig. 4.41 Using column major order results in activating, reading and precharging an SDRAM row for every read issued from the CPU

To assess the performance (speed) of the row-major and column-major matrix reads, we use the code in Listing 4.8. For each test, the code first sets the PC8 pin and reads the timer TIM3 counter value (this is the start of the test). After the test, we reset the PC8 pin and read the timer TIM3 counter value (this is the start of the test). By setting and resetting the PC8 pin, we can measure the duration of each test using an oscilloscope. The timer TIM3 runs at 1MHz (1 us resolution). Hence, we can estimate the duration of each test simply by reading the timer counter before and after the test.

```
// Row-major order access:
HAL_GPIO_WritePin(GPIOC, GPIO_PIN_8, GPIO_PIN_SET);
timer_val_start = __HAL_TIM_GET_COUNTER(&TIM3Handle);
SDRAM_mat_row_access_test();
HAL_GPIO_WritePin(GPIOC, GPIO_PIN_8, GPIO_PIN_RESET);
timer_val_end = __HAL_TIM_GET_COUNTER(&TIM3Handle);
if (timer_val_end > timer_val_start)
    elapsed_rows = timer_val_end - timer_val_start;
else
    elapsed_rows = timer_val_end + (65536-timer_val_start);

// Column-major order access:
HAL_GPIO_WritePin(GPIOC, GPIO_PIN_8, GPIO_PIN_SET);
timer_val_start = __HAL_TIM_GET_COUNTER(&TIM3Handle);
SDRAM_mat_col_access_test();
timer_val_end = __HAL_TIM_GET_COUNTER(&TIM3Handle);
HAL_GPIO_WritePin(GPIOC, GPIO_PIN_8, GPIO_PIN_RESET);
if (timer_val_end > timer_val_start)
    elapsed_cols = timer_val_end - timer_val_start;
else
    elapsed_cols = timer_val_end + (65536-timer_val_start);
```

Listing 4.8 Code used to test the speed of row-major and column-major matrix read from the SDRAM

Figure 4.42 shows the oscilloscope trace for the signal on the GPIOC pin. It shows that the row-major order read lasts for about 2.3 ms, while the column-major order read lasts for about 10 ms. Using the timer counter, we estimate the duration of the row-major order read to 2365 us and the duration of the column-major order read to 9816 us. Both measurements show that the row-major order read is about four times faster than the column-major order read, which is in accordance with the rough estimation from Figs. 4.40 and 4.41.

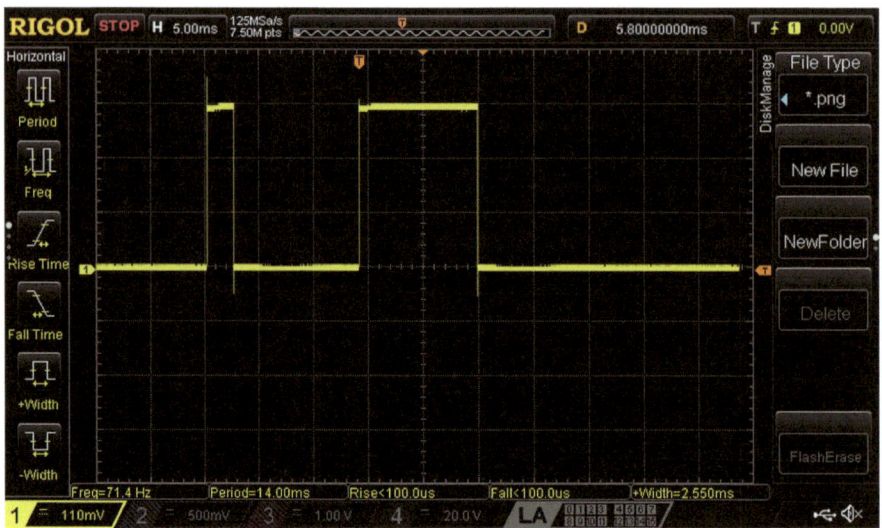

Fig. 4.42 Oscilloscope trace on the GPIOC pin 8. The row-major order matrix read lasts for about 2.5 ms while the column-major order matrix read lasts for more than 10 ms

4.9.3.1   Using DMA to Transfer Data from an External SDRAM to the Internal SRAM

Direct memory access (DMA) is used to provide high-speed data transfer between peripherals and memory and between memory and memory without any CPU action (except DMA controller initialization and DMA transfer request in case of memory-to-memory transfer). As already described in Sect. 3.5.2, the DMA controller in STM32 microcontrollers (actually, there are two DMA controllers, DMA1 and DMA2, respectively) have 16 streams in total (8 for each DMA controller), each dedicated to managing memory access requests from one or more peripherals. Each stream can have up to 8 channels (requests) in total. Each DMA controller has an arbiter for handling the priority between DMA requests. According to the STM32F69I reference manual, the memory-to-memory mode in DMA is a mode that doesn't need any triggering request from a peripheral, and it will happen just after the stream enable bit is set. Also, according to the STM32F69I reference manual, only the DMA2 could handle memory-to-memory data transfers. The stream can be enabled just by setting the Enable bit (EN) in the DMA SxCR register. Then, the stream immediately fills the FIFO up to the threshold level. When the threshold level is reached, the FIFO contents are drained and stored in the destination.

Before using the DMA2 controller to transfer data from one memory region to another, we must configure (initialize) the DMA2 controller as described in Sect. 3.5.2. When configuring the DMA controller we:

1. Select a stream that we wish to use. Any available stream in the DMA2 controller can be used for memory-to-memory transfers.
2. Select a channel; this is irrelevant for memory-to-memory transfers because a peripheral device does not trigger the DMA transfer through a channel. Instead, it is triggered by setting the EN bit in the DMA SxCR register.
3. Set a priority for a selected DMA stream.
4. Set the number of data to be transferred (it can be any value from 1 to 65535).
5. Set the source and destination transfer width (byte, half-word, word).
6. Set the source and destination addresses.
7. Select whether the source and destination addresses should be incremented during the transfer. For memory-to-memory transfers, both addresses should be incremented during the transfer.
8. Select whether the burst transfers of 4, 8 or 16 beats should be used during the transfer.

Programming DMA is relatively easy. Recall from Sect. 3.5.2 that each stream can be controlled using four registers: memory address register, peripheral address register, number of data register, and configuration register. Once set, DMA takes care of memory address increment without disturbing the CPU. Now that it is clear how the DMA works from a theoretical point of view, we can use the HAL library to configure and use a DMA controller. The HAL library abstracts most of the underlying hardware details. The DMA controller is abstracted in HAL with a C structure DMA_HandleTypeDef. Let us describe more in-depth only the two most important fields of this structure:

- Instance: this is the pointer to the DMA Stream descriptor we will use. For example, DMA2_Stream1 indicates the first stream of DMA2. The stream descriptor is a C structure that contains all DMA stream registers. The reference to the Instance structure points to the actual peripheral address. For example, the DMA2_Stream1 is defined in HAL as a pointer to the stream descriptor structure, and it holds the register base address for DMA2 Stream1 registers.
- Init: is an instance of the C structure DMA_InitTypeDef, which is used to configure the DMA Stream and channel.

DMA_InitTypeDef is defined in the following way:

```
typedef struct
{
    uint32_t Channel;
    uint32_t Direction;
    uint32_t PeriphInc;
    uint32_t MemInc;
    uint32_t PeriphDataAlignment;
    uint32_t MemDataAlignment;
    uint32_t Mode;
    uint32_t Priority;
    uint32_t FIFOMode;
    uint32_t FIFOThreshold;
```

```
13    uint32_t MemBurst;
      uint32_t PeriphBurst;
15  } DMA_InitTypeDef;
```

Listing 4.9 DMA `DMA_InitTypeDef` C structure

Let us briefly describe the C `DMA_InitTypeDef` structure:

- `Channel`: Specifies the channel used for the specified stream. It can assume the values DMA_CHANNEL_0, DMA_CHANNEL_1 up to DMA_CHANNEL_7. The peripherals are bound to streams and channels during the MCU design, so we should consult the datasheet for our microcontroller to see the stream/channel bound to the peripheral we want to use with DMA.
- `Direction`: Specifies if the data will be transferred from memory-to-peripheral, memory-to-memory or peripheral-to-memory.
- `PeriphInc`: Specifies whether the Peripheral address register should be incremented or not during the DMA transfer. Recall that a DMA controller has one peripheral port used to specify the address of the peripheral register involved in the DMA transfer. Since a DMA transfer usually involves several bytes, the DMA can be configured to increment the peripheral register for every transmitted byte.
- `MemInc`: Specifies whether the memory address register should be incremented or not during the DMA transfer.
- `PeriphDataAlignment`: Specifies the Peripheral data width. Transfer data sizes of the peripheral and memory are fully programmable through this field and the next one. The DMA controller is designed to automatically perform data alignment when source and destination data sizes differ.
- `MemDataAlignment`: Specifies the Memory data width.
- `Mode`: the DMA controller has two working modes: normal and circular. In normal mode, the DMA sends the specified amount of data from the source to the destination port and stops the activities. It must be re-activated again to do another transfer. In circular mode, it automatically resets the transfer counter at the end of transmission and starts transmitting again from the first byte of the source buffer.
- `Priority`: Specifies the software priority for the DMA Stream. The priority allows the internal arbiter in the DMA controller to rule concurrent requests.
- `FIFOMode`: Specifies if the stream uses the FIFO buffer. Recall that each stream has an independent 4-word (4 * 32 bits) FIFO. The FIFO temporarily stores data coming from the source before transmitting it to the destination. The FIFO introduces one important advantage: it reduces S(D)RAM access time by supporting burst transactions. The FMC SDRAM controller used in our case issues four READ commands in a row, thus reading four consecutive words from an active SDRAM row. Using the FIFO allows for storing these four words efficiently before they are sent to SRAM.
- `FIFOThreshold`: Specifies the FIFO threshold level. The FIFO will be drained to the destination when this threshold is achieved.

- `MemBurst`: Specifies the amount of data to be transferred to/from memory in a single non-interruptible transaction.
- `PeriphBurst`: Specifies the amount of data to be transferred to/from peripheral (or memory for mem-to-mem DMA transfers) in a single non-interruptible transaction.

All HAL functions related to DMA manipulation are designed so that they accept as the first parameter an instance of the C structure `DMA_HandleTypeDef`. To initialise the DMA Stream, we first set all desired parameters in the `DMA_InitTypeDef` structure and then use the HAL function `HAL_DMA_Init (DMA_HandleTypeDef *hdma)`. The following code illustrates configuring and initialising the DMA2 Stream 1 for memory-to-memory transfers using FIFO and burst of length 4:

```
HAL_StatusTypeDef DMA2_SDRAM_Config(DMA_HandleTypeDef* DmaHandle)
{
  /* Enable DMA2 clock */
  __HAL_RCC_DMA2_CLK_ENABLE();

  /* Select the DMA Stream to be used #*/
  DmaHandle->Instance = DMA2_Stream1;

  /* Set the DMA Parameters */
  /* DMA_CHANNEL_0                          */
  DmaHandle->Init.Channel = DMA_CHANNEL_0;
  /* M2M transfer mode                      */
  DmaHandle->Init.Direction = DMA_MEMORY_TO_MEMORY;
  /* Peripheral increment mode Enable */
  DmaHandle->Init.PeriphInc = DMA_PINC_ENABLE;
  /* Memory increment mode Enable           */
  DmaHandle->Init.MemInc = DMA_MINC_ENABLE;
  /* Peripheral data alignment : Word */
  DmaHandle->Init.PeriphDataAlignment = DMA_PDATAALIGN_WORD;
  /* memory data alignment : Word           */
  DmaHandle->Init.MemDataAlignment = DMA_MDATAALIGN_WORD;
  /* Normal DMA mode                        */
  DmaHandle->Init.Mode = DMA_NORMAL;
  /* priority level : high                  */
  DmaHandle->Init.Priority = DMA_PRIORITY_HIGH;
  /* FIFO mode enabled                      */
  DmaHandle->Init.FIFOMode = DMA_FIFOMODE_ENABLE;
  /* FIFO threshold: full                   */
  DmaHandle->Init.FIFOThreshold = DMA_FIFO_THRESHOLD_FULL;
  /* Memory burst                           */
  DmaHandle->Init.MemBurst = DMA_MBURST_INC4;
  /* Peripheral burst                       */
  DmaHandle->Init.PeriphBurst = DMA_PBURST_INC4;

  /* Initialize the DMA stream */
  if (HAL_DMA_Init(DmaHandle) != HAL_OK)
  {
    /* Initialization Error */
    return HAL_ERROR;
  }

  /* Configure NVIC for DMA transfer complete/error interrupts */
```

```
43   HAL_NVIC_SetPriority(DMA2_Stream1_IRQn, 0, 0);

45   /* Enable the DMA STREAM global Interrupt */
     HAL_NVIC_EnableIRQ(DMA2_Stream1_IRQn);

47
     return HAL_OK;
49   }
```

Listing 4.10 DMA2 controller configuration and initialization

In the above code, the DMA2 Stream 1 is configured to automatically increment the source and destination addresses for each transmitted word. Remember that we are transferring the block of contiguous memory words in the matrix from the external SDRAM to the matrix in the internal SRAM. Hence, the addresses in the source and destination block should increase for each word transferred. Also, we enable the FIFO, set the FIFO threshold level to full and enable the burst transfers of size 4. In such a way, the DMA controller will read four words from SDRAM to the FIFO and then transfer them to SRAM.

After DMA initialisation, we should set the priority for the DMA2 Stream 1 interrupt and enable the interrupt request generated by DMA2 Stream 1. The DMA2 Controller will then assert an interrupt request whenever the DMA 2 Stream 1 completes the DMA transfer. Hence, we should also implement the minimal interrupt handler for the DMA2 Stream 1 as follows:

```
1   void DMA2_Stream1_IRQHandler(void)
    {
3     /* Check the interrupt and clear flag */
      HAL_DMA_IRQHandler(&DMA2_SDRAM_Handle);
5   }
```

Listing 4.11 DMA2 stream 1 interrupt handler

Now we should create a DMA handle variable and initialize the DMA2 Stream 1 by passing reference to the DMA handle into the DMA2_SDRAM_Config function:

```
1   DMA_HandleTypeDef       DMA2_SDRAM_Handle;

3   ...

5   // Configure DMA2 for SDRAM:
7   if (DMA2_SDRAM_Config(&DMA2_SDRAM_Handle) != HAL_OK) {
      Error_Handler();
9   }
```

Listing 4.12 DMA2 stream 1 handle and its configuration

To initiate the memory-to-memory DMA transfer, we use HAL_DMA_Start (DMA_HandleTypeDef *hdma, uint32_t SrcAddress, uint32_t DstAddress, uint32_t DataLength). The arguments are the pointer to the DMA handle, the source address, the destination address and the number of data to transfer. This function sets the EN bit in DMA SxCR, which in turn triggers the DMA controller to start the transfer. Hence, to transfer a matrix from the external SDRAM to SRAM, we implement and use the following function:

```
void SDRAM_DMA_mat_row_access_test(void){
  volatile uint32_t address;

  for (int k = 0; k < N; k++)
  {
    HAL_DMA_Start(&DMA2_SDRAM_Handle,
                  (uint32_t) PA3_SDRAM_DEVICE_ADDR_RW,
                  (uint32_t) matrixB,
                  MAT_ROWS * SDRAM_COLS);
    HAL_DMA_PollForTransfer(&DMA2_SDRAM_Handle,
                            HAL_DMA_FULL_TRANSFER,
                            HAL_MAX_DELAY);
  }
}
```

Listing 4.13 Matrix transfer using DMA

The function `HAL_DMA_PollForTransfer()` waits for DMA transfer to
complete. Otherwise, the CPU would continue to execute the program and would
not bother with DMA transfer (which would be the desired way), but in our case,
we are going to measure the time required to transfer the matrix from SDRAM to
SRAM using DMA; hence, we should wait for DMA to terminate. The function
`HAL_DMA_PollForTransfer()` is used here for the sake of simplicity, but it
is strongly recommended to use the DMA interrupt handler instead. Finally, we can
add the DMA matrix transfer to a set of the previous performance tests in Listing 4.8
as follows:

```
// Row-major order access:
HAL_GPIO_WritePin(GPIOC, GPIO_PIN_8, GPIO_PIN_SET);
timer_val_start = __HAL_TIM_GET_COUNTER(&TIM3Handle);
SDRAM_mat_row_access_test();
HAL_GPIO_WritePin(GPIOC, GPIO_PIN_8, GPIO_PIN_RESET);
timer_val_end = __HAL_TIM_GET_COUNTER(&TIM3Handle);
if (timer_val_end > timer_val_start)
  elapsed_rows = timer_val_end - timer_val_start;
else
  elapsed_rows = timer_val_end + (65536-timer_val_start);

// Column-major order access:
HAL_GPIO_WritePin(GPIOC, GPIO_PIN_8, GPIO_PIN_SET);
timer_val_start = __HAL_TIM_GET_COUNTER(&TIM3Handle);
SDRAM_mat_col_access_test();
timer_val_end = __HAL_TIM_GET_COUNTER(&TIM3Handle);
HAL_GPIO_WritePin(GPIOC, GPIO_PIN_8, GPIO_PIN_RESET);
if (timer_val_end > timer_val_start)
  elapsed_cols = timer_val_end - timer_val_start;
else
  elapsed_cols = timer_val_end + (65536-timer_val_start);

// DMA transfer:
HAL_GPIO_WritePin(GPIOC, GPIO_PIN_8, GPIO_PIN_SET);
timer_val_start = __HAL_TIM_GET_COUNTER(&TIM3Handle);
SDRAM_DMA_mat_row_access_test();
timer_val_end = __HAL_TIM_GET_COUNTER(&TIM3Handle);
HAL_GPIO_WritePin(GPIOC, GPIO_PIN_8, GPIO_PIN_RESET);
if (timer_val_end > timer_val_start)
  elapsed_cols = timer_val_end - timer_val_start;
else
  elapsed_cols = timer_val_end + (65536-timer_val_start);
```

Listing 4.14 Code used to test the speed of row-major, column-major and DMA matrix read from
the SDRAM

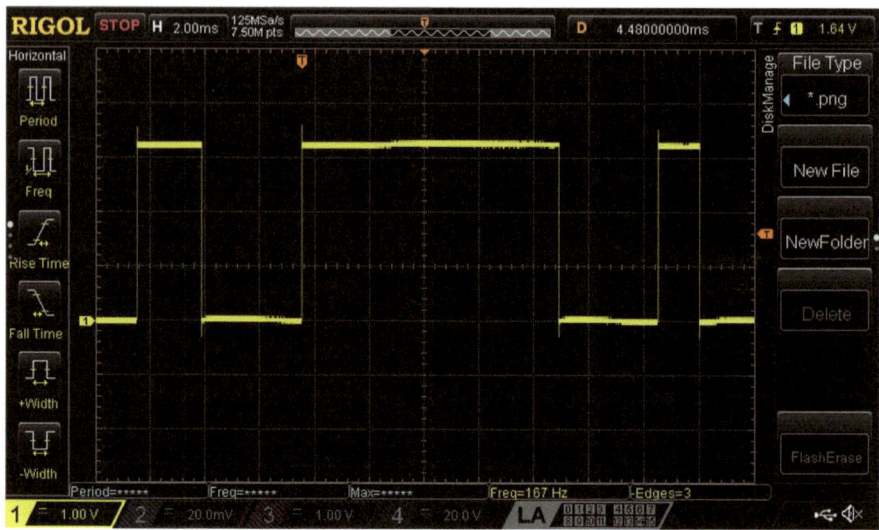

Fig. 4.43 Oscilloscope trace on the PC8 pin. The row-major order matrix read lasts for about 2.5 ms, the column-major order matrix read lasts forabout 10 ms while DMA transfer lasts for about 1.5 ms

When executing the DMA performance test, we observe from Fig. 4.43 that the time required to transfer the matrix from the external SDRAM to the internal SRAM is only about 1500 us. Why is the DMA controller faster than the CPU, considering that the same amount of data is being transferred from/to the same devices in both cases?

```
J_LOOP:
  ; address = PA3_SDRAM_DEVICE_ADDR_RW  + ((i*SDRAM_COLS + j)↩
    <<2);
  add.w    r1, r3, #0          ; r1 <- r3
  ldr      r2, [pc, #60]       ; r2 <- 0xC0008000 (SDRAM ↩
    address)
  add.w    r2, r2, r1, lsl #2  ; r2 <- r2+(r1*4) LOAD FROM ↩
    SDRAM
  ; matrixB[i][j] = *(uint32_t*)address;
  ldr      r0, [r2, #0]        ; r0 <- M_SDRAM[r2]
  ldr      r2, [pc, #52]       ; r2 <- matB base address
  str.w    r0, [r2, r1, lsl #2] ; matB[i][j] <- r0  STORE TO ↩
    SRAM
  ; for(int j=0; j<SDRAM_COLS; j++) {
  adds     r3, #1             ; inc r3 (r3 holds j)
  cmp      r3, #255           ; if j <= 255
  ble.n    J_LOOP             ;      loop back
```

Listing 4.15 Assembly code corresponding to the instructions created by the compiler for the innermost loop in Listing 4.6. There are 11 instructions executed in each iteration of the innermost loop; hence 11 instructions are executed for transferring one word from SDRAM to SRAM. The first four instructions are used to calculate the address in SDRAM. Then, four instructions are used to read the word from SDRAM and write it to SRAM, and finally, the last three instructions increments the innermost loop counter, compare it to 255 and loop if not equal

Well, the answer lies in the fact that the DMA controller does not execute instructions. For each word transferred, the CPU fetches the `LDR` instruction (load register with word), executes it (it loads the data from SDRAM to an internal register), fetches the `STR` instruction (store register as word), and finally executes it (it stores the data from the internal register to SRAM). Besides LDR and STR instructions, in each loop iteration, the CPU executes a bunch of other instructions required to calculate the address in SDRAM, increment and compare the loop index, etc. (see Listing 4.15). The DMA controller only transfers data from SDRAM (in bursts!) and forwards them to SRAM (in bursts!) without fetching and executing the load/store instructions! Besides offloading the CPU, this is another benefit of utilizing DMA controllers.

4.10 Double Data Rate SDRAM

How can we further speed up memory transfers? The solution is to access two adjacent columns simultaneously with one READ/WRITE command. So, instead of reading/writing one 8-bit memory word (column), we can read/write two adjacent 8-bit memory words (columns). But with that solution, a new challenge arises. How to transfer two 8-bit words in the same amount of time as one 8-bit word? One solution would be to have a bus twice as wide. Thus, instead of the 8-bit data bus (DQ[7:0]), the SDRAM device would have had a 16-bit data bus (DQ[15:0]). However, this could be challenging because more wires mean more noise on the data bus and worse data/signal integrity. The second solution would be to have a twice as fast bus. But this is also challenging because higher frequency means worse data/signal integrity and higher power consumption. The better solution is to **transfer data at both clock edges to double data bus bandwidth without a corresponding increase in clock frequency or in data bus width**.

In SDRAM devices, each time a column read command is issued, the control logic determines the duration of the data burst, and each column is moved separately from the sense amplifiers through the I/O logic to the external data bus. However, the separate control of each column limits the operating data rate of the SDRAM device. **In Double Data Rate (DDR) SDRAM devices, two adjacent columns are moved in parallel from the sense amplifiers to the output data register**, and the data is then pipelined through a multiplexor to the external data bus. The feature to access two columns at a time is referred to as **2N-prefetch**. Figure 4.44 illustrates the simplified block diagram of a DDR SDRAM device with four independent banks. We can see that the internal structure is similar to the internal structure of an SDRAM device except for the IO block. The memory arrays and banks used in DDR SDRAMs are the same as in SDRAMs. The name "double data rate" refers to the fact that a DDR SDRAM with a certain clock frequency achieves nearly twice the bandwidth of an SDRAM running at the same clock frequency due to this double pumping. Double data rate SDRAM is a significant improvement of SDRAM. DDR SDRAMs have been used in computer systems' memory since 2001.

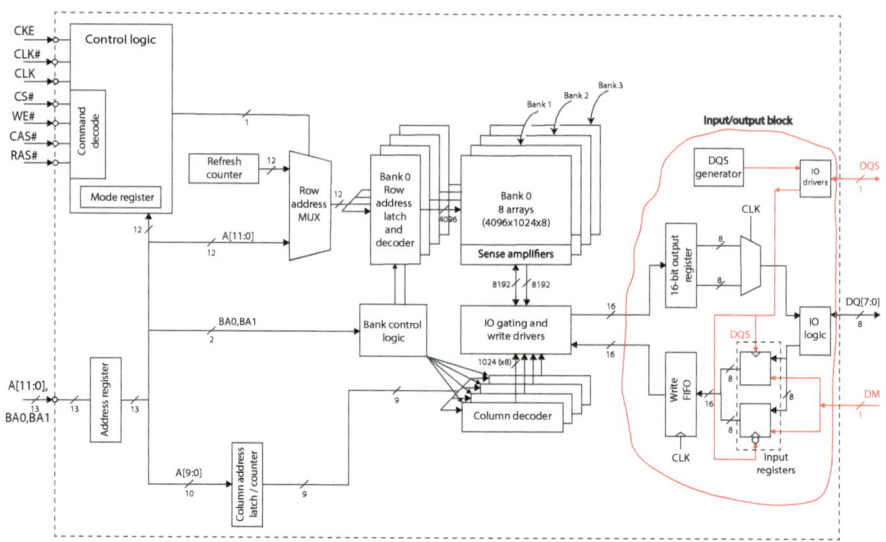

Fig. 4.44 Simplified block diagram of a DDR SDRAM device with four banks

The main difference in the internal organization of DDR SDRAM over SDRAMs is an improved I/O block. The I/O block of an 8-bit DDR SDRAM device from Fig. 4.44 now consists of a 16-bit output register, a 2/1 multiplexor, a DQS generator, two 8-bit input registers, a write FIFO and IO logic. Figure 4.44 shows that, in the case of the READ access, given the width of the external data bus (DQ) as 8-bit, 16 bits are moved from the sense amplifiers to the output register, and the 16 bits are then pipelined through the multiplexor to the external data pins. The clock signal controls the select input of the multiplexor. In the case of the WRITE access, two 8-bit data are stored successively (one after the other) in two 8-bit input registers and then transferred together into a 16-bit write FIFO. From there, data is transferred to the sense amplifiers through IO gating and write drivers. Besides, DDR SDRAMs have two new control signals: data strobe (DQS) and data mask (DM). In the following subsections, we are going to describe the operation of the IO block during the READ and WRITE accesses and the role of DQS and DM in more detail.

The downside of the 2N-prefetch architecture is that short column bursts are no longer possible. In DDR SDRAM devices, a minimum burst length of 2 columns of data is accessed per column read command.

4.10.1 Functional Description

The DDR SDRAM uses a double data rate architecture to achieve high-speed operation. The double data rate architecture is essentially a 2N-prefetch architecture with an I/O block designed to transfer two data words per clock cycle at the I/O pins. A single read or write access for the DDR SDRAM effectively consists of a single 2N-bit-wide, one-clock cycle data transfer at the internal DRAM core and two corresponding N-bit-wide, one-half clock cycle data transfers at the I/O pins.

The DDR SDRAM operates from a **differential clock**. Differential clock employs two complementary clock signals, CLK and CLK#. In general, a clock signal can be regarded as a binary signal whose duty cycle is nominally 50%. As we know, the clock signal is used to synchronize and capture data at its rising or falling edges. In DDR SDRAMs, data are synchronized and captured at both clock edges. But clocks are notoriously bad at having 50% duty cycles at high frequencies. As a rule of thumb, high frequency is generally considered to be above 100MHz. So, the reason for having two separate clocks is to allow for more precise alignment of the rising edges of the clock with the data. The crossing of CLK going HIGH and CLK# going LOW is referred to as the positive edge of CLK. Commands (address and control signals) are registered at every positive edge of CLK.

Read and write accesses to the DDR SDRAM are burst-oriented. Accesses start at a selected location and continue for the BL number of locations in a sequence. Similarly to SDRAMs, accesses begin with the registration of an ACTIVE command, which may then be followed by a READ or WRITE command. The address bits registered coincident with the ACTIVE command are used to select the bank and row to be accessed. The address bits registered coincident with the READ or WRITE command are used to select the bank and the starting column location for the burst access. The DDR SDRAM provides for programmable READ or WRITE burst lengths of 2, 4, or 8 locations.

4.10.1.1   Read

Figure 4.45 illustrates the Operation of the I/O block during the READ access to an 8-bit DDR SDRAM. First, 16 bits (two adjacent 8-bit columns) are transferred from the sense amplifiers to the 16-bit output register as the consequence of the READ command. Then, when CLK is HIGH, the first 8-bit word is transferred through the multiplexor onto the I/O pins; when the CLK signal is LOW, the second 8-bit word is transferred through the multiplexor onto the IO pins. In such a way, two 8-bit words from the DRAM array are transferred in one clock cycle. A bidirectional data strobe (DQS) signal is transmitted, along with data, for use in data capture at the memory controller. The DQS generator generates the DQS signal and synchronizes it with the memory controller's global clock. Hence, the **DQS signal is edge-aligned with data for READs**.

4.10.1.2   Write

Figure 4.46 illustrates the operation of the I/O block during the WRITE access to an 8-bit DDR SDRAM. Two 8-bit words are successively transferred from the data bus into the input registers. Two input registers form a DDR input pair. A bidirectional data strobe **(DQS) signal is now transmitted by the memory controller**, along with data, for use in data capture at DDR SDRAM. The first 8-bit word is captured into the first data input register at the positive edge of DQS, while the second 8-bit word is captured into the second input register at the negative edge of DQS. Hence, **input data is registered on both edges of DQS**, and **DQS signal is center-aligned with**

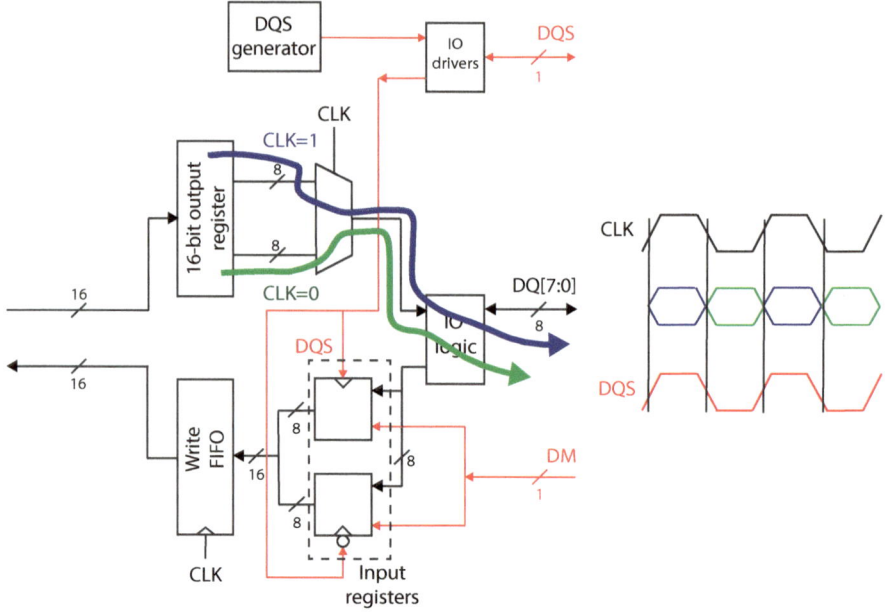

Fig. 4.45 Operation of the IO block during READ

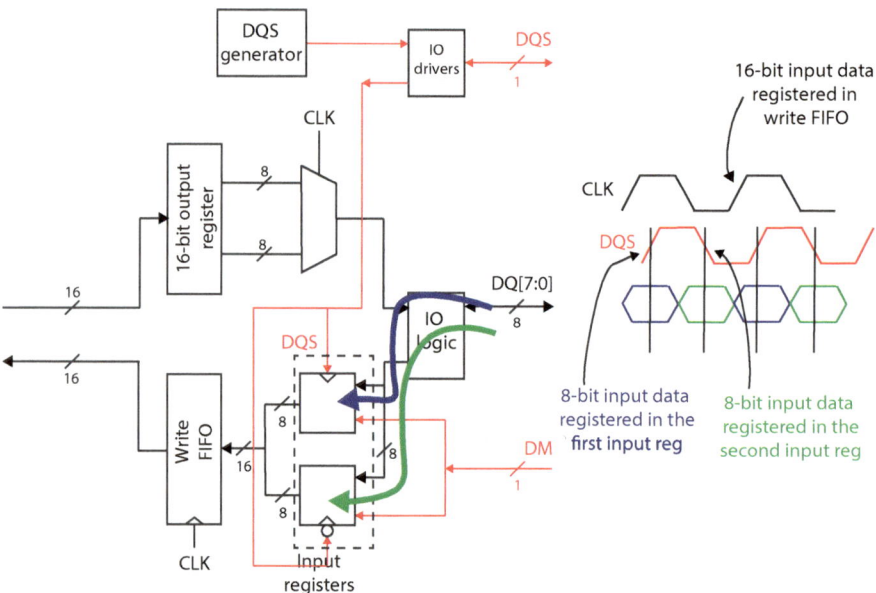

Fig. 4.46 Operation of the IO block during WRITE

data for WRITEs. Then, the 16-bit data is transferred into the write FIFO at the positive edge of the CLK signal and written to the sense amplifiers and the DRAM array during the PRECHARGE command.

4.10.2 DDR SDRAM Timing Diagrams

Within the realms of computer memory, understanding DDR SDRAM timings and timing diagrams is paramount for optimizing system performance, ensuring data integrity, and unlocking the full potential of memory modules. DDR SDRAM timing diagrams represent the intricate timing relationships governing the operation of DDR synchronous dynamic random-access memory. These diagrams visually depict the various timing parameters, such as CAS latency, tRCD, tRP, commands, data, and addresses, along with their respective timing intervals and relationships. By presenting these timings in a graphical format, DDR SDRAM timing diagrams provide a clear and concise means of understanding the precise sequence of events that occur during memory read, write, and refresh operations. They offer invaluable insights into the synchronization and coordination required for efficient data transfer within the memory module and between the memory module and the rest of the system, facilitating optimal performance tuning and troubleshooting in computer memory subsystems.

4.10.2.1 Read Bursts

Figure 4.47 shows the timing for a read burst with CL = 2 and BL = 4. During READ bursts, the valid data-out element from the starting column address is available following the CL after the READ command. Each subsequent data-out element is valid at the next positive or negative clock edge (i.e., at the next crossing of CLK and CLK#). The DDR SDRAM drives DQS along with output data. The initial LOW state on DQS is known as the *read preamble*; the LOW state coincident with the last data-out element is known as the *read postamble*. Upon completion of a read burst, assuming no other commands have been initiated, the DQ will go High-Z.

Data from any READ burst may be concatenated with data from a subsequent READ command. In such a way, a continuous flow of data can be maintained. The first data element from the new burst will follow the last element of a completed

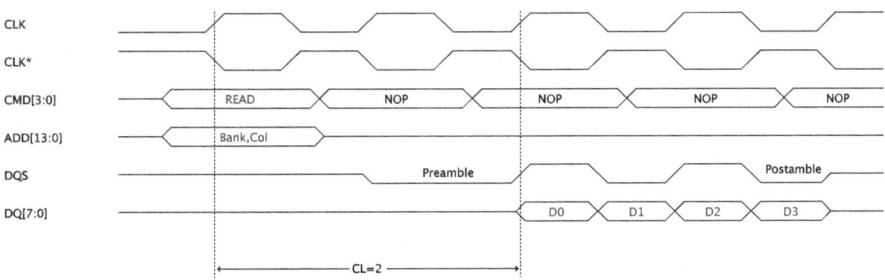

Fig. 4.47 The DDR READ burst with CL = 2 and BL = 4

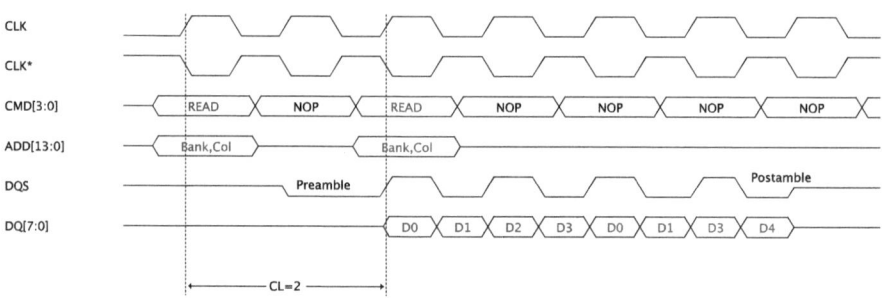

Fig. 4.48 Two consecutive DDR READ bursts with CL = 2 and BL = 4

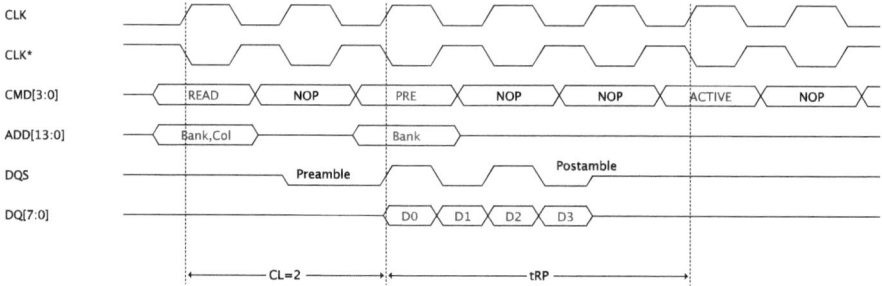

Fig. 4.49 DDR READ to PRECHARGE

burst if the new READ command is issued x cycles after the first READ command, where x equals the number of desired data element pairs (recall that the 2N-prefetch architecture requires pairs). This is shown in Fig. 4.47.

A PRECHARGE command may follow a READ burst to the same bank. The PRECHARGE command should be issued x cycles after the READ command, where x equals the number of desired data element pairs ($x = BL/2$). This is shown in Fig. 4.49. Following the PRECHARGE command, a subsequent command to the same bank cannot be issued until both t_{RAS} and t_{RP} have been met (Fig. 4.48).

4.10.2.2   Write Bursts

Figure 4.50 shows the timing for a WRITE burst with BL = 4. Input data appearing on the DQ is written to the memory array subject to the data mask (DM) input coincident with the data. The DQS and DM signals are now transmitted by the memory controller, along with data. If the DM signal is registered LOW, the corresponding input data is written to memory. If the DM signal is registered HIGH, the corresponding input data is ignored, and a WRITE is not executed to that column location. During WRITE bursts, the first valid input data element is registered on the first rising edge of DQS following the WRITE command. Subsequent data elements are registered on the successive edges of DQS. The LOW state on DQS between the WRITE command and the first rising edge is known as the *write preamble*, and the LOW state on DQS following the last input data element is known as the *write postamble*. The first

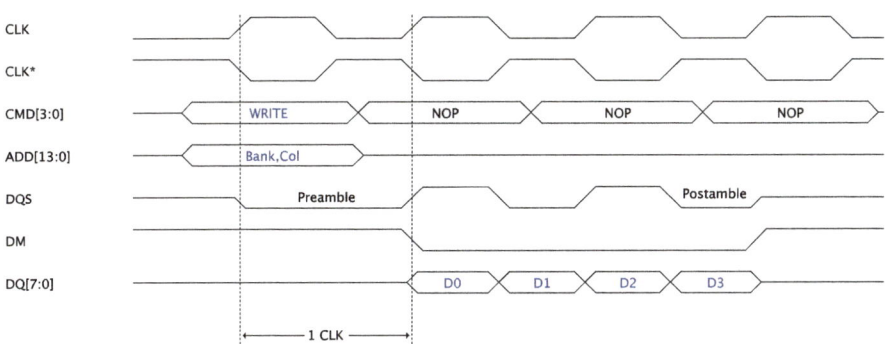

Fig. 4.50 The DDR WRITE burst with BL = 4

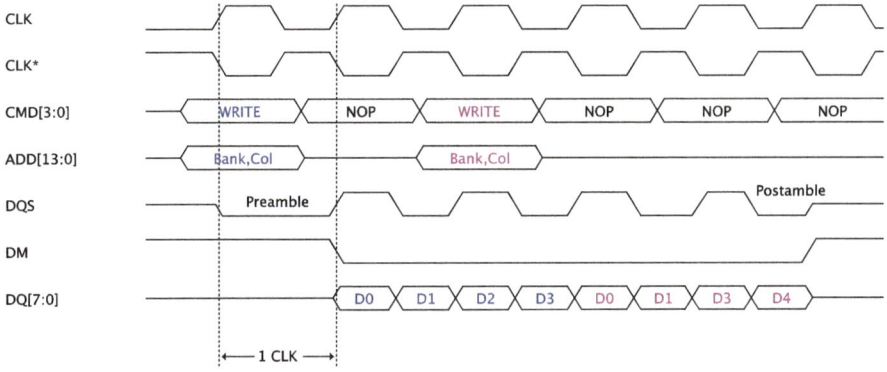

Fig. 4.51 Two DDR WRITE bursts with BL = 4

input data element following the WRITE command, along with its DQS, should be valid on the data bus one clock period after the WRITE command. Actually, most modern DDR SDRAMs specify this time between the WRITE command and the first corresponding rising edge of DQS from 75% to 125% of one clock cycle. In all of the WRITE diagrams, this time is one clock cycle.

Data for any WRITE burst may be concatenated with a subsequent WRITE command. The new WRITE command should be issued x cycles after the first WRITE command, where x equals the number of desired data element pairs. Figure 4.51 illustrates two concatenated bursts with BL = 4.

A PRECHARGE command to the same bank may follow a WRITE burst, as shown in Fig. 4.52. There is a time period, **write recovery time** (t_{WR}), associated with the WRITE-to-PRECHARGE command sequence. Only the data-in pairs registered prior to the t_{WR} period are written to the internal array. After the PRECHARGE command, a subsequent command to the same bank cannot be issued until t_{RP} is met.

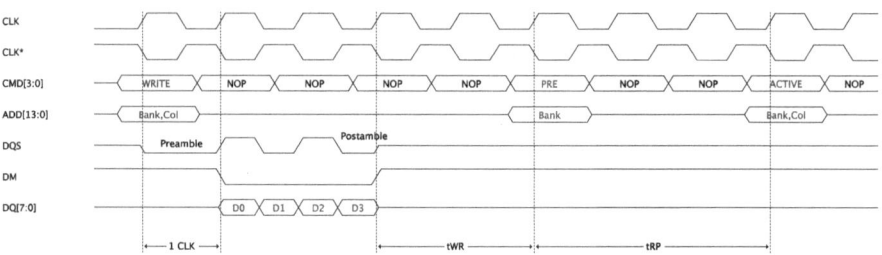

Fig. 4.52 DDR WRITE to PRECHARGE

4.10.3   Address Mapping

Now that we are familiar with the basic operations in SDRAMs, we can move forward
and see how an address from the CPU should be mapped into SDRAM's bank, row,
and column address. The memory controller performs the address mapping. Suppose
we are addressing a DDR SDRAM chip that consists of 8 banks, and each bank has
eight DRAM arrays of size 4096 rows by 1024 columns. To address such a DDR
SDRAM chip, we need 12 bits for the row address, three bits for the bank address,
and 10 bits for the column address.

Figure 4.53 shows the naive way of address mapping, where the top address bits
are used to address the bank, the middle 14 bits are used to address the row, and
the last 10 bits select the column. The main problem of such naive address mapping
is that consecutive rows are in the same bank; hence, there is no bank interleaving.
In the case of consecutive memory transfers consisting of more than one row, the
currently open row should first be precharged before the new row is open.

The better way of address mapping would be to take advantage of bank interleaving, such that consecutive rows are in different banks. In this way, we can open a
new row before the currently accessed row is precharged. We say that the precharge
time is masked. Figure 4.54 shows the address mapping, where **bank interleaving**
is used. Now, the top address bits select the row, while the middle address bits select
the bank. Each time the end of a row is reached, the same row in a different bank is
accessed (Fig. 4.55).

3	12	10
Bank	Row	Column

Fig. 4.53 Naive address mapping

12	3	10
Row	Bank	Column

Fig. 4.54 Bank interleaving

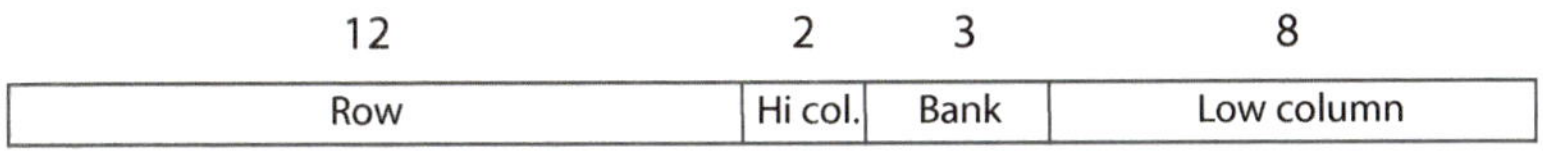

12	2	3	8
Row	Hi col.	Bank	Low column

Fig. 4.55 Cache block interleaving

Table 4.2 Summary of important timings in SDRAMs

Name	Symbol	Description
CAS latency	CL	The number of cycles between sending a column address to the memory and the beginning of the data in response to a READ command. This is the number of cycles it takes to read the first bit of memory from a DRAM with the correct row already open. CL is an exact number that must be agreed on between the memory controller and the memory
Row Address to Column Address Delay	t_{RCD}	The minimum number of clock cycles required between opening a row and issuing a READ/WRITE command. The time to read the first bit of memory from an SDRAM without an active row is $t_{RCD} + \mathrm{CL}$
Row Precharge Time	t_{RP}	The minimum number of clock cycles required between issuing the precharge command and opening the next row. The time to read the first bit of memory from an SDRAM with the wrong row open is $t_{RP} + t_{RCD} + \mathrm{CL}$
Row Active Time	t_{RAS}	The minimum number of clock cycles required between a row active command and issuing the precharge command. This is the time needed to refresh the row internally, and overlaps with t_{RCD}. In SDRAM modules, it is usually $t_{RCD} + \mathrm{CL}$

The third way of address mapping would be to take into account the cache memory. Typically, the cache block is 64 bytes in size. In reality, memory reads or writes are rarely random due to the locality of reference. If a cache is used to support the locality of references, the CPU will access consecutive cache blocks. Hence, the cache misses will occur on the consecutive 64 bytes in memory. For example, if a cache block is stored in the last 64 bytes of a row, the cache miss on the next cache block would require precharging the row and opening a new one. In the case where consecutive cache blocks are stored in different banks, a row precharge would not be required. Thus, it would be better to put consecutive cache blocks into different banks—this is called **cache block interleaving**. Figure shows the address mapping, where cache block interleaving is used. Now, the column bits are split into two parts. Low column bits select the word within the cache block. The remaining high column bits address the cache block in different banks.

4.10.4 Memory Timings: A Summary

So far, we have learned that each memory operation is associated with one or more memory timings that should be met in order to perform these operations correctly. Table 4.2 summarizes the most important memory timings.

CL, t_{RCD}, and t_{RP} are for most modern SDRAMs, typically around 13 ns, and have not changed significantly since the SDRAMs were first introduced. Actually, the DRAM cell and array process technologies have not significantly changed over the decades, and only the techniques to speed up memory transfers have been (e.g., synchronous interface, bank interleaving, etc.). The next subsection covers the techniques to speed up memory transfers in DDR SDRAMs.

4.10.5 DDR Versions

To boost the performance of DDR SDRAMs, DDR SDRAMs have been further improved. Due to its nature (data is stored as a charge) and the process technology used to implement DRAM cells, the DRAM core (DRAM arrays) has mostly stayed the same over the decades, and its speed of operation remains relatively low. In SDRAMs, the clock rate used to transfer data on the data bus equals the clock rate used to transfer data between internal latches, sense amplifiers, and input/output data registers. The following improvements aim to speed up memory transfers by employing larger prefetch or by increasing the frequency on the data bus (and not the frequency of the SDRAM core). These subsequent improved versions of DDR SDRAM are numbered sequentially: DDR2, DDR3, and DDR4.

DDR SDRAMs have 2N-prefetch, and the typical frequencies of the SDRAM core and the data bus are 133, 167, and 200 Mhz. In DDR2 SDRAM devices, the number of columns prefetched is 4. Hence, **DDR2 employs 4N-prefetch**. Besides, the DDR2 internal clock runs at half the DDR2 external bus clock rate. DDR2 offers data bus clock rates of 266 MHz, 333 MHz, and 400 MHz. DDR2 also lowers power by dropping the voltage from 2.5 v (DDR) to 1.8 v. **DDR3 increased the prefetch to 8N**. DDR3 bus clock rate is four times faster than DDR3 internal clock rate. DDR3 also drops the voltage to 1.5 v and has a maximum data-bus clock speed of 800 MHz. **DDR4 also employs 8N-prefetch** but drops the voltage to 1 to 1.2 v and has a maximum data-bus clock rate of 1600 MHz. DDR4 bus clock rate is four times faster than DDR4 internal clock rate.

DDR5, or Double Data Rate 5, is the latest generation of DDR SDRAM technology. It represents a significant advancement over its predecessor, DDR4, offering improvements in speed, capacity, and efficiency. DDR5 offers significantly higher data transfer rates compared to DDR4 memory. 16N-prefetch is a new feature introduced with DDR5 memory, offering higher memory bandwidth and improved performance compared to DDR4 memory with 8N prefetch. With 16N prefetch, DDR5 memory doubles the data throughput compared to 8N prefetch in DDR4. While DDR4 memory modules typically operate at a voltage of 1.2 v, DDR5 memory modules operate at a reduced voltage of around 1.1 V. The lower operating voltage of DDR5 memory helps reduce power consumption and heat generation, which is especially important in high-performance computing systems, servers, and data centers where power efficiency is critical. Lower voltage operation can also improve system stability and reliability by reducing the risk of electrical overstress and component degradation. DDR5 also has higher frequencies than DDR4, with a bus clock rate of 3200–4200 MHz.

4.11   DIMM Modules

The capacities of one DDR3 SDRAM chip are 1, 2, 4, and 8 Gbits, while the capacities of one DDR4 SDRAM chip are 4, 8, 16, and 32 Gbits. We can connect two or more chips together to increase memory capacity and bandwidth, as illustrated in Fig. 4.56. Each chip in Fig. 4.56 is the DDR SDRAM chip from Fig. 4.44, containing four banks, each of size 4096x1024x8 bits. Hence, one DDR SDRAM chip is 16Mx8 bits in size. Both chips in Fig. 4.56 share the memory, the control (DQS and DM), and the command bus (CS#, WE#, RAS#, and CAS#); hence, both chips are accessed simultaneously. A set of DRAM chips connected to the same chip select (CS#) signal, which are therefore accessed simultaneously, is referred to as **a rank**. The chips in Fig. 4.56 form a DDR SDRAM of size 16Mx16 bits. Hence, by connecting two DRAM chips as in Fig. 4.56, we have increased the capacity and the data bus bandwidth, as now 16 data bits are transferred simultaneously.

We can further increase DRAM's size and bandwidth by connecting more than two chips in one rank. Figure 4.57 illustrates a rank composed of four DDR SDRAM chips of size 16Mx8 bits. Again, all four DDR SDRAM chips share the same CS# signal and are accessed simultaneously. The rank is of size 16Mx32 bits, as now 32 data bits are transferred simultaneously.

We can even form two independent ranks. In such a way, we can interleave the accesses to both ranks (similarly to bank interleaving) and mask latencies. While accessing one rank, we can activate a row in another rank or refresh another rank. Figure 4.58 illustrates two independent ranks, Rank0 and Rank1. For each rank, there is a separate CS# signal: CS0# for Rank0, and CS1# for Rank1. Now, both ranks share the same data bus, as only one rank can be read or written simultaneously.

In modern computer systems, DRAM chips are combined on a printed circuit board designed for use in personal computers, workstations, and servers. The memory chips are placed on both sides of the printed circuit board. Typically, there are eight

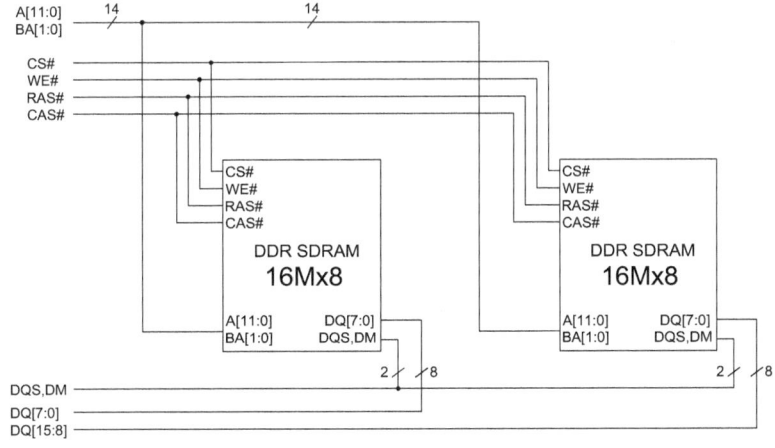

Fig. 4.56 A rank composed of two DRAM 16Mx8 chips

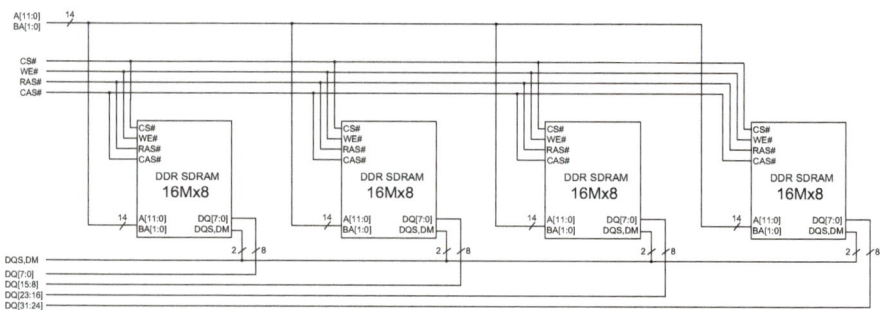

Fig. 4.57 A rank composed of four DRAM 16Mx8 chips

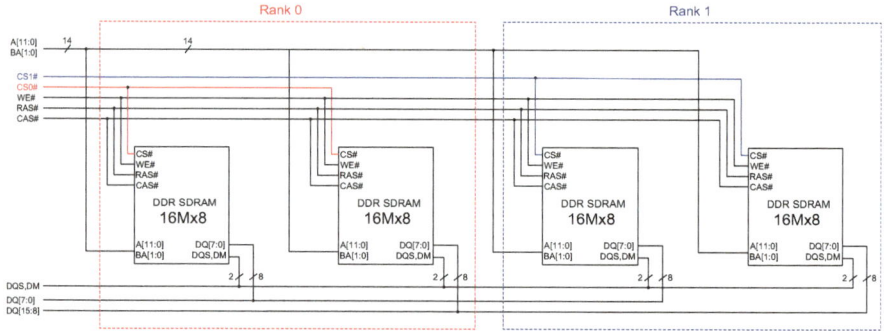

Fig. 4.58 Two ranks each containing two DRAM 16Mx8 chips

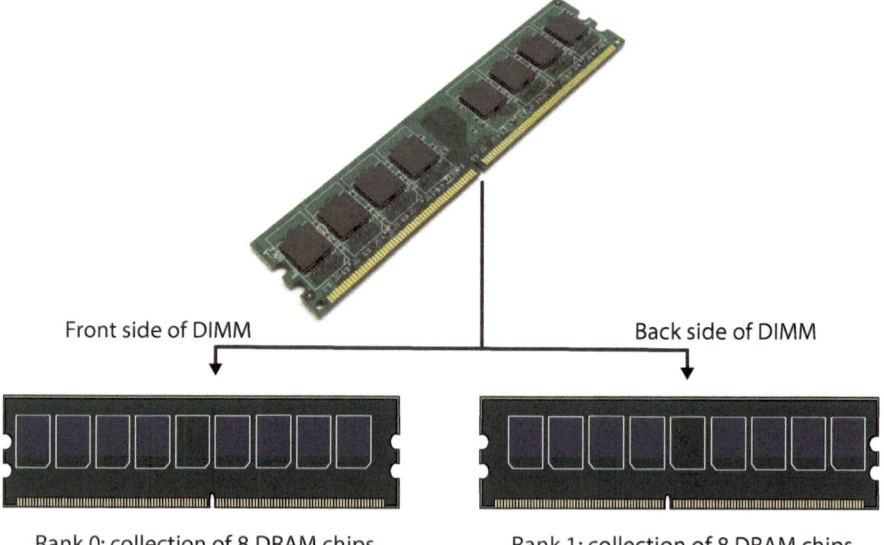

Front side of DIMM   Back side of DIMM

Rank 0: collection of 8 DRAM chips   Rank 1: collection of 8 DRAM chips

Fig. 4.59 A DIMM module

Table 4.3 Comparison of DDR SDRAM generations and DIMMs

Generation		Chip		Data bus		Timings	DIMM			
DRAM	DRAM name	Clock (Mhz)	Prefetch	Clock (MHz)	MT/s	CL-t_{RCD}-t_{RP}	t_{CL} (ns)	MB/s	Voltage	DIMM name
DDR	DDR-266	133	2N	133	266	2.5-3-3	18.8	2128	2.5	PC-2100
DDR	DDR-300	150		150	300			2400		PC-2400
DDR	DDR-400	200		200	400	3-3-3	15	3200		PC-3200
DDR2	DDR2-533	133	4N	266	533	4-4-4	15	4264	1.8	PC2-4300
DDR2	DDR2-667	166		333	667	5-5-5		5336		PC2-5300
DDR2	DDR2-800	200		400	800	6-6-6		6400		PC2-6400
DDR3	DDR3-1066	133	8N	533	1066	7-7-7	13.12	8528	1.5	PC3-8500
DDR3	DDR3-1333	166		666	1333	9-9-9	13.5	10664		PC3-10700
DDR3	DDR3-1600	200		800	1600	11-11-11	13.75	12800		PC3-12800
DDR4	DDR4-2400	300	8N	1200	2400	18-18-18	13.5	19200	1.2	PC4-19200
DDR4	DDR4-2666	333		1333	2666	20-20-20	13.6	21333		PC4-21333
DDR4	DDR4-3200	400		1600	3200	22-22-22	13.75	25600		PC4-25600
DDR5	DDR5-6400	400	16N	3200	6400	52-52-52	16	51200	1.1	PC5-51200

(8) memory chips placed on one side of the printed circuit boards. A printed circuit board containing memory chips on both sides is referred to as **dual in-line memory module (DIMM).** For instance, the 64-bit data bus for DIMM requires eight 8-bit chips addressed in parallel. The DRAM chips on one side of the DIMM module form one rank: they share the same chip select (CS#) signal and are, therefore, accessed simultaneously. Figure 4.59 illustrates a DIMM module and its two ranks, Rank 0 and Rank 1. In practice, all DRAM chips on DIMM share all of the other command and control signals, and only the chip select pins for each rank are separate. Each side of a DIMM, containing eight 8-bit DRAM chips, is one rank, and each rank has a 64-bit-wide data bus.

Manufacturers use the rather confusing labeling of SDRAM chips and DIMM modules. When DDR SDRAMs are packaged as DIMMs, they are confusingly labeled by the peak DIMM bandwidth. For example, when DDR SDRAMs with a clock frequency of 133 MHz are packed as a DIMM, the DIMM name becomes PC2100. The name comes from 133MHz x 2(DDR) x 8 bytes (eight 8-bit DRAM chips in a rank) equals 2100 MB/sec. Also, confusing names are used to label the DRAM chips. DRAM chips are labeled with the number of bits per second rather than their clock rate, so a 133 MHz DDR SDRAM chip is called a DDR266. Table 4.3 shows the relationships among internal and data-bus clock rates, prefetch, transfers per second per chip, chip names, DIMM bandwidth, DIMM supply voltage, and DIMM names.

DDR, DDR2, DDR3, DDR4, and DDR5 memories are classified according to the maximum speed at which they can work, as well as their timings. The important memory timings of commercial memory chips are usually given as triple:

$$CL - t_{RCD} - t_{RP},$$

where CL, t_{RCD}, and t_{RP} are given in data-bus clock cycles. For example, a DDR3-1333 chip can be described as 9-9-9, meaning that CL equals nine bus clock cycles,

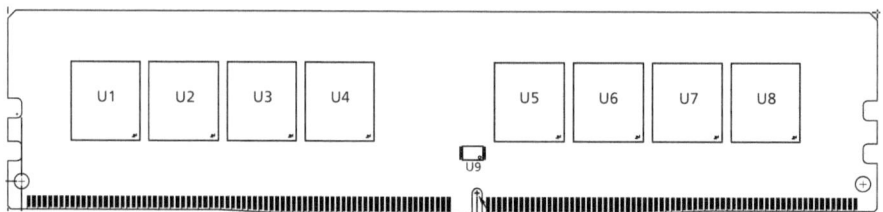

Fig. 4.60 288-Pin Micron (1G x 64 bit) DDR4 SDRAM DIMM—Front side

t_{RCD} equals nine bus clock cycles, and t_{RP} equals nine bus clock cycles. As the bus clock rate of a DDR3-1333 chip is 667MHz, all timings equal 13.5 ns.

DDR5 memory modules support higher capacities compared to DDR4, allowing for larger memory configurations in computer systems. DDR5 modules are expected to offer capacities of up to 128 GB per module initially, with the potential for even higher capacities as the technology matures. Compared to DDR4, DDR5 modules further reduce memory voltage to 1.1 V, thus reducing power consumption. DDR5 DIMMs also introduce a new voltage regulator module (VRM) on the memory module itself, allowing for more precise power delivery control and potentially reducing overall system power consumption. DDR5 memory includes features to enhance reliability and data integrity, such as improved error correction capabilities and support for on-die termination (ODT) to minimize signal reflections and improve signal integrity. The initial availability of DDR5 modules began in 2021, with mass adoption expected to ramp up in subsequent years as the technology matures and prices become more competitive.

Unlike the previous DDR version, each DDR5 DIMM has two independent channels. Earlier DIMM generations featured only a single channel and one Command/Address bus controlling the whole memory module with its 64 data lines. Both subchannels on a DDR5 DIMM each have their own Command/Address bus, controlling 32 data lines, resulting in a total number of 64 data lines. The reduced data bus width is compensated by a doubled minimum burst length of 16 (also because of 16N-prefetch), which preserves the minimum access size of 64 bytes, which matches the cache line size used by modern microprocessors.

4.11.1 Micron DDR4 DIMM Module

DDR4 DIMMs have a standardized structure and connector layout designed to facilitate easy installation and compatibility with a wide range of computer systems. Figures 4.60 and 4.61 show the front and back sides of a Micron DDR4 DIMM module. Each side contains one rank composed of eight DDR4 DRAM chips (U1-U4, U5-U8, U10-U17).

The small chip located alongside the memory chips on the front side of a DDR4 DIMM module is a serial presence detect (SPD) EEPROM (Electrically Erasable Programmable Read-Only Memory). The SPD EEPROM is essential in providing

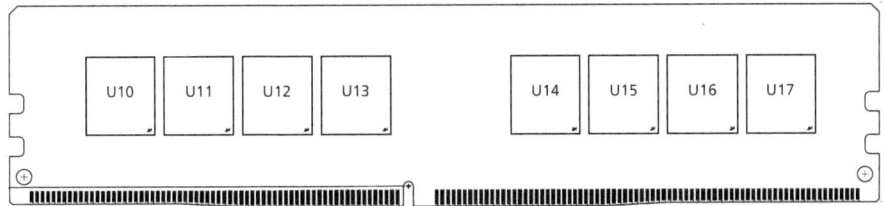

Fig. 4.61 288-Pin Micron (1G x 64 bit) DDR4 SDRAM DIMM—Back side

vital information about the DDR4 DIMM module to the system BIOS during initialization and configuration. The SPD EEPROM stores crucial information about the DDR4 DIMM module, including its manufacturer, model, capacity, speed rating, timing parameters, and other relevant details. During the system boot-up process, the BIOS reads the information stored in the SPD EEPROM to identify the DDR4 DIMM module and configure it appropriately for optimal performance. This information helps the system BIOS determine the supported operating frequencies, timings, and voltage settings for the DDR4 DIMM module, ensuring compatibility and stability. The presence of the SPD EEPROM allows DDR4 DIMM modules to support plug-and-play functionality, enabling the system to detect and configure memory modules automatically without requiring manual intervention. By reading the information stored in the SPD EEPROM, the system BIOS can determine the characteristics of the DDR4 DIMM module and adjust its settings accordingly, simplifying the installation and setup process for end users.

DDR4 DIMMs have a row of gold-plated connectors along the bottom edge of the module, which are used to connect the DIMM to the memory slot on the motherboard. These connectors are arranged in a specific pattern and configuration to ensure proper alignment and electrical connection with the memory slot on the motherboard. The number and arrangement of connectors may vary depending on the DIMM's form factor and capacity. Common configurations include 288-pin DIMMs for desktop and server applications.

DDR4 DIMMs feature a small notch located near the middle of the connector row on the bottom edge of the module. The notch is offset slightly from the center of the DIMM and is used to ensure correct orientation and alignment when inserting the DIMM into the memory slot. The position and size of the notch are standardized across DDR4 DIMMs to prevent compatibility issues and ensure proper installation.

Figure 4.62 shows the functional diagram of the DDR4 DIMM module. Each rank has its own chip select signal.

4.12 Memory Channels

We have learned that multiple banks and multiple ranks enable concurrent DRAM accesses. Multiple ranks can be further used to form a **channel**, but only one rank can be activated at a time. **Multiple independent channels** serve the same pur-

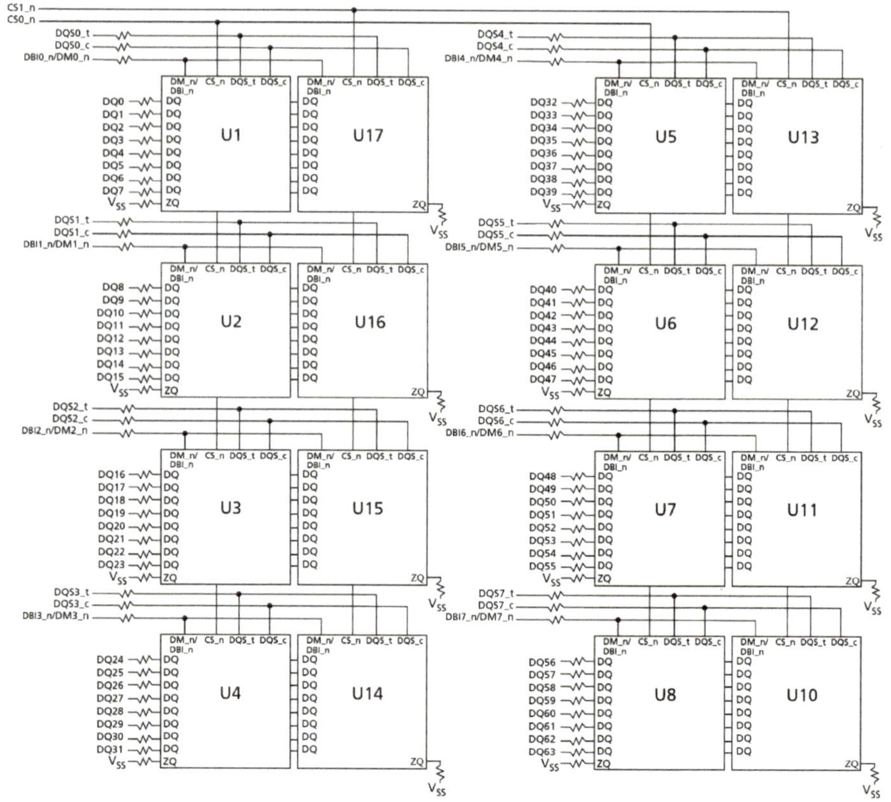

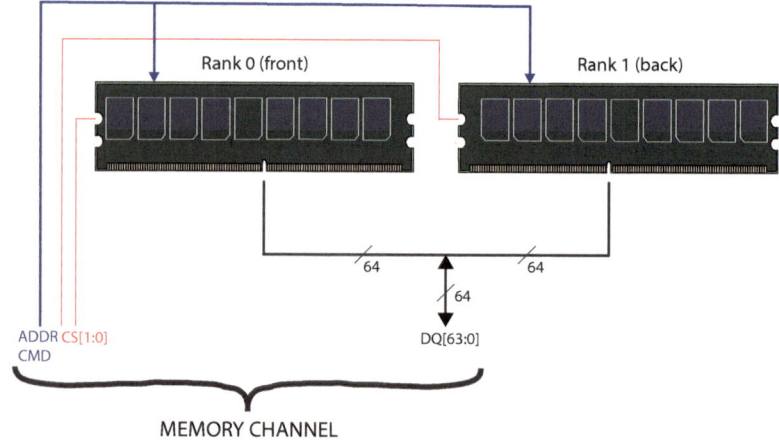

Fig. 4.62 288-Pin Micron (1G x 64 bit) DDR4 SDRAM DIMM functional block diagram

Fig. 4.63 A memory channel

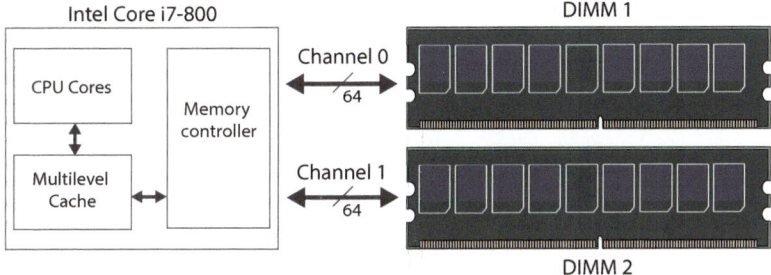

Fig. 4.64 A dual channel configuration supported by Intel Core i7-800. DIMM 1 and DIMM 2 should be identical in capacity, speed and CAS latency

Fig. 4.65 Color codes of channels on PC motherboard

pose as multiple banks or ranks, but they are even better because they **have separate data buses**. In such a way, bus bandwidth is increased. The advantage of running two or four channels is that they will provide the same capacity as a larger single-channel while doubling and quadrupling the amount of memory bandwidth. Of course, multiple channels bring a few disadvantages: more board wires and more pins (on the memory controller) are required. Multiple-channel architecture is a technology implemented on motherboards by the motherboard manufacturer and does not apply to memory modules. Also, a memory controller (which is a part of the chipset) must support multiple-channel architecture. Theoretically, dual-channel configurations double the memory bandwidth when compared to single-channel configurations.

Figure 4.63 illustrates one channel formed from two ranks on the same DIMM module. Indeed, in multi-channel architectures, one channel is formed from at least one DIMM module. In today's desktop computers, up to two DIMM modules can be used to form one channel.

Most of today's computer systems support dual-channel configuration. Dual-channel-enabled memory controllers in a PC system architecture use two 64-bit data channels. For example, the Intel Core i7-800 series supported dual-channel configuration, as illustrated in Fig. 4.64.

Figure 4.65 shows a part of a motherboard that supports two memory channels. The motherboard has four DIMM sockets. To distinguish the channel's sockets on the motherboard, the sockets are color-coded. The motherboards use two colors. The colored pair of sockets is a dual-channel set. A **matching pair of DIMMs** are two

Fig. 4.66 A matching pair of DIMMs form two channels

DIMMs that **are identical in capacity, speed, and CAS latency**. A matching pair should be used in both memory channels, i.e., a matching pair of DIMMs should be installed on the same color sockets. Another matching pair then goes into the remaining two sockets. Figure 4.66 shows two identical DIMM modules (a matching pair) inserted into the same-color sockets (red), forming two identical memory channels, A and B. Ideally, all DIMM modules should be identical in a system, or else we may end up with some memory being potentially downclocked to the lowest common denominator.

Intel Core i7-900 series DDR3 uses a triple-channel architecture, while modern high-end processors like the Intel Core i9 and AMD Ryzen Threadripper series support quad-channel memory. The quad-channel architecture can be used only when all four DIMM memory modules (or a multiple of four) are identical in capacity and speed and are placed in the same-color quad-channel sockets. When two DIMM memory modules are installed, the architecture will operate in a dual-channel mode; when three memory modules are installed, the architecture will operate in a triple-channel mode. On motherboards supporting quad-channel configuration, a similar color-coding scheme is used for dual-channel DIMM sockets. A same-color quadruple is a quad-channel set. A matching DIMM module quadruple (i.e., four DIMMs that are identical in capacity, speed, and CAS latency) should be installed on the same color sockets.

4.12.1   Case Study: Intel i7-860 Memory

At the beginning of the chapter, we have introduced the i7-860 and its memory hierarchy. This system is again illustrated in Fig. 4.67. Now, we are going to describe the system with its real memory components and the case of an L3 miss.

The i7-860 supports up to two 64-bit memory channels, each consisting of a separate set of DDR3 1066/1333 DIMMs, each of which can transfer in parallel. The i7-860 supports up to two DIMMs per channel and a total of up to 16 GB of memory.

In the case of L3 miss, both 64-bit memory channels are used simultaneously as one 128-bit channel (since there is only one memory controller, and the same address of the missing block in L3 is sent on both channels) to fill the missing block in L3.

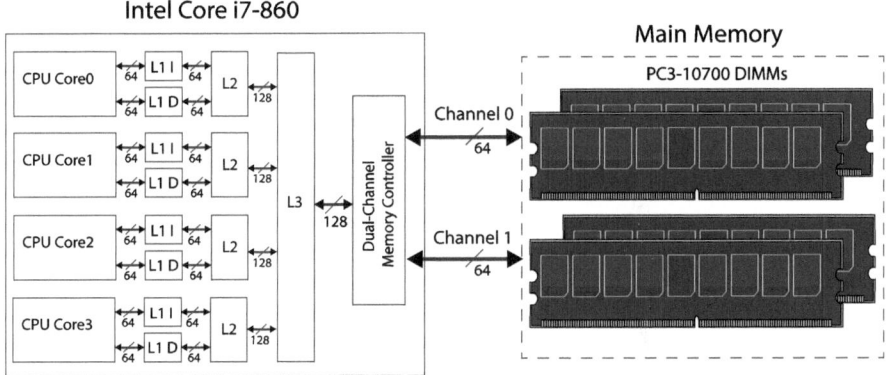

Fig. 4.67 Intel i7-860 memory

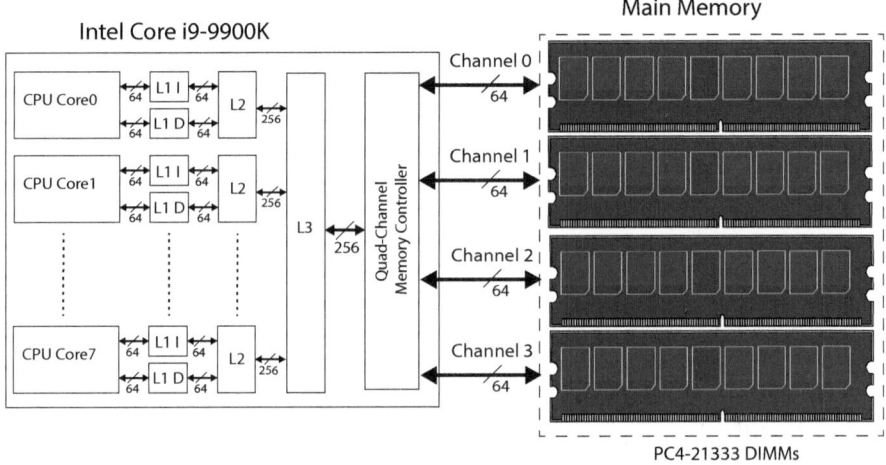

Fig. 4.68 Intel i9-9900K memory

Using DDR3-1333 (DIMM PC3-10700), the i7-860 has a peak memory bandwidth of just over 21 GB/sec. Thus, the memory controller fills the 64-byte cache block at a rate of 16 bytes (124 bits) per memory clock cycle.

If we assume the peak memory bandwidth, a 64-byte block is transferred at the rate of 21GB/s, which equals three ns. Of course, we cannot assume that the missing block is transferred at the peak memory bandwidth. At best, we can assume that the row in SDRAMs containing the missing block is open. Thus, we have to add the CAS latency (CL), which equals 13.5 ns for DDR3-1333 chips. Thus, the missing block in L3 can be filled in 16.5 ns. The i7-860 runs at 2.8 GHz, which means that one CPU cycle equals 0.36 ns. Thus, the missing block in L3 will be available no prior than in 47 CPU cycles. In the case that the row containing the missing block is not open and all rows in that bank are precharged, we should add at least t_{RCD} to the above block access time. As t_{RCD} also equals 13.5 ns, the block is transferred in 29 ns or 81 CPU cycles. Finally, if we have to precharge a row before opening the row containing the missing block, the block will be transferred in 42.5 ns or 119 CPU cycles.

4.12.2   Case Study: i9-9900K Memory

Figure 4.68 illustrates the Intel i9-9900K system. Intel i7-9900K is an out-of-order execution processor that includes eight cores. The L1 and L2 caches are separate for each core, while the L3 cache is shared among the cores on a chip. The L1 cache is the 32 KB, eight-way set-associative cache. The L2 cache is the 256 KB, four-way set-associative cache. Finally, the L3 cache is the 16 MB, 16-way set-associative cache. The i9-9900K supports up to four 64-bit memory channels, each consisting of a separate set of DDR4-2666 DIMMs (PC4-21333), and each of which can transfer in parallel, thus the peak memory bandwidth is 41.6 GB/s. The i9-9900K supports up to two DIMMs per channel and a total of up to 128 GB of memory.

Caches

5

CHAPTER GOALS

Have you ever wondered how your computer manages to access frequently used data with lightning speed, seemingly defying the limitations of its main memory? The answer lies in a clever and essential component of modern computer systems: caches. These hidden heroes are crucial in optimizing memory access and enhancing system performance by storing frequently accessed data and instructions in a high-speed memory buffer. In this chapter, we will unravel the mysteries of caches, exploring their architecture, operation, and impact on system performance.

From this chapter, you should gain a basic understanding of the design and operation of caches, including:

- Understand the fundamental concept of caches and their role in computer systems.
- Explore the motivation behind cache design and its impact on system performance.
- Learn about the principles of locality and how they influence cache behavior.
- Examine the organization and architecture of caches, including cache levels, associativity, and cache line size.
- Understand the trade-offs involved in cache design, such as capacity versus latency and hit rate versus miss rate.
- Learn about cache replacement policies and their impact on cache performance.

P. Bulić, *Understanding Computer Organization*, Undergraduate Topics in Computer Science, https://doi.org/10.1007/978-3-031-58075-8_5

- Explore the concept of cache hierarchies and the benefits of multilevel cache designs.
- Understand the organization and management of L1, L2, and L3 caches in modern processors.
- Investigate cache design considerations in different processor architectures.

5.1 Introduction

In computer systems, where performance is paramount and data access times can make or break the user experience, caches are indispensable tools for optimizing memory access and enhancing system responsiveness. As a crucial component of the memory hierarchy, caches play a pivotal role in bridging the gap between high-speed processors and slower main memory, facilitating efficient data access and manipulation in modern computing environments.

At its core, a cache is a specialized form of high-speed memory that stores frequently accessed data and instructions, enabling rapid retrieval and reducing the latency of memory accesses. By keeping a subset of frequently used data closer to the processor, caches mitigate the performance bottleneck associated with accessing data from slower, larger main memory, thereby improving overall system performance and responsiveness.

The concept of caching revolves around the principle of locality, which refers to the tendency of programs to access a relatively small portion of memory repeatedly or to access nearby memory locations in close succession. Caches exploit this principle by storing recently accessed data and instructions in fast, on-chip memory, allowing the processor to access them quickly without fetching them from slower main memory.

Throughout this exploration of caches in computer systems, we will delve into the fundamental principles, architectures, and optimizations that underpin their operation and effectiveness. From the basics of cache organization and management to advanced techniques for cache coherence and consistency, we will uncover the intricacies of caching and its profound impact on system performance and efficiency.

5.2 Memory Hierarchy

The fastest components of computer systems are central processing units. They execute instructions and process data, and we expect them to do it as fast as possible. However, we must remember that central processing units have instructions and operands in the main memory. One of the main goals we try to achieve when producing main memories is the data density per unit area in the silicon chip. We have learned that SDRAMs are fundamental components of computer memory, offering substantial storage capacity at a relatively low cost. However, SDRAMs come with a significant drawback: slowness. This inherent sluggishness can lead to performance bottlenecks, as the CPU often spends valuable clock cycles waiting for data retrieval from SDRAM. The slowness comes from a slightly different semiconductor technology we use to implement the SDRAM main memories than the CPU. The semiconductor technology used to implement main memories (i.e., DRAM cells) allows for a high density of memory cells. Unfortunately, it dramatically affects the memory's speed, and the main memories are ten or hundred times slower than central processing units. The main memory could also be made with faster technology (SRAM cells), but the price would be too high and unacceptable.

Therefore, we are faced with the following important challenge: how can we provide central processing units with fast access to instructions and operands while maintaining a large amount of cheap memory in the system? The solution to this challenge lies in the so-called **memory hierarchy**. The memory hierarchy is a tiered arrangement of different memory types, with each level offering a trade-off between speed and capacity. At the top of this hierarchy are the fastest but smallest memory components, like processor registers, while SDRAM occupies a lower tier. In the memory hierarchy, *we use several different memories in the system* (in terms of price and speed). Central processing units should only access small, fast memories that store instructions and operands and guarantee short access times. These small and fast memories are called **caches**. If the instructions or operands needed by the CPU are not in this small and fast memory, then the instructions and operands should be loaded from the larger and slower memories into these small and fast ones. But wait, if CPUs are allowed to access only these small and fast memories, how do we know what instructions and operands to store in them? Remember that modern CPUs are based on the Von Neumann architecture. A Von Neumann CPU has a PC that increases by 1 every time an instruction is fetched from memory and, hence, points to the next instruction to be fetched. This means that, except for jump instructions, we are very good at predicting which instructions the CPU will need. This is called **spatial locality**, which says that if we are accessing a memory word with address A, it is very likely that the next memory access will be to a memory word with address A+1. In addition, we often use loops in programs, so we often access the same set of memory words over a period of time. This is called **temporal locality**. Temporal and spatial locality are fundamental principles in Von Neumman's architecture that describe memory access patterns within a program's execution. These principles are crucial for understanding and optimizing the performance of computer systems, particularly memory hierarchies.

Temporal locality refers to the tendency of a program to access the same memory locations repeatedly over a short period of time. In other words, if a program accesses a specific memory location, it will likely reaccess the same location soon. The most common reason for temporal locality is program loops. For instance, in a loop, operands or instructions within the loop body are frequently accessed in subsequent iterations. Besides, data items read from or written to memory remain relevant for a significant portion of the program's execution. Reusing previously fetched data eliminates the need to reload it from slower memory levels, resulting in improved performance. Small and fast memories exploit temporal locality by storing recently

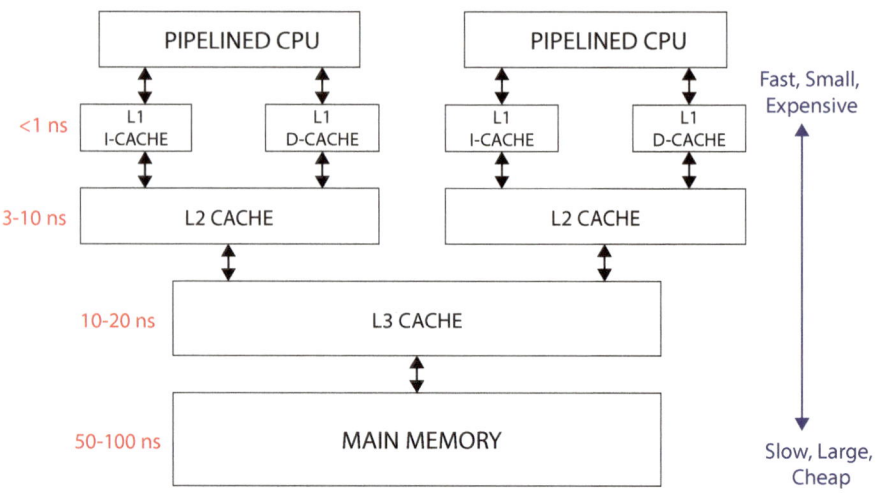

Fig. 5.1 Memory hierarchy in modern computer systems

accessed data and instructions. When a program accesses a location, it is copied into the small and fast memory, making future access to that location much faster.

Spatial locality refers to the tendency of a program to access memory locations close to each other. The most common reason for temporal locality is using a program counter to address instructions. Small and fast memories exploit spatial locality, also. We do not store only the requested item (instruction or operand) in the small and fast memories but also nearby items. If the program accesses one location, data from nearby locations is readily available in the small and fast memory.

Figure 5.1 shows the typical memory hierarchy in modern computer systems. Today, computer systems contain several pipelined CPU cores. Each CPU core has two high-speed on-chip level one (L1) caches with instructions and operands, respectively. Recall from the computer architecture course that a pipelined CPU can access operands (in the memory stage) and instructions (in the instruction fetch stage) simultaneously; hence, we need two separate caches, one for instructions and the other for operands. These L1 caches are made using the fastest and most expensive semiconductor technology. They are usually very small (only 16 or 32 kB) and contain a small subset of instructions and data a CPU core needs. If the requested operands or instruction is found in the cache, we say a **cache hit** occurred. Cache hits are desirable and essential for achieving high-performance computing because they reduce the time spent waiting for data to arrive from slower memory levels. Because of temporal and spatial locality, the rate of cache hit in L1 caches is very high (usually higher than 95%). When the data or instruction requested by the CPU is not found in the cache memory, we say that a **cache miss** occurred. If a cache miss occurs in the L1 cache, the requested data or instructions are copied from the slightly slower and slightly bigger on-chip L2 cache into the L1 cache (in the case of the L2 cache hit). If an L2 cache miss occurs, the requested data or instructions are copied from the slower and bigger L3 cache into the L2 cache (in the case of the L3 cache hit). Finally, if an L3 cache miss occurs, the requested data or instructions are copied from the very slow and large off-chip main memory in the L3 cache. Note that several CPU cores share the same L3 cache.

Cache memories are a simple solution that we brought to computer science from everyday life. Still, at the same time, they represent one of the most astonishing and indispensable technical improvements in modern computer systems, quietly revolutionising how our computers work behind the scenes. Cache memories are high-speed, small-sized memory buffers that play a critical role in bridging the gap between the blazing speed of a CPU and the relatively sluggish access times of main memory (SDRAM). As we delve into cache memories, we'll discover their remarkable ability to enhance system performance by strategically storing frequently used data and instructions, optimising memory access patterns, and keeping our computing experiences smooth and efficient. In this exploration of cache memories, we'll unveil the inner workings of these vital components, shedding light on how they improve the responsiveness of your devices and the efficiency of your software.

Caches, together with the memory hierarchy, are at the heart of computer performance, ensuring that applications run swiftly and efficiently. In this chapter, we will delve deeper into the memory hierarchy and caching, understanding how these architectural elements form the backbone of modern computing systems. So, let us start our journey through the intricacies of cache memories, where speed meets intelligence to deliver a seamless computing experience.

5.3 Cache Structure and Organisation

Without locality, the cache would be useless because the probability of the operands or instructions needed by the CPU for being in the cache would be far too small. However, since locality exists, the cache can contain a relatively small subset of the content of the main memory (or a slower memory level of the memory hierarchy). Having only a subset of the main memory in the cache requires a special cache structure that is slightly more complex than the main memory structure. Besides keeping the subsets of the main memory, the cache should also keep the information that tells us which subsets are in the cache. Therefore, the cache consists of two parts: a **control** part and a **data** part.

The data part contains **cache lines** or **blocks**. A cache block is a fundamental unit of storage within a cache memory. A cache block is nothing but 2^b consecutive (contiguous) memory words whose addresses differ only in the lower b bits. These blocks are organized within the cache's data part and play a crucial role in improving memory access times. Cache blocks are designed to exploit spatial locality, which is the tendency of programs to access nearby memory locations sequentially. By storing contiguous data in cache blocks, the cache takes advantage of this behaviour, reducing cache misses when accessing nearby data. Cache blocks are used to store a fixed-size portion of data retrieved from the main memory or a slower level of the memory hierarchy. Each cache block contains the contiguous data or instructions fetched from the main memory. The common cache block size in modern computers is 64 bytes. Therefore, a cache block contains 64 adjacent memory words whose addresses differ only in the lower six bits. Cache blocks are the smallest data units that can be transferred between the cache and main memory or between two levels in the memory hierarchy. When data is fetched from or written to the main memory, it is done in entire cache blocks, even if only a portion of the block is needed.

The control part contains metadata that is used to manage and identify the contents of the cache. Metadata for each block includes a **cache tag** and a **valid bit**. The cache tag is the address of the block in the main memory. It stores information about which memory address range is stored in that cache block and is used to identify the block. The cache tag is essential for determining if a requested memory location is present in the cache. Cache tags are used to compare the requested memory address with the stored tags. Hence, we use cache tags to determine cache hit or miss. Besides a cache tag, each block metadata contains a valid bit. The valid bit indicates whether the data in the cache block is valid and can be used. If the valid bit is set, the data is valid and can be used for memory access; otherwise, it's not. The valid bit helps ensure cache coherency.

Figure 5.2 presents a simple cache structure and its operation. The cache in Fig. 5.2 contains only eight blocks, each storing eight 8-bit words. This cache is used only for explanation, and there are no such small and simple caches in reality. The CPU address in this example is 16-bit long, and the three least significant bits determine the offset of the memory word within a block. As there are eight blocks in the cache, the next three address bits (bits 5 to 3) determine the index of the block (placement of the block) within the cache. The remaining ten most significant bits determine the address of the block in the main memory. When the CPU makes a memory access request, the cache controller extracts the tag portion of the memory address and compares it with the cache tag in the cache block determined by the memory address. The cache controller performs this comparison to determine whether the requested data is present in the cache. If the tag extracted from the memory address matches the tag stored for a specific cache block and the valid bit is set (indicating valid data), it is a **cache hit**. The requested data is then fetched from the cache, providing faster access to the CPU. If the tag does not match or the valid bit is not set (indicating an invalid cache block), it is a **cache miss**. In this case, the cache controller must fetch the data from a slower memory level to satisfy the CPU's request. The retrieved block is placed in the cache for future accesses. As each block from the main memory is mapped to exactly one cache block determined with the three bits (5 to 2) in the address, we say that the cache is directly mapped. In **direct-mapped** cache, there is a one-to-one correspondence between main memory blocks and cache blocks, making it one of the simplest forms of cache organisation.

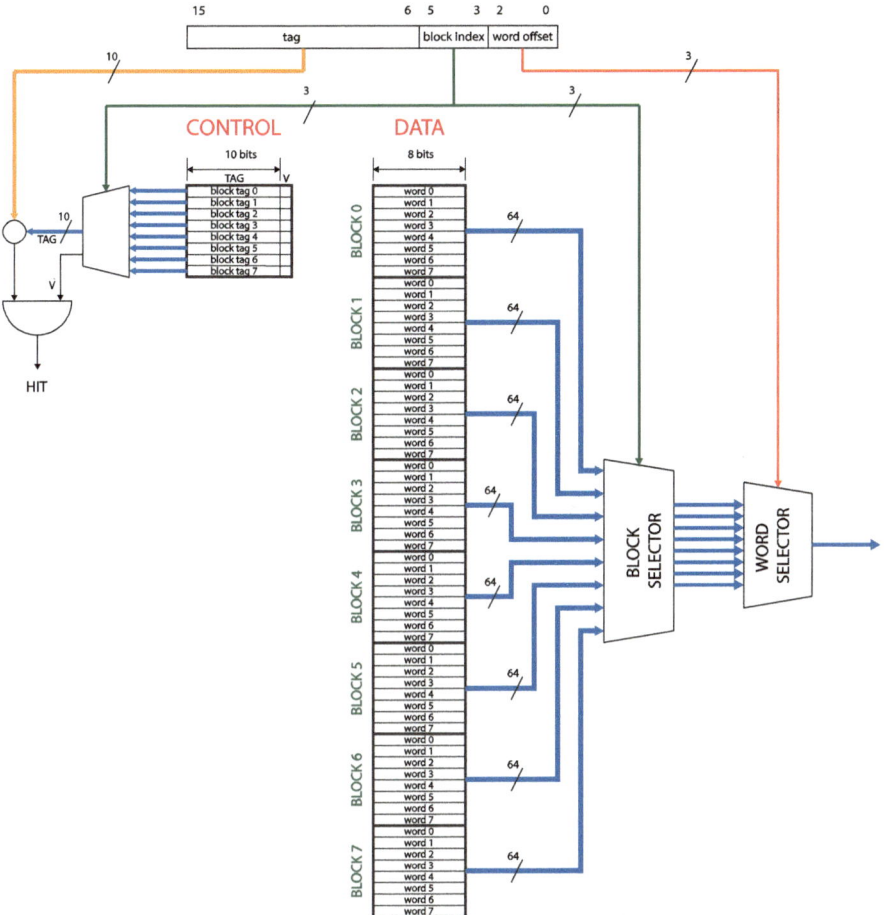

Fig. 5.2 A simple direct-mapped cache with eight blocks, each storing eight 8-bit words

5.4 Direct Mapped Cache

A direct-mapped cache is one of the simplest forms of cache memory organization used in computer systems. In a direct-mapped cache, each block of main memory is mapped to a specific cache line. Direct-mapped caches are characterized by their straightforward structure and quick access times but may suffer from conflicts when multiple memory blocks map to the same cache line. To delve deeper into the details of a direct-mapped cache, we'll explore its structure, how memory addresses are mapped to cache lines, and the strategies employed for efficient data retrieval. Understanding direct-mapped caches is vital to grasping the fundamental principles of cache memory in computer architecture.

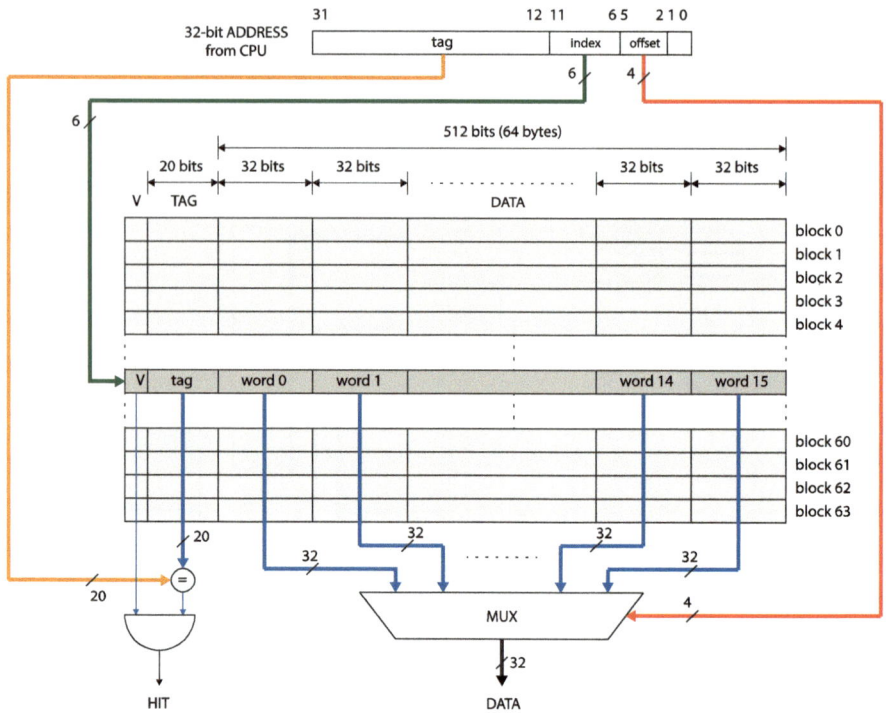

Fig. 5.3 A 4 kB cache that contains 64 blocks with 16 32-bit words per block

Figure 5.3 presents the 4 kB (2^{12}) cache that contains 64 (2^6) cache lines (blocks) (2^6) with 16 (2^4) words per block. The address size is 32 bits, and the data word size is also 32 bits (2^2B). Hence, the cache size is $2^6 \times 2^4 \times 2^2$B $= 2^{12}$B $= 4$kB. To access data in the cache, the CPU provides a 32-bit memory address. The cache controller looks up this address to determine if the requested data is present in any of the cache lines. The memory address provided by the CPU is divided into three main parts: the **tag**, the **index**, and the **offset**. Let's break down how these components work in our 4 KB cache with 64 cache lines (blocks), each having 16 words of 32 bits (four bytes).

The **offset** specifies the word within a cache line (block) that is being accessed. We need four bits to represent the offset since we have 16 words in a cache line. Since each memory word is 32 bits (4 bytes), the two least significant bits in the address are ignored (always zero).

The **index** determines which cache line (block) within the cache is being accessed. As our cache contains 64 cache lines, we need six bits to index all the blocks ($2^6 = 64$). Accessing a cache line is often called **indexing**. Index selects the word from the cache line (block) using a 16-to-1 multiplexor. In practice, to eliminate the multiplexor in real chips, caches use a separate SRAM for data and a smaller SRAM for tags. The **index** and **offset** fields form the address bits for the data SRAM. In our case, the data SRAM is 32 bits wide and has 1024 32-bit data words.

The **tag** contains the upper bits of the memory address, uniquely identifying a particular memory block in the main memory. To find the tag size, we subtract the combined size of the index and offset from the total address size. Since our cache contains 64 blocks (2^6), each block holds 16 words (2^4), and each word in the cache line is four bytes (2^2), we need 20 bits for the tag (address size – index size – offset size – word size = 32 bits – 6 bits – 4 bits –2 bits = 20 bits for the tag).

5.4.1   Read Operations in Direct-Mapped Caches

When the CPU needs to read or write data, it provides a memory address. The memory address is split into the tag, index, and offset components based on their respective sizes. Then, the search is performed in the cache. The index is used to select the specific cache line (block) in the cache that corresponds to the requested memory address. The offset field selects the searched word from the cache line. Simultaneously, the tag from the address is compared to the tag stored in that cache line. If the tags match, it's a cache hit. The cache line contains the requested data, and the data word already present at the data lines can be quickly accessed. If the tags don't match, it's a cache miss. The required memory block from the next level in the memory hierarchy is loaded into the cache line determined by the index bits. The cache line's tag is updated with the new tag from the address, making it ready for future access. The cache miss often triggers an exception, and the corresponding interrupt handler triggers the task switch. Hence, the CPU continues the execution of another task and does not wait for a block to be filled into the cache line.

5.4.2   Handling Writes in Direct-Mapped Caches

While reading data from a direct-mapped cache is a simple and fast operation, special consideration should be given to the write operation. The main question in write operations is where data should be written: in the cache or the main memory? This is an important issue because caches are local (private) to CPUs (recall that only L3 is shared among various CPUs). If the most recent copy of the data is held in L1 or L2 caches, other CPUs may not be aware that the most recent copy exists outside the main memory. This is known as the problem of **cache consistency**. Cache consistency requires that all copies of a particular piece of data, stored in different levels of the memory hierarchy (e.g., cache memory and main memory), must remain synchronized and reflect the same content. Ensuring cache consistency is a complex task, as it requires efficient mechanisms for tracking and managing data states and synchronizing access across multiple cache levels. However, it is vital for the reliability, predictability, and performance of modern computer systems, particularly in multi-core and multi-processor environments.

In a direct-mapped cache, write strategies refer to the methods employed for handling write operations, specifically how writes are managed in the cache. There are **two primary strategies for handling writes** in a direct-mapped cache: **write-through** and **write-back**.

In a **write-through cache**, when the CPU performs a write operation, the data is written not only to the cache but also immediately to the corresponding location in upper levels and in main memory. This strategy ensures that main memory always contains the most up-to-date data, making it consistent with the data in the cache. The advantages of a write-through cache are data consistency and simple and predictable behaviour. The disadvantages of a write-through cache are write latency (as writing to both the cache and main memory can introduce some delay) and increased memory traffic.

In a **write-back cache**, when a write operation is executed, the data is written only to the cache, and the corresponding location in upper levels and the main memory is updated later, usually when the cache line is evicted or replaced. In this strategy, the cache maintains a **dirty bit** for each cache line to track modified data. A write-back cache has lower write latency and reduces memory traffic. The main disadvantage is complexity, as managing the dirty bit and ensuring that modified data is eventually written back to the main memory adds complexity to cache management.

The choice between these write strategies depends on the specific requirements of the system and the trade-offs it is willing to make. Write-through is simpler and ensures data consistency but can introduce write latency. Write-back offers better performance but requires more sophisticated cache management to maintain data consistency. Some systems use a combination of both strategies, allowing the cache to operate in a write-back mode while periodically ensuring that modified data is written back to the main memory to maintain consistency. This approach combines the advantages of both strategies but adds complexity to cache management.

5.5   Set Associative Cache

As we have seen, each main memory block maps to exactly one cache line in a direct-mapped cache. The index bits in the memory address select the cache line to which a memory block is mapped. In the direct-mapped cache from Fig. 5.3, six index bits select one of 64 cache lines. Recall that the tag field represents the address of a memory block that maps to a cache line selected by the index field. **Therefore, there are precisely 2^{tag} memory block that maps to the same cache line!** In the direct-mapped cache from Fig. 5.3, there are 1M (2^{20}) memory blocks that map to the same cache line. This can lead to **cache conflicts**, causing one block to replace another prematurely. To mitigate cache conflicts, we use **set-associative caches** that have a slightly different structure. In set-associative caches, we **group multiple cache lines into one set**. Here, **a memory block maps to any available cache line in exactly one set**. As a result, there are fewer conflicts because multiple blocks can map to the same set but not the same line within that set. The name set-associative cache derives from the way it organizes cache lines into sets and associates those sets

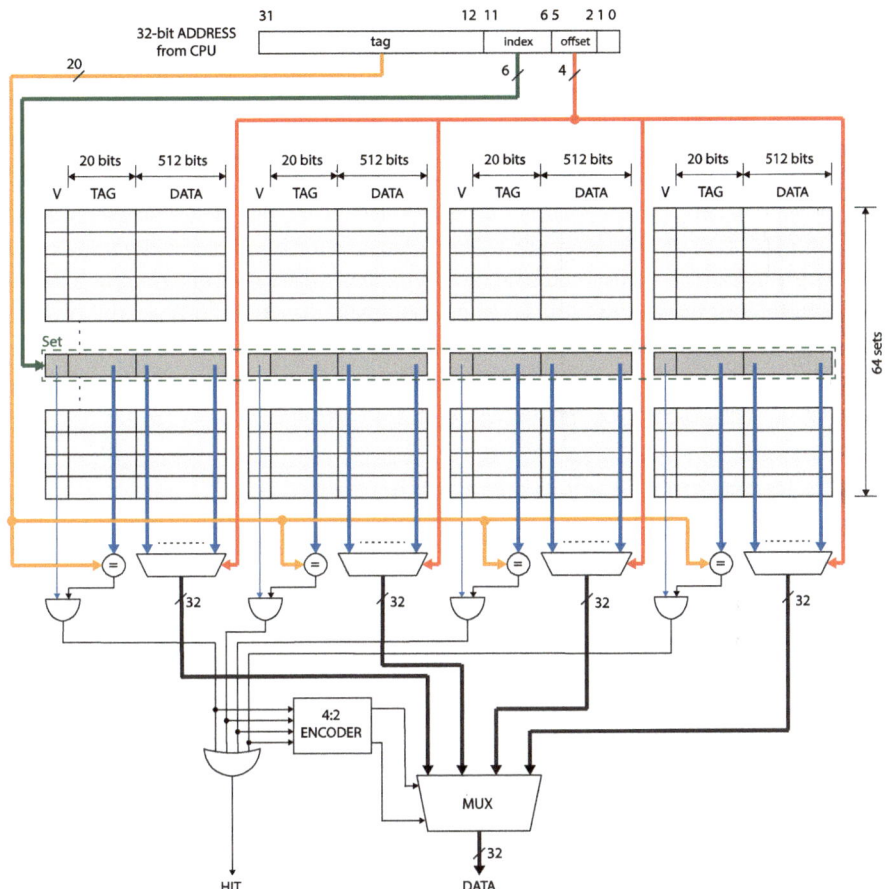

Fig. 5.4 A 16 kB four-way set-associative cache

with specific blocks in the main memory. In a set-associative cache, the term **N-way set-associative cache** refers to the associativity level of the cache. Specifically, an N-way set-associative cache **means that each set in the cache contains N cache lines**.

A 16 kB four-way set-associative cache is depicted in Fig. 5.4. Locating a block in a 4-way set-associative cache involves a process similar to locating a block in a direct-mapped cache. When a memory address is provided to the cache, the cache controller interprets the address to determine the set index, tag, and word offset. The address is divided into three parts. The tag bits uniquely identify the memory block in the main memory. It is used to check if the data in a cache line is the same as the requested data. The index field specifies to which set in the cache the requested memory block is mapped. Finally, offset identifies the location of the desired memory word within a cache line. The cache in Fig. 5.4 contains 64 (2^6)

sets. Each set contains four cache lines of 16 32-bit data words. Therefore, the size of the cache is $64\ sets \times 4\ lines\ per\ set \times 16 \times 4 = 16$ kB. In the selected set, the memory word pointed by the offset bits is selected by four multiplexors in each cache line. Simultaneously, the tag fields of all four cache lines in the selected set are compared to the tag field from the memory address. A cache hit occurs if there is a match and the valid bit is set. Otherwise, a cache miss occurs. The four comparators in Fig. 5.4 determine which element of the selected set (if any) matches the tag. The four outputs of the comparators are 4:2 encoded, and the two-bit code is used to select the data from one of the four indexed blocks using a 4-to-1 multiplexor. For example, suppose the tag field in the third cache line equals the tag field from the provided address. In that case, the 4:2 encoder encodes the output of four comparators as '10' (which index of the third cache line in the selected set) and the 32-bit word from the third cache line is selected using the output 4-to-1 multiplexor.

5.5.1 Replacing a Block in a Set-Associative Cache

Recall that in a direct-mapped cache, the newly requested block can be mapped to exactly one cache line when a miss occurs. In this case, the content of this cache line is replaced with a new memory block. In an N-way set-associative cache, N cache lines in a set can hold a newly requested memory block. If all cache lines in a set are full, we face the challenge of which cache line to replace. In other words, all cache lines in a set are candidates for replacement, and we must choose the most appropriate one. Set-associative caches use the **Least Recently Used (LRU)** algorithm to determine which cache line within a set to replace when a new block is brought into the set. The LRU algorithm ensures that the cache line that hasn't been accessed for the longest time is replaced. The key idea behind the LRU algorithm is to **evict the cache line that hasn't been accessed for the longest time**, under the assumption that **less recently used data is less likely to be needed in the near future**. The LRU algorithm ensures that the cache remains populated with the most relevant data for the workload, potentially reducing cache misses and improving overall cache performance.

The LRU algorithm maintains a history of the recent access patterns for each cache set. The cache controller updates the LRU information based on the cache line accessed during a cache hit. Initially, all cache lines in a set are considered equally recently used. When a cache line in a set is accessed (either for a read or a write), it is marked as the most recently used. The other cache lines in the set are adjusted to reflect their relative access times. When a new block needs to be brought into a set that is already full, the cache controller selects the cache line that is marked as the "least recently used" for eviction. However, implementing the LRU algorithm can be relatively complex and requires additional hardware to track access history accurately.

5.5.2   Choosing the Associativity Level

Increasing the associativity (the set size) increases the size of a set-associative cache and reduces the miss ratio. The size of a set in a set-associative cache, often referred to as the associativity level, directly impacts cache hits and the overall cache performance. Hence, choosing the right set size is a crucial design decision. As we increase the set size (i.e., the associativity level), we generally reduce the likelihood of cache conflicts. Multiple memory blocks can map to the same set but not to the same cache line within that set. This reduces the probability of cache misses caused by contention for the same cache line. Conversely, decreasing the set size (reducing associativity) makes the cache more similar to a direct-mapped cache. This can lead to more cache conflicts, potentially causing cache misses when multiple memory blocks contend for the same set.

However, larger set sizes increase the complexity of the cache design and the hardware requirements. As associativity increases (e.g., from 8-way to 16-way set-associative cache), the complexity and hardware requirements escalate further. Hence, higher associativity requires more power and space. A smaller set size with lower associativity may be favoured in cost-sensitive or power-constrained environments. There's often a trade-off between reducing cache conflicts (by increasing associativity) and managing hardware complexity. The balance between cache performance and complexity must be carefully considered. Extremely high associativity levels may provide diminishing returns while adding substantial complexity.

Here are some common associativity levels that are often used in various caches within a modern computer system. A 4-way set-associative cache contains four cache lines in each set. This is a common choice for L1 instruction and data caches, offering better performance than direct-mapped caches while maintaining manageable hardware complexity. An 8-way set-associative cache provides even higher associativity, making it well-suited for L2 (level 2) caches. This level of associativity further reduces cache conflicts. In some cases, particularly in shared caches for multi-core processors or in last-level caches (L3), a 16-way set-associative cache is used to accommodate the needs of multiple cores and diverse workloads.

Finally, if we increase the size of a set while maintaining the same cache size, we must reduce the number of sets. Eventually, there will be only one set left in the cache. This is a **fully associative cache**. It has the highest level of associativity, where any memory block can map to any cache line. Fully associative caches provide the greatest flexibility but are limited to only 8 or 16 cache lines due to the high hardware complexity. They are often used in scenarios where only a few cache lines are required to store highly frequently accessed data. We will see their usage in the next chapter.

5.6   Cache Controller

Along with a cache, a CPU always contains a cache controller. A cache controller is a hardware component responsible for managing the operation of a cache memory subsystem within a computer system in a largely invisible way to the program. It plays a crucial role in coordinating data transfers between the cache memory and the main memory, as well as controlling the cache's behavior and policies. Here are some key functions and responsibilities of a cache controller:

1. **Cache Management**: The cache controller is responsible for managing the contents of the cache memory, including the storage and retrieval of data blocks from the main memory, as well as the replacement and eviction of cache lines when the cache becomes full. It automatically writes code or data from the main memory into the cache. When the core requests instructions or data from a particular address, but there is a cache miss, the request must be passed to the next level of the memory hierarchy, an L2 cache, or external memory. It also causes a cache linefill—the contents of a piece of main memory are transferred into the cache. Simultaneously to cache line fill, the requested data or instructions are streamed directly to the core. Hence, the core need not wait for the linefill to complete before using the data. This data is immediately supplied to the core pipeline while the cache hardware and external bus interface read the rest of the cache line in the background.
2. **Cache Look-up**: When it receives a request from the core, it checks to see whether the requested address is to be found in the cache. This is known as a cache look-up. It does this by comparing a subset of the address bits of the request with tag values associated with lines in the cache. If there is a match, known as a hit, and the line is marked valid, then the read or write occurs using the cache memory.
3. **Cache Prefetching**: The cache controller may implement prefetching techniques to anticipate future memory accesses and proactively fetch data into the cache before it is actually requested by the CPU, thereby reducing memory access latency and improving overall performance.
4. **Cache Control Policies**: The cache controller defines and enforces various cache control policies, such as the cache replacement policy (e.g., Least Recently Used—LRU, Random Replacement) and cache write policies (e.g., Write-Through, Write-Back), to optimize cache performance and behavior.
5. **Cache Coherency**: In multi-core or multi-processor systems, cache coherence is essential to ensure that all caches have a consistent view of memory. Occasionally, data and instructions in the cache and data in external memory might be different; this is because the processor can update the cache contents, which still need to be written back to the main memory. This is a problem known as cache coherency. This can be a particular problem when you have multiple cores or an external DMA controller. The cache controller implements cache coherence protocols to maintain data consistency across multiple cache levels and processor cores.
6. **Cache Synchronization**: In systems with multiple cache controllers or cache hierarchies, the cache controller may coordinate cache synchronization operations, such as cache flushes and cache invalidations, to maintain cache coherence and ensure data consistency.

5.7   Case Study: Cache in STM32F7 and STM32H7 Series Devices

The STM32F7 Series and STM32H7 Series devices feature a Cortex-M7 core with an advanced cache system designed to improve performance in embedded applications. They include two 16 Kbytes L1 caches for the instructions and the data. An L1 cache stores a set of data or instructions near the CPU, so the CPU does not have to keep fetching the same repeatedly used data.

The L1 caches on all Cortex-M7 cores are divided into lines of 32 bytes. Both instruction and data caches have 512 lines. Each line is tagged with an address. The data cache is a 4-way set associative (four lines per set), and the instruction cache is a 2-way set associative. Set-associative caches offer a balance between the simplicity of direct-mapped caches and the flexibility of fully associative caches. They provide better performance than direct-mapped caches by reducing the likelihood of cache conflicts (where multiple memory addresses map to the same cache line) while still maintaining a relatively simple cache structure compared to fully associative caches.

The STM32F7 Series and STM32H7 Series devices provide control and configuration registers for managing the operation of the cache. These registers allow the software to enable or disable the cache and invalidate cache lines.

5.8   Case Study: Cache in Processors with ARMv8-A Architecture

Processors that implement the ARMv8-A Architecture contain two or more levels of cache. For example, the Cortex-A53 and Cortex-A57 processors normally contain two or more levels of cache, that is, a small L1 instruction and data cache and a larger, unified L2 cache, which is shared between multiple cores. Additionally, clusters of cores can share an external L3 cache implemented as an external SRAM memory block. The in-core cache controller checks all instruction fetches and data reads or writes in the cache. However, we can mark some parts of memory, such as those containing peripheral devices, for example, as non-cacheable. Cortex-A57 contains a two-way set associative 32KB L1 cache with a 64-byte cache line length. Hence, there are 256 cache sets.

Virtual Memory

6

CHAPTER GOALS

Have you ever wondered how your computer manages to run multiple programs simultaneously, juggling vast amounts of data and instructions with apparent ease? The answer lies in a sophisticated concept known as virtual memory. While your computer's physical memory (RAM) has finite capacity, virtual memory extends its capabilities by intelligently managing memory resources. In this chapter, we will embark on a journey to explore the intricate world of virtual memory, uncovering its underlying principles, mechanisms, and benefits.

Upon completion of this chapter, you will be able to:

- Understand the fundamental concept of virtual memory and its role in modern computing systems.
- Understand the motivation behind virtual memory and its benefits for multitasking, memory management, and system stability.
- Examine the mechanisms used for virtual-to-physical address translation in virtual memory systems.
- Learn about page tables, TLBs (Translation Lookaside Buffers), and the process of address translation during memory accesses.
- Explore the concept of paging and its role in virtual memory management.
- Explore real-world examples showcasing the use of virtual memory in diverse processors.
- Examine the interaction between caches and virtual memory systems.
- Understand how caches interact with virtual address translation and page replacement algorithms.
- Learn about cache virtualization techniques and optimizations for improving virtual memory performance.

© The Author(s), under exclusive license to Springer Nature Switzerland AG 2024
P. Bulić, *Understanding Computer Organization*, Undergraduate Topics in Computer
Science, https://doi.org/10.1007/978-3-031-58075-8_6

6.1 Introduction

In the previous sections, we have seen that the physical SDRAM memory is limited. All processes running on a computer system share this limited amount of physical memory. Moreover, the amount of physical memory varies among different computer systems, even if they have the same processor architecture. Ideally, we would have a computer system with a memory size that is precisely equal to the size of the CPU's memory space. For example, for a 64-bit CPU, we would like to have 2^{64} physical memory words present in the system. Theoretically, we could add additional SDRAM memory to a computer system, but this is often limited by the available density of memory chips, the capacity of the memory controller, and, of course, the high price of the SDRAM chips. To overcome this limitation, we could make processes think they have unlimited physical memory. This technique is called **virtual memory**.

Virtual memory enables efficient multitasking by allowing multiple programs to run simultaneously, each with its own virtual address space. Any program running on a system with virtual memory presumes that it is the sole program running and can generally use any memory location within the whole CPU's address space. In such a way, programs are isolated from each other, preventing one program from directly accessing the memory space of another program.

Besides, virtual memory creates a hierarchy of memory storage, with the fastest and most expensive storage being the physical SDRAM and slower, more extensive storage, like a hard drive disk (HDD) or solid-state drive (SDD), serving as secondary storage. In such a way, it allows the execution of programs larger than the physical memory by swapping portions of these programs between the smaller SDRAM and the bigger HDD or SDD, as needed.

In summary, virtual memory is a sophisticated memory management system that abstracts and optimizes the use of physical memory resources, enhancing computer systems' overall performance and capabilities. It provides a crucial layer of abstraction that allows programs to run in a seemingly limitless memory space, improving system responsiveness and supporting the execution of complex and resource-intensive tasks. Let's embark on a journey to explore the remarkable world of virtual memory, understanding how it transforms the way our computers work and, ultimately, how it enhances our digital.

6.2 The Benefits and Downsides of Virtual Memory

Virtual memory is a fundamental concept in computer systems that enables them to manage and utilise the available physical memory resources efficiently, extending the capabilities of a computer far beyond its physical constraints. It is a powerful and versatile technology that plays a pivotal role in modern operating systems, ensuring the seamless execution of diverse and resource-intensive applications. In this virtual realm, the computer appears to have an almost limitless amount of memory, allowing it to multitask, run large programs, and maintain system stability. **Virtual memory is a memory management technique used in computer systems to give applica-**

tions an illusion of a vast and contiguous memory space, even when the physical SDRAM available may be limited. It creates an abstraction layer that separates the address spaces used by applications from the actual physical memory.

The main reasons to have virtual memory are:

1. **Program positional independence:** Programs (processes) can be designed and executed as if they have access to an unlimited, contiguous block of memory, even if the physical memory is limited. Programs being positionally independent means that the memory addresses used by a program during its execution are not fixed or hardcoded. Instead, the program uses virtual addresses, and the operating system is responsible for mapping these virtual addresses to physical addresses. Positional independence enables dynamic loading, allowing parts of a program to be loaded into memory only when needed. This flexibility enhances resource utilization by loading portions of programs on-demand, reducing the initial memory footprint. Positional independence simplifies memory management for both the operating system and developers. It allows the OS to load programs at different memory locations without requiring modifications to the program's code. This abstraction makes it easier to develop and run programs without being constrained by the actual size of the physical memory or their position in the physical memory.

2. **Isolation and Protection:** Virtual memory enables the isolation of processes from each other. Each process is allocated its own virtual address space, providing the illusion that it has exclusive access to the entire memory. This isolation prevents one process from interfering with the memory contents of another, enhancing system stability and security.

3. **Address space abstraction:** Virtual memory abstracts the physical memory (RAM) from the applications. Each process works with virtual addresses, and the operating system manages the mapping between virtual and physical addresses. This abstraction simplifies application development by allowing programmers to work with a larger, more flexible address space.

4. **Memory protection:** Virtual memory systems provide memory protection mechanisms. Each process has its own set of permissions for accessing different parts of its virtual address space. Unauthorized access attempts result in memory protection faults, preventing processes from corrupting each other's data.

5. **Code Sharing:** Multiple instances of the same program can share the same code in memory since the virtual addresses are independent of the physical memory locations. This promotes efficient use of memory and helps reduce redundancy.

6. **Large Program Support:** Virtual memory enables the execution of programs that exceed the size of available physical RAM. It achieves this by moving parts of the program in and out of storage, ensuring that the active portions remain in physical memory.

7. **Effective Use of Storage:** When physical memory is exhausted, less frequently used portions of programs are temporarily moved to disk, freeing up space for other processes. This prevents programs from being limited by the size of the physical memory and optimizes overall system performance. In essence, virtual memory is the architectural marvel that empowers modern computer systems to

handle complex tasks, run multiple applications concurrently, and transcend the confines of physical memory, thereby providing a seamless and efficient computing experience.

There are also several downsides of virtual memory. Implementing virtual memory requires complex logic and resources. Each time a program accesses memory, **the virtual address must be translated into a physical address**, which is a complex and time-consuming operation. If the data being requested does not reside in main memory (SDRAM), it must be retrieved from secondary storage (HDD or SSD). To help alleviate the performance penalties of translating virtual addresses to physical addresses, dedicated **memory management unit** (MMU) hardware should be implemented within a processor.

6.3 Memory Management Unit

The memory management unit (MMU) is a piece of hardware that translates virtual addresses issued by a CPU core into physical addresses. Each processor core capable of running a multiuser and multitasking operating system has an MMU. Before enabling the virtual memory, MMU should be properly configured by software running at an appropriate privilege level. Figure 6.1 shows an MMU within the microprocessor. The CPU core executes programs with their own virtual address space. Hence, each time an instruction is fetched, or the CPU executes LOAD/STORE instructions, virtual addresses are issued from the CPU core. These virtual addresses are fed into the MMU, which translates them into physical addresses. The memory controller then uses these physical addresses to access the main memory (SDRAM).

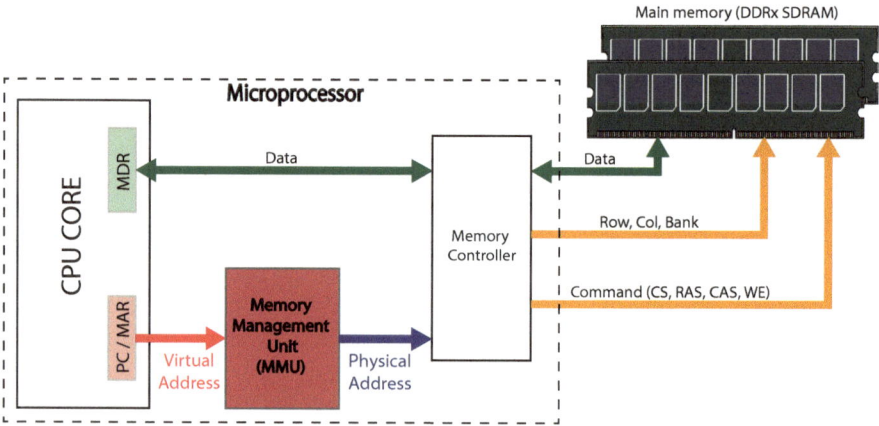

Fig. 6.1 A memory management unit (MMU) within the microprocessor translates virtual addresses into physical addresses

Each virtual address A_v can be in the main memory or on the HDD/SSD disk. If the virtual address A_v is in the main memory, the MMU translates it to a physical address A_p, and the CPU accesses a word in the main memory at the address A_p. But if A_v is on the disk, the MMU initiates a request to transfer data from the disk to the main memory. When the transfer is done, A_v is translated to A_p, and the CPU makes the access.

Unlike the cache, where the requested block from the main memory is transferred to the cache by hardware, with the MMU, the requested data block is transferred from the disk to the main memory by software. It usually works by raising an interrupt, and the interrupt service program transfers a data block from the disk to the main memory. The slowness of the software solution is negligible compared to the time it takes to transfer a block of data from the disk to the main memory.

6.4   Virtual Address Translation

A Memory Management Unit (MMU) is a hardware component in a computer system responsible for translating virtual addresses used by programs into physical addresses in the system's main memory. This translation is crucial for the proper functioning of modern operating systems and applications.

When a program runs, it operates assuming it has access to a contiguous virtual memory. This virtual memory space is typically much larger than the physical memory (SDRAM) available in the system. When the program accesses memory, it uses virtual addresses. The MMU then translates these addresses into physical addresses corresponding to specific physical memory locations (RAM).

All memory accesses issued by software use virtual addresses, requiring the MMU to translate the virtual address to a physical address for each access. **The MMU translates the virtual addresses of contiguous blocks in virtual memory rather than individual memory locations. Translating individual memory locations would be highly inefficient because it would require a separate translation for each memory access.** By translating the virtual address of an entire block at once, the MMU reduces the overhead associated with address translation. A virtual memory block is called a **page**. **MMUs typically work by dividing the virtual address space into pages of equal size**. Pages are typically relatively large (e.g., 4 kB) compared to individual memory locations. This granularity strikes a balance between efficient address translation and memory management. The process of address translation using pages is called **paging**. There are several benefits of paging:

1. **Efficiency:** If each memory location were translated individually, the overhead of managing and storing translation mappings would be significantly higher.
2. **Spatial locality:** Recall that programs often exhibit spatial locality, meaning they tend to access memory locations near each other. The MMU leverages this spatial locality by translating entire pages to improve performance. Once a page is accessed, it's likely that subsequent memory accesses within the same page will occur, minimizing the need for additional translations.

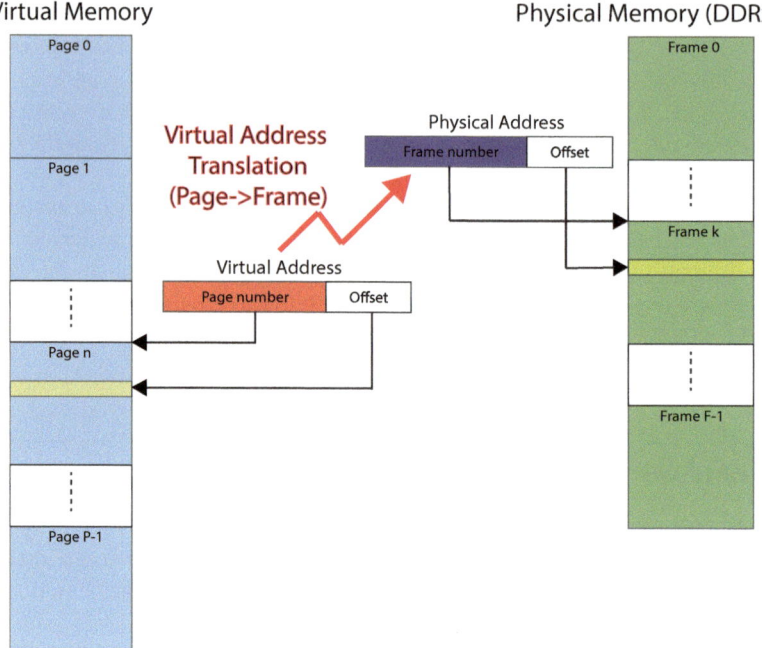

Fig. 6.2 Virtual address translation using paging

3. **Hardware implementation:** Implementing address translation for individual memory locations would require more complex hardware and incur higher latency for each memory access. Translating entire pages allows the MMU to utilize simpler structures, resulting in faster and more efficient address translation.

Overall, translating virtual addresses at the page level rather than for each memory location provides a good balance between efficiency and hardware complexity, making it the preferred approach in modern computer systems.

The basic idea of paging is simple and presented in Fig. 6.2. Virtual memory is divided into blocks of equal size called **pages**. Physical memory is also divided into blocks of equal size, called **page frames**. Pages and page frames are the same size. Typical sizes of pages and page frames are 4 kB, 2 MB, or 4 MB (always a power of 2). **Any page from virtual memory can be transferred to any of the frames in physical memory**. With a known size of the virtual and physical address space, the number of pages and frames is fixed and known in advance. **The MMU translates the virtual address of the page (called the page number) to the physical address of the page frame (called the frame number)**. Each virtual address consists of the page number and the memory word offset within the page. Similarly, each physical address consists of a frame number and a memory word offset within the page frame.

Suppose a computer system with a CPU that uses n-bit virtual addresses. The virtual address space in such a system is of size 2^n memory words. Suppose this

Fig. 6.3 Virtual address translation

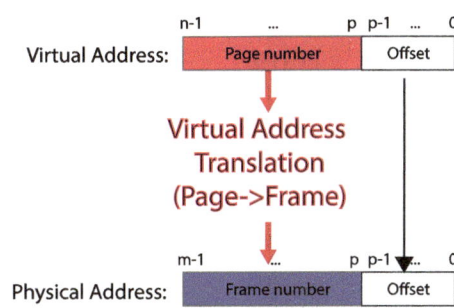

system contains the main memory (physical address space) of size 2^m memory words. Suppose the pages and frames are of size 2^p memory words. Then, the virtual address space contains 2^{n-p} pages, and the physical memory contains 2^{m-p} frames. Figure 6.3 shows virtual address translation for such a system.

The virtual address contains the p-bit offset within a page and the $(n-p)$-bit page number. The page number is the address of a page in the virtual address space, while the offset is the address of a memory word within the page. The physical address contains the p-bit offset within a frame and the $(m-p)$-bit frame number. The frame number is the address of a frame in the main memory, while the offset is the address of a memory word within the frame. The virtual address space is often larger than the physical address space, hence $n > m$, but this is not a requirement in general. Virtual and physical addresses are often of the same width despite virtual address space being much larger than physical address space.

The MMU translates the $(n-p)$-bit page number to $(m-p)$-bit frame number, while the offset in the physical address is the same as the offset from the virtual address. **The physical address is formed simply by concatenating the frame number with the offset from the virtual address**. The MMU performs this translation using data structures called **page tables**. A page table contains mappings between page numbers and frame numbers. The following section will describe page tables and their usage in virtual address translation in detail.

The number of bits in the page offset field determines the page size. The number of pages in the virtual address space need not match the number of page frames in the main memory. The larger number of pages gives a program the illusion of an unbound amount of memory it can use. Recall that the main memory usually does not hold all the pages a program uses. Indeed, the main memory holds only a portion of all pages that belong to a program. Other pages are stored on disk to save space in the main memory and to allow other programs to use the main memory. When a program tries to access a page that is not in the main memory, a so-called **page fault** occurs. A page fault will trigger an interrupt, and the interrupt service routine will transfer the requested page from the secondary storage (HDD or SSD) into the main memory. Often, there is no free frame in the main memory for the requested page, so before transferring a page from the disk, the interrupt service program will transfer the least used frame from the main memory to the disk to free a frame for a new page. This operation is called **page swapping**. Page swapping is a mechanism used in virtual memory systems to manage memory resources efficiently. Due to a very slow transfer between the main memory and secondary storage, the page fault takes millions of CPU cycles to process.

The choice of page size in a virtual memory system can significantly affect system performance. Sizes from 4 kB ($p = 12$) to 4 MB ($p = 22$) are typical today. Larger page sizes typically result in fewer page faults because each page contains more memory locations. With larger pages, there's a higher likelihood that subsequent memory accesses will be within the same page, reducing the number of page faults. This can improve overall system performance by reducing the overhead associated with handling page faults. On the contrary, smaller page sizes allow for more efficient memory utilization. With smaller pages, the operating system can fit smaller memory allocations more precisely, reducing internal fragmentation. Internal fragmentation refers to the wasted space within a page due to the allocation of memory in fixed-size pages that may not be fully utilized by the data stored within them. This inefficiency arises when the allocated memory space within a page is larger than what is actually needed to store the data. In other words, internal fragmentation occurs when a page (or page frame) contains allocated memory space along with unused space that other programs cannot utilize. This unused space represents wasted memory resources.

6.5   One-Level Paging

The MMU uses data structures called **page tables** to translate a virtual address to a physical address. The translation process involves consulting the page table to determine the mapping between the virtual address and the corresponding physical address. **Page tables are stored in the main memory**. The operating system initializes and maintains the page table, which stores the mappings between virtual pages and physical frames. Entries in the page table are called **page descriptors**. **Each page descriptor corresponds to a virtual page and contains the corresponding physical frame number where the page is located in physical memory**. The MMU will read the corresponding descriptor from the page table to translate a page number to a frame number. Hence, the MMU is connected to the main memory, as presented in Fig. 6.4. The MMU will address the main memory for two reasons: to read a page descriptor from the page table or to physically address the memory word requested by the CPU.

Figure 6.5 shows the virtual address translation using paging for a computer system with 32-bit virtual and physical addresses where pages are of size 4 kB. This type of paging is often referred to as **one-level paging**. The virtual address space contains 2^{32} one-byte memory words (4 GB in total); hence, there are $4\,GB/4\,kB = 1\,M$ pages. An in-memory page table contains 1M (1048576) 32-bit page descriptors (one per page) that map virtual addresses into physical addresses. The size of the page table is 4 MB (1M descriptors $\times$ 32 bit). To perform the translation, the MMU uses the page number extracted from the virtual address (20 uppermost bits) to index into the page table. The descriptor corresponding to the virtual page is then retrieved.

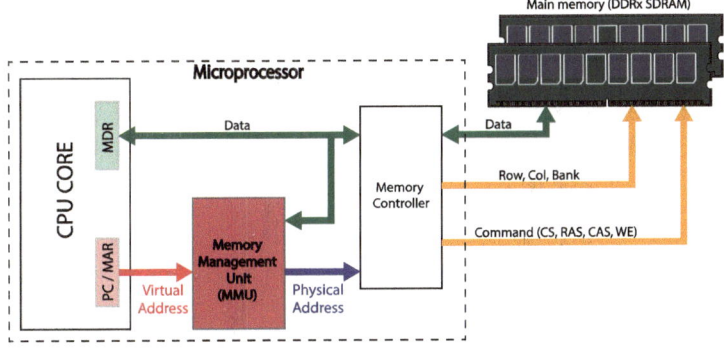

Fig. 6.4 A memory management unit (MMU) within the microprocessor should have access to the main memory in order to read page descriptors from the page table

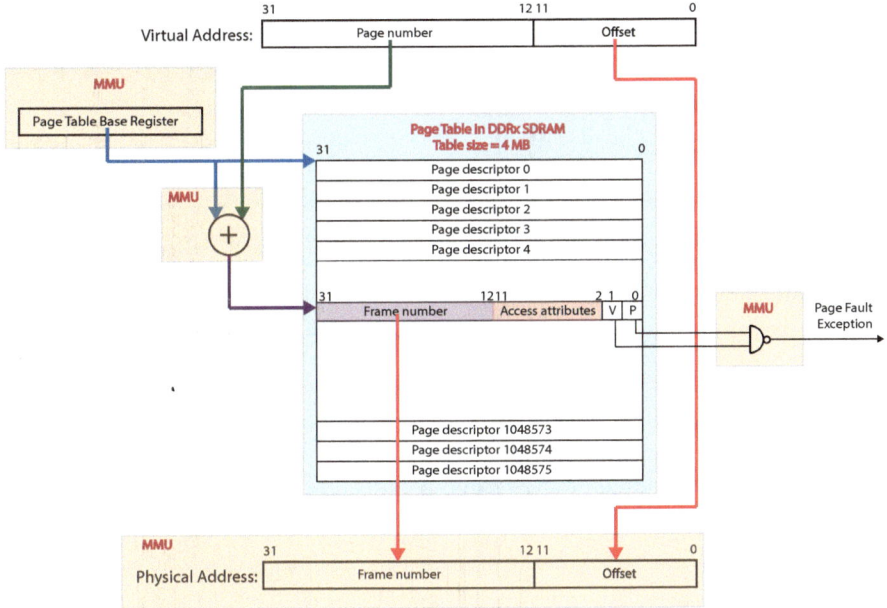

Fig. 6.5 Virtual address translation with one-level paging

The MMU contains a register (Page Table Base Register) that holds the page table base address and directly accesses the page table in the main memory. The MMU forms the address of the page descriptor that corresponds to the requested page by summing the page number from the virtual address and the page table base address from the Page Table Base Register. Each page descriptor contains the 20-bit frame number where the corresponding page resides in physical memory. The MMU reads the descriptor and combines the 20-bit frame number with the 12-bit offset portion of the virtual address to generate the 32-bit physical address, which is then used to access the memory location. **Hence, one access to the main memory is required to read the descriptor and form the physical address used to access the requested memory word**.

Before enabling the MMU, the operating system must appropriately set up the page table, and the OS must tell the MMU where the page table resides in memory by setting its base address in the MMU's Page Table Base Register. Once the MMU is configured, it will translate all virtual addresses issued by software to physical addresses. The physical addresses are given to the memory controller that issues commands and row/bank/column addresses to the SDRAM chips (or DIMM modules). Each page descriptor in the page table contains:

1. **Valid bit (V)**: Indicates whether the page descriptor is valid. MMU uses this bit to detect page faults and generate the page fault interrupt request. Upon reset, all page descriptors are invalid. When a page is transferred to the main memory for the first time, the operating system initializes the corresponding page descriptor and sets the valid bit. When the valid bit is set, the MMU can translate the virtual address using the descriptor. Otherwise, the page fault is generated and the OS transfers the page to the main memory.
2. **Present bit (P)**: Indicates whether the corresponding virtual page is currently present in physical memory. If this bit is set, the frame number in the descriptor is valid and can be used to form the physical address. Otherwise, the page is currently not present in the main memory (it has been previously swapped to secondary storage), and the MMU generates the page fault.
3. **Frame Number**: Specifies the physical address of the frame number where the corresponding virtual page is located in the physical memory. When the valid and present bits indicate that the descriptor is valid and the page is present in the main memory, the 20-bit FN is used to construct the physical address.
4. **Access Attributes**: Define the access permissions for the virtual page, such as read, write, or execute permissions. These bits control the level of access allowed to the page. Depending on the system architecture, additional metadata, such as dirty bits (indicating whether the page has been modified), may be included in this field.

6.6 Two-Level Paging

More sophisticated memory management techniques, such as **multi-level paging**, may be preferred. Here, we will describe the two-level paging technique presented in Fig. 6.6.

One large page table (used in one-level paging) is split into several smaller page tables in two-level paging. These smaller page tables are then structured as a hierarchical tree data structure consisting of two levels:

1. a **level-2 page table** (also referred to as page directory): at the top level of the tree is the level-2 page table (page directory). The level-2 page table is an array of

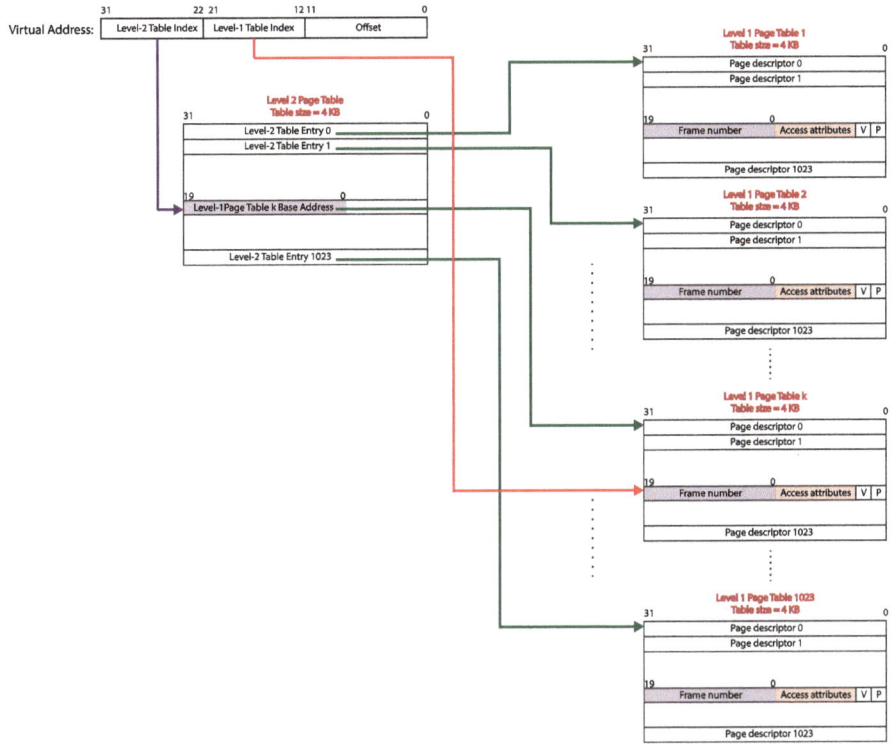

Fig. 6.6 Two level paging

entries, each corresponding to a portion of the virtual address space and directly pointing to one of the level-1 page tables. The size of the page directory is typically determined by the number of the level-1 page tables. The number of level-1 page tables is determined by the maximum size of the virtual address space and the desired size of pages.

2. **level-1 page tables**: the level-1 page tables contain entries for individual pages (page descriptors) within the corresponding portion of the virtual address space. The size of the level-1 page table is, in general, determined by the number of descriptors it contains, but the most common size of the level-1 page table equals the size of one page. **In this way, each level-1 page table is regarded in the operating system as one page that can reside in one frame in the main memory and can be swapped between the main memory and the secondary storage**.

This organization allows for the efficient management of large virtual address spaces by dividing the address translation process into two stages. As each level-1 page table is the same size as one page, we can manage level-1 page tables as we manage page tables, i.e., they can reside in the main memory or the secondary storage. This means the operating system can swap out portions of the bottom-level page tables to secondary storage (such as a hard disk) when not actively in use and swap them back into main memory when needed. Overall, the operating system manages bottom-level page tables as integral components of the virtual memory system, ensuring efficient address translation and memory management for processes running on the system. **This significantly reduces the physical memory required to store page tables**.

Figure 6.6 presents two-level paging for the system from Fig. 6.5. The large page table from Fig. 6.5, containing 1M descriptors, is split into 1024 level-1 page tables, each containing 1024 descriptors ($1024 \times 1024 = 1M$). Then, the level-2 page table is added as a tree root, where each of the 1024 entries directly points to one level-1 page table. The page descriptors in level-1 page tables are 32-bit long, meaning that each level-1 page table is of size 4 kB. Also, each entry in the level-2 page table is 32-bit long, meaning that level-1 page table has also the size of 4 kB. The operating system need to hold in the main memory the level-2 page table and at least one level-1 page table, which is accessed by the MMU. The other level-1 page tables can reside in the secondary storage until needed by the MMU.

The hierarchical tree structure of two-level paging allows for efficient address translation and memory management, particularly in systems with large virtual address spaces. It reduces the memory overhead of managing large page tables while providing scalability and flexibility to accommodate varying memory allocation patterns and page sizes. Let us now describe the virtual address translation in two-level paging presented in Fig. 6.7.

Virtual address translation involves two stages in two-level paging: top-level translation using the level-2 page table and bottom-level translation using the level-1 page table. When a program accesses memory using a virtual address, the MMU extracts the level-2 page table index and the level-1 page table index from the virtual address. The level-2 page table index identifies the entry in the top-level level-2 page table, and the level-1 page table index identifies the entry in the bottom-level level-1 page table.

As presented in Fig. 6.7, the MMU uses the 10-bit level-2 page table index to index into the top-level level-2 page table, retrieving the corresponding level-2 page table entry. The level-2 page table entry contains a 20-bit pointer to the bottom-level page table responsible for the portion of the virtual address space specified by the level-2 page table. Please note that level-1 page tables are 4 KB large and are kept in 4 KB physical frames in the main memory; hence, the last 12 bits of the page table address are always 0.

The MMU then uses the level-1 page table index to index into the bottom-level page table. It retrieves the corresponding page descriptor, which contains the physical frame number where the requested page resides in physical memory. The MMU combines the frame number from the page descriptor with the offset portion of the virtual address to generate the physical address. The offset portion of the virtual address specifies the byte offset within the frame. Finally, the physical address generated by the MMU is used to access the memory location in physical memory.

If the MMU encounters an invalid level-2 table entry or page descriptor during the translation process, or P bits of the level-2 table entry or page descriptor are cleared,

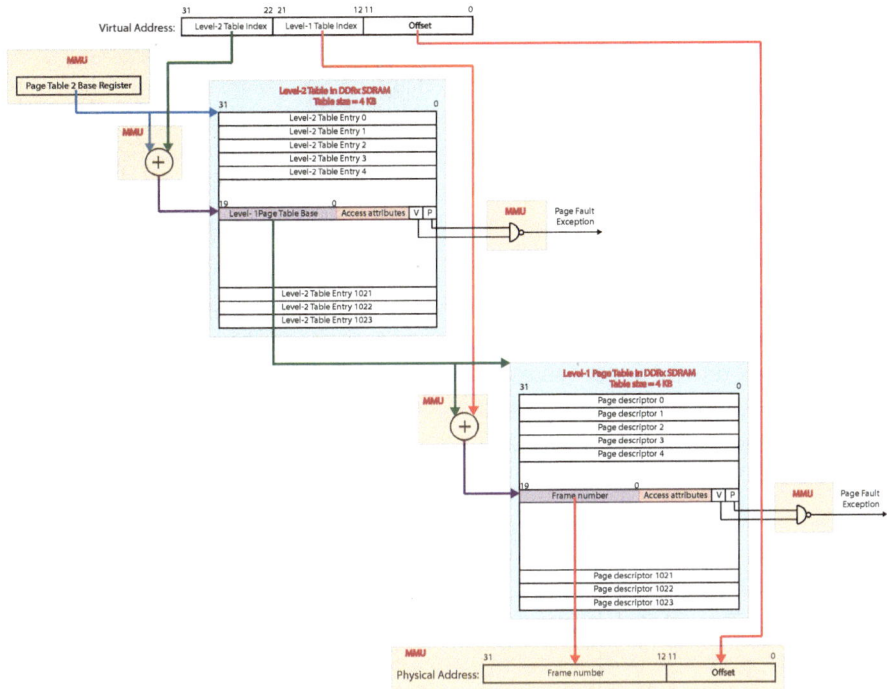

Fig. 6.7 Virtual address translation with two-level paging

indicating that the corresponding portion of the virtual address space is not currently in physical memory, a page fault occurs. The operating system intervenes to load the required page or level-1 page table from secondary storage into physical memory and updates the level-2 page table or page table accordingly.

Overall, virtual address translation in two-level paging involves hierarchical lookup of page directory and page table entries to map virtual addresses to physical addresses, allowing programs to access memory transparently while efficiently managing large virtual address spaces. Unfortunately, two-level paging prolongs the time needed to translate virtual addresses. This is because the MMU needs to access the main memory twice: firstly, to retrieve the level-2 page table entry, and secondly, to retrieve the page descriptor from the level-1 page table. The access time in a two-level paging system refers to the time taken by the CPU's Memory Management Unit (MMU) to translate a virtual address into a physical address and access the corresponding memory location.

Modern computer systems use even more levels to translate virtual addresses. For example, Intel uses 5-level paging, and ARMv8 uses 4-level paging, further prolonging the access time. Recall that accessing page tables in main memory can be slow due to the long access times of SDRAM-based main memory. Therefore, minimizing virtual address translation latency is crucial for maintaining system performance in memory-intensive applications. The next section describes fast and efficient address translation techniques to improve system performance and responsiveness.

6.7   Translation Lookaside Buffers

On a computer with virtual memory, every time the CPU accesses memory, it is necessary to translate the virtual address to the physical address. Since the content of the page table is required for virtual address translation, it would be ideal from the point of view of translation speed if the tables were in the MMU. Unfortunately, this is not feasible due to their sizes. The tables are so large that they are stored in main memory or even in secondary storage when the MMU does not need them. In an N-level paging system, the MMU must traverse N levels of page tables to find the corresponding physical address. Each level of page table access requires accessing main memory, which has longer access times compared to faster cache memory. This increased number of memory accesses adds significant latency to the translation process. The above number of accesses applies to virtual address translations that do not result in a page fault. When a page fault occurs, the number of accesses and, thus, the access time increases significantly. Systems with frequent and multi-level address translation would suffer from long memory access time, resulting in poor overall system performance and responsiveness. Moreover, accessing the main memory consumes more power compared to accessing the CPU cache. The system with virtual memory would need to perform frequent memory accesses, leading to increased power consumption and reduced energy efficiency.

Obviously, if we don't have some mechanism to accelerate the virtual address translation, a computer with virtual memory will be much slower and less efficient than one without virtual memory. Therefore, we must have some built-in mechanism for accelerating virtual address translation within the CPU (MMU). This mechanism is basically a small and fast cache that stores the most recently used page descriptors. We have seen that the page descriptor describes a translation for several thousand memory words (e.g., 4 KB) within one page. This means that the MMU uses the same descriptor for a long time due to spatial and temporal locality. Therefore, the obvious solution to speed up the virtual address translation is to store the descriptor that the MMU is currently using in a fast cache inside the MMU. This way, the MMU will avoid accessing the main memory each time when translating the virtual page address.

These special caches, which contain only descriptors from the page tables, are called **Translation Lookaside Buffers (TLB)**. All computers with virtual memory have a TLB; without it, the virtual memory would be useless. The TLB is similar to the instruction and operand caches we met in the previous chapter. The only difference is that the TLB stores descriptors rather than instructions or operands.

TLBs rely on spatial locality to improve their effectiveness in caching translations. TLBs cache translations for recently accessed virtual addresses and their corresponding physical addresses. When a virtual address is translated, the resulting physical address and other associated information (i.e., page descriptor) are stored in the TLB.

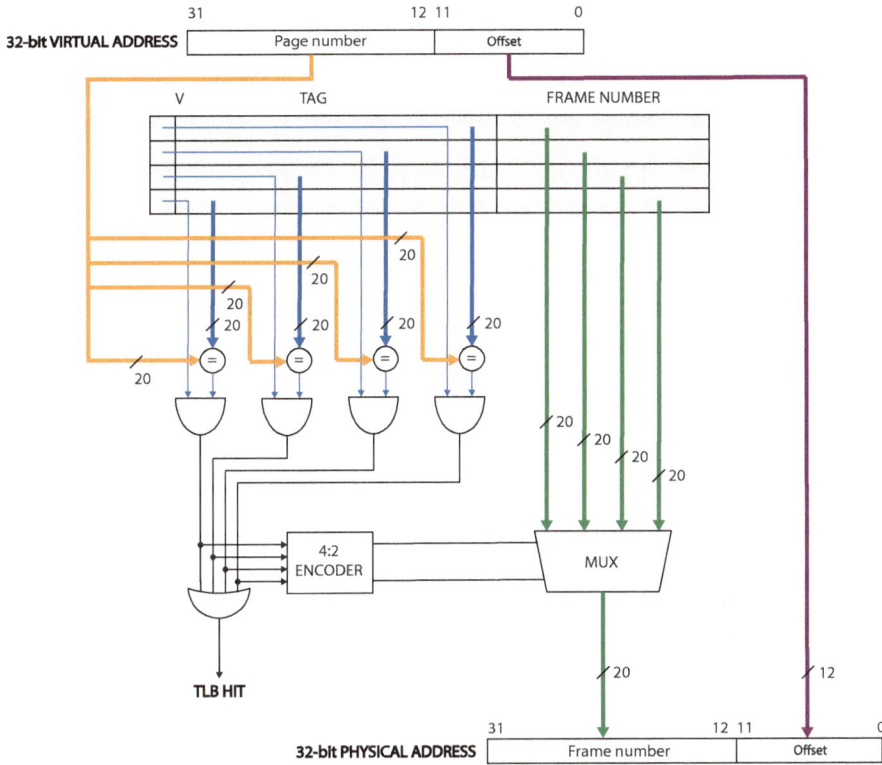

Fig. 6.8 Fully associative TLB with four entries

When a subsequent memory access references the same virtual address or nearby virtual addresses (i.e., the same page), the TLB can provide the translation directly without the need to access the page tables in the main memory. By caching translations based on spatial locality, TLBs can significantly reduce the latency of address translation and improve the overall performance of memory access operations. This is particularly beneficial in systems with large virtual address spaces or memory-intensive applications, where efficient translation of virtual addresses to physical addresses is essential for maintaining system performance and responsiveness.

There is another significant difference between instruction/operand caches and TLBs. Since the virtual page number changes significantly more slowly than the block addresses in the instruction/operation cache, we can achieve a very high hit probability with a much smaller TLB. Therefore, TLBs are often implemented as small, fully associative (or set-associative) translation caches.

The Translation Lookaside Buffer is a hardware cache used in memory management units (MMUs) to store recently used virtual-to-physical address translations. TLBs are typically organized as associative caches. The TLB consists of a finite number of entries, each storing a virtual-to-physical address translation. The number of entries in a TLB can vary depending on the specific architecture and design constraints. Figure 6.8 presents the organization of a fully associative TLB with four entries (TLB stores four descriptors). Again, we assume 32-bit virtual and physical addresses and a 4 KB page size. To keep the description simple, we focus only on the read operation.

Each tag entry in TLB holds the virtual page number used to look up translations in the TLB. In the fully associative TLB, every TLB tag is compared against the virtual page number since the required descriptor can be stored anywhere in TLB. This comparison determines if the translation is present in the TLB. Each data entry in TLB contains the physical frame number (and additional metadata from the page descriptor if necessary). If the valid bit of the matching tag is set, we have a TLB hit, and the associated frame number is read from the TLB entry to form the physical address. Hit in TLB always means the page is valid and present in the main memory.

In modern computer systems, TLBs can hold 16-512 entries, and the hit time is typically one clock cycle. For example, the TLB size in ARM Cortex A8 is 32, while in ARM Cortex A73, the TLB size is 48 entries. Designers often use a wide variety of associativities in TLBs. Some systems use fully associative TLBs because the fully associative mapping results in a lower miss rate. Since TLBs are small, the cost to implement the fully associative mapping in hardware is reasonable. Other systems may use large TLBs with small associativity.

TLBs cache recent translations of virtual addresses to physical addresses. When a virtual address needs to be translated, the MMU first checks the TLB. If the translation is found in the TLB (a TLB hit), the translation can be performed much faster than if it had to access the page tables in memory. TLBs speed up memory access by reducing the latency of address translation. Therefore, TLBs are an essential component of MMUs in modern computer systems, providing fast and efficient address translation to improve system performance and responsiveness.

6.7.1   Multilevel Translation Lookaside Buffers

Multilevel Translation Lookaside Buffers are a feature found in modern processors from both Intel and AMD. Multilevel TLBs enhance translation efficiency by organizing TLBs into multiple levels, similar to multilevel caches. Intel and AMD CPUs nowadays are built with multiple TLBs, for example, a small L1 TLB (potentially fully associative) that is extremely fast and a larger and slower L2 TLB. Here's a brief overview of each level:

1. **L1 TLB**: The L1 TLB, also known as the primary TLB, is typically small and fast, residing on the processor core itself. It stores a subset of frequently used virtual-to-physical address translations, allowing for fast access to recently accessed memory locations. The L1 TLB is usually fully associative, providing quick access to translations with minimal latency. Moreover, there are usually two L1 TLBs, one for pages containing data (DTLB) and the other for pages containing instructions (ITLB).

2. **L1 TLB**: The L2 TLB serves as a secondary level of translation caching, supplementing the L1 TLB. It is larger in size compared to the L1 TLB, allowing it to store a larger number of translations and accommodate a broader range of memory access patterns. The L2 TLB is usually shared among multiple processor cores, providing a centralized cache for translation entries.

For instance, AMD Opteron Family 15H has a fully associative L1 DTLB with 32 entries for 4 KB pages and an 8-way associative L2 TLB with 1024 entries for 4 KB pages. Similarly, Intel Core i7 has a 4-way set associative L1 DTLB with 64 entries for 4 KB pages and a 4-way set associative L2 TLB with 512 entries for 4 KB pages.

6.8   Integrating Caches and Virtual Memory

Until now, we have discussed caches and virtual memory separately, but in every computer system, caches and virtual memory coexist and form a memory hierarchy. Therefore, when we have a cache and virtual memory in the computer system, the following question arises: should we access the cache with a physical or virtual address? Let's take a closer look at both options. First, recall two basic operations when accessing the cache:

1. **Cache Tagging**: refers to the process of comparing a portion of a memory address against a cache tag to identify the memory block stored in that cache line. During a cache tagging, the cache controller compares the tag of the requested memory address with the tags of the cache lines to determine if the requested data is present in the cache.
2. **Cache Indexing**: refers to the process of determining the location within the cache where a particular memory block should be stored or looked up. The cache index is derived from part of the memory address of the data being accessed. This portion of the memory address is used to select the specific cache block or set where the data should be stored or retrieved.

6.8.1   Physically Indexed and Physically Tagged (PIPT) Cache

When a CPU generates a physical address, access to the cache precedes access to the main memory. Data is checked in the cache using the tag and block index (directly mapped cache) or set index (set-associative) bits. Such a cache where the tag and index bits are generated from a physical address is called a **Physically Indexed and Physically Tagged (PIPT)** cache. But in today's computer systems, the CPU generates virtual addresses. When using a PIPT cache, the virtual address needs to be converted to its corresponding physical address before the PIPT cache can be searched for data. Therefore, accessing a PIPT cache requires translating the virtual address to a physical address before indexing the cache. This translation introduces additional latency that can impact overall memory access time and system performance.

6.8.2 Virtually Indexed and Virtually Tagged (VIVT) Cache

An immediate but naive solution is a **Virtually Indexed and Virtually Tagged (VIVT)** cache. VIVT cache directly checks the data in the cache and fetches it without translating the virtual address to a physical address, reducing the hit time significantly. VIVT caches are simpler to implement compared to PIPT cache organizations because they use virtual addresses for both indexing and tagging. There is no need for complex address translation or additional hardware support for physical addresses, which reduces hardware complexity and cost. However, there are challenges with VIVT caches. One challenge is that **virtual addresses may alias**, meaning different virtual addresses can map to the same cache index. This happens when two different programs use two different virtual addresses for the same physical address (e.g. when both programs access the same OS subroutine). These duplicate addresses, called aliases, could result in two copies of the same data in a cache; if one is modified, the other will have the wrong value. This can lead to cache conflicts and reduced cache performance. With a PIPT cache, this wouldn't happen because the accesses would first be translated to the same physical cache block.

6.8.3 Virtually Indexed and Physically Tagged (VIPT) Cache

Today's computer systems employ a compromise approach, such as **Virtually Indexed and Physically Tagged (VIPT)** caches, which use virtual indexing and physical tagging to balance performance and complexity. The VIPT cache uses tag bits from a physical address and an index from a virtual address. These caches work simultaneously with TLBs. Figure 6.9 shows such a solution where TLB and VIPT cache work together and simultaneously.

The cache is indexed using the port of the virtual address that does not translate (i.e., virtual offset), and the tag (part of the physical address) is obtained from the VIPT cache. The cache indexing (data access and tag access) starts immediately when the CPU issues the virtual address, and there is no need to translate the virtual address for cache indexing. As the cache is indexed with the part of the virtual address that does not change during the address translation, this is indeed equal to indexing the cache with the physical address.

Simultaneously with indexing, the TLB is searched with a virtual page number, and a physical address is obtained from the TLB. Finally, the tag obtained from the VIPT cache during cache indexing is compared against the physical address from the TLB. If they both are the same, then it is a cache hit; else, it is a cache miss.

Since TLB is smaller in size than cache, TLB's access time will be lesser than cache's access time. Hence, the physical address needed for cache tagging is obtained from the TLB in time to compare it against the tag from the VIPT cache. The VIPT cache takes the same time as the VIVT cache during a hit and solves the problems

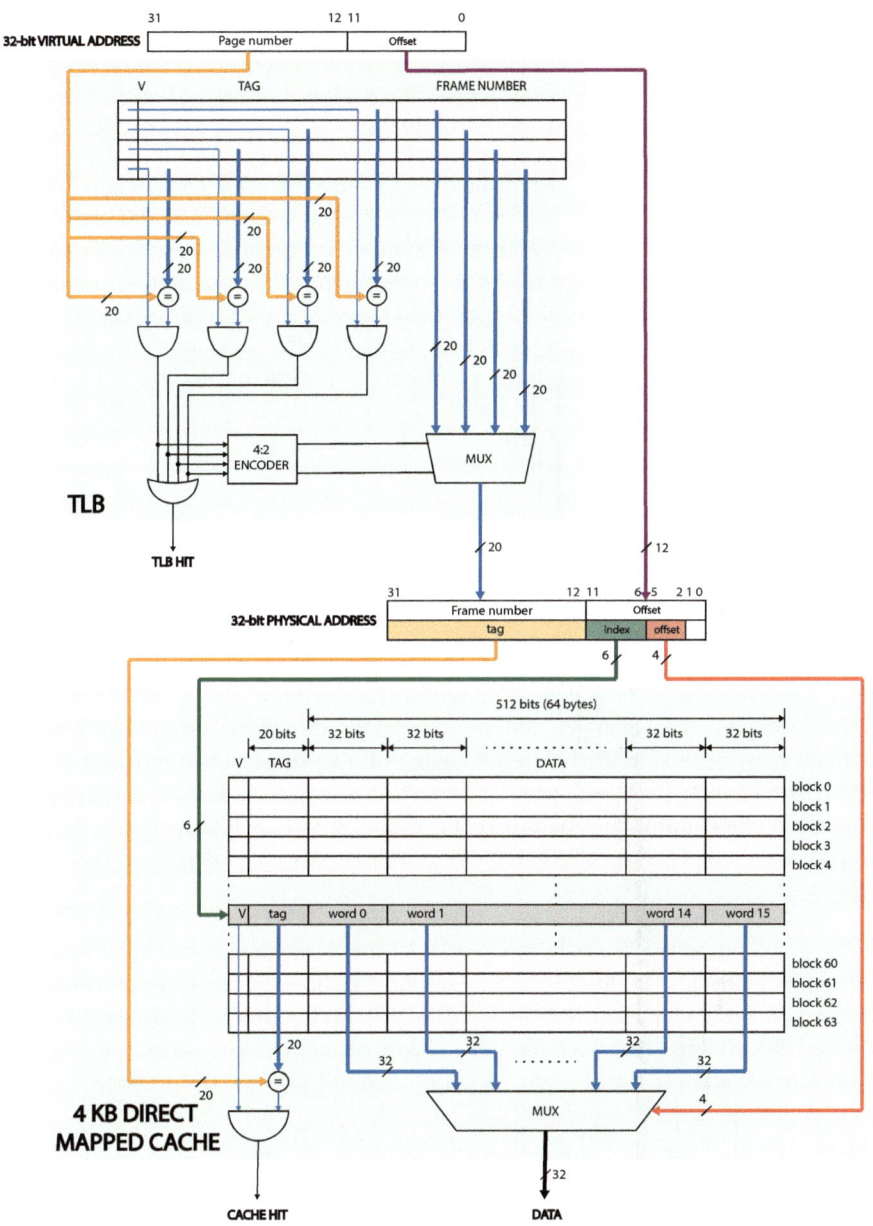

Fig. 6.9 Directly mapped VIPT cache

of the VIVT cache. Since the TLB is accessed in parallel with the cache, the flags and tags can be checked simultaneously. The VIPT cache uses part of the physical address as an index, and since every memory access in the system will correspond to a unique physical address, data for multiple processes can coexist in the cache.

However, there is a small challenge with VPIT caches. As the VIPT cache is indexed with the offset, the size of the directly mapped VIPT cache cannot exceed the size of a page. In Fig. 6.9, we assume a 4 KB page size and a 12-bit offset (quite common in today's systems) used for cache indexing. Hence, the maximum size of the directly mapped VIPT cache is 4 KB. To overcome this limitation, set-associative VIPT caches are commonly used. Figure 6.10 presents such a solution using a 4-way set-associative VIPT cache. In the set-associative VIPT cache, the offset from the virtual address is used to index the set in the cache (and not the block as in the directly mapped cache). A set, in turn, contains several blocks (four blocks in a set in the VIPT cache presented in Fig. 6.10).

6.9   Case Study: AMD64 5-Level Paging

The AMD64 virtual address translation mechanism enables system software to create separate address spaces for each process or application. These address spaces are known as virtual address spaces. AMD64 uses the paging mechanism to selectively map individual pages in the virtual address space using a set of hierarchical page tables. The paging mechanism and the page tables are used to provide each process with its own private region of physical memory for storing its code and data. Besides, the AMD64 paging provides a process protection mechanism. Processes are protected from each other simply by isolating them within the virtual address space. This way, the process cannot access physical memory that is not assigned to its virtual address space.

The AMD64 architecture allows the translation of 64-bit virtual addresses into 52-bit physical addresses, although processor implementations can support smaller virtual and physical address spaces. Figure 6.11 shows an overview of the page-translation hierarchy used in AMD64. As this figure shows, a virtual address is divided into fields, each of which is used as an offset in a translation table. To complete the address translation, a walk through all table entries referenced by the virtual address fields is required. The 12 lowest-order virtual address bits are used as the byte offset into the 4 KB physical page. In a 5-level paging with 4 KB pages, the translation is performed by dividing the virtual address into seven fields. The virtual address fields are as follows:

1. Bits 63:57 are a sign extension of bit 56, as required for canonical address forms. Bit patterns that are valid addresses are called *canonical addresses*. The x86-64 and AMD64 architecture divide canonical addresses into low and high groups. Low canonical addresses range from 0x0000 0000 0000 0000 to 0x0000 7FFF FFFF FFFF. High canonical addresses range from 0xFFFF 8000 0000 0000 to 0xFFFF FFFF FFFF FFFF.

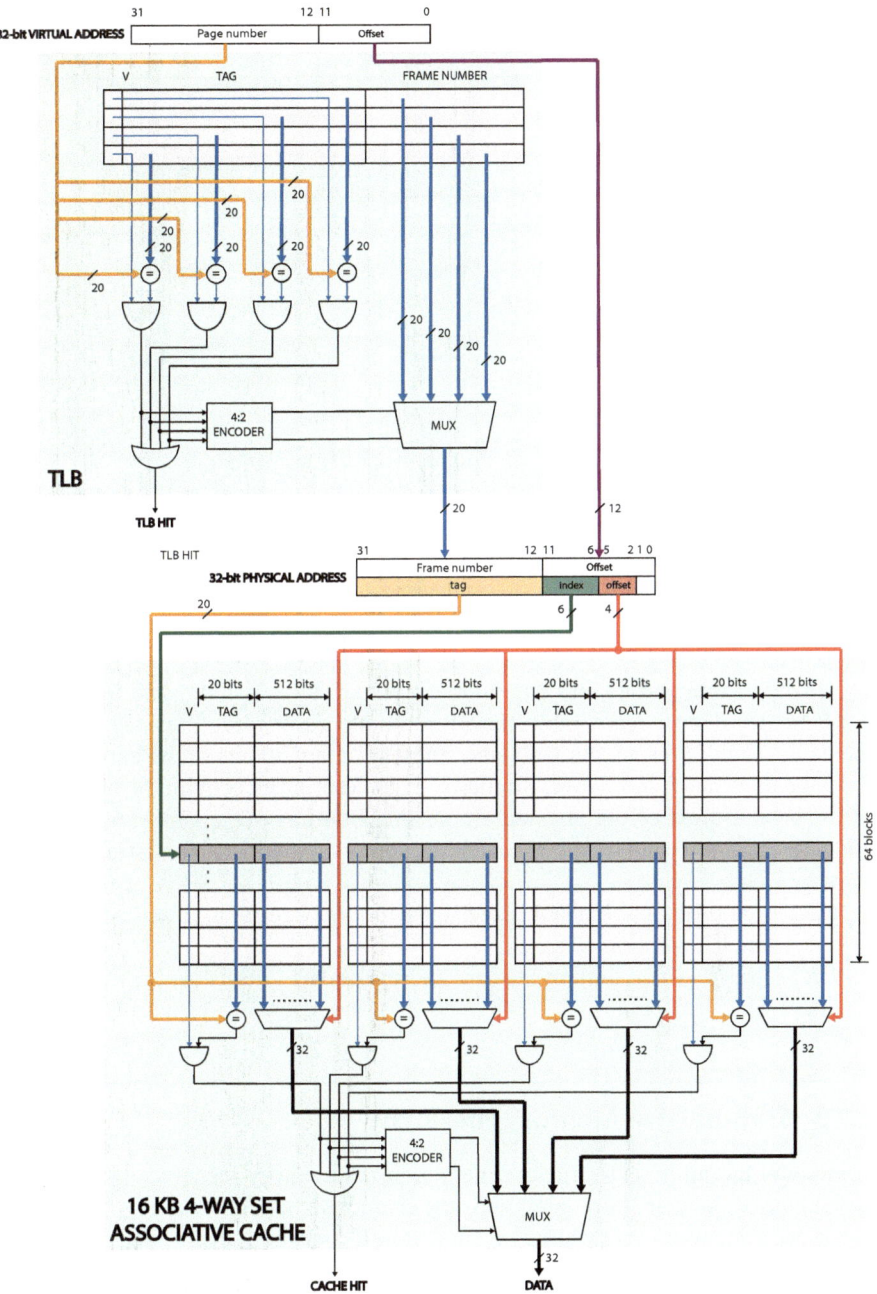

Fig. 6.10 4-way set-associative VIPT cache

Virtual Address:

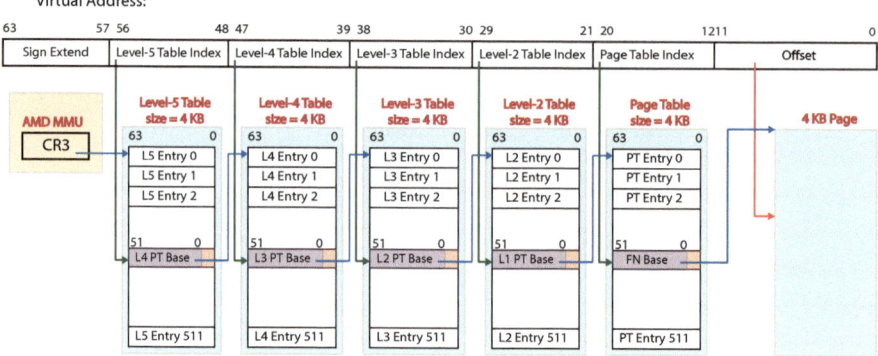

Fig. 6.11 AMD64 5-level paging

2. Bits 56:48 provide the index into the 512-entry level-5 table.
3. Bits 47:39 provide the index into the 512-entry level-4 table.
4. Bits 38:30 provide the index into the 512-entry level-3 table.
5. Bits 29:21 provide the index into the 512-entry level-2 table.
6. Bits 20:12 provide the index into the 512-entry page table.
7. Bits 11:0 provide the byte offset into the page and physical frame.

All page table entries and page descriptors are 64 bits long, and all tables are 4 KB in size. Hence, each table can fit into a physical frame. Besides control and meta bits (Valid, Present, etc.), all table entries contain 40 bits of the base address of the page table at the next level. Similarly, page descriptors contain the upper 40 bits of the base address of the physical frame. AMD64 has two control registers (CR2 and CR3) that are used by the virtual address translation mechanism (i.e., MMU):

1. **CR2**—the processor loads this register with the page-fault virtual address when a page-fault exception occurs.
2. **CR3**—this register contains the base address of the highest-level page-translation table.

6.10   Summary of Memory Hierarchy

A memory hierarchy is a structure that organizes memory resources in a computer system based on their speed, size, and cost characteristics. The memory hierarchy in computer systems exists to optimize the trade-offs between speed, capacity, and cost associated with accessing data. In memory hierarchy, we tend to put faster, smaller, and more expensive memory near the CPU and slower, larger, and considerably cheaper memory further from the CPU. The locality of memory references is the main concept that enables the implementation of a memory hierarchy. Locality of reference refers to the tendency of programs to access a relatively small portion of memory at any given time. Locality of memory reference enables the efficient use

Table 6.1 Typical levels of memory hierarchy

Name	CPU registers	Cache (L1, L2, L3)	Main memory	Secondary (disk) storage
Level	1	2	3	4
Implementation technology	CMOS	CMOS SRAM	CMOS SDRAM	NAND Flash
Location	Within CPU	On CPU chip	Off CPU chip, DIMM modules	Separate device
Access time (ns)	0.1 ns	0.3–10	15–150	25000–100000

of memory by storing frequently accessed data and instructions close to the CPU in the small, fast, and expensive memory, hence reducing memory access latency.

Table 6.1 summarizes the typical levels of the memory hierarchy. The typical levels in the hierarchy slow down and get cheaper and larger as we move away from the CPU. Registers are the smallest and fastest storage locations within the CPU. They hold data and instructions that are currently being processed by the CPU. Registers have the fastest access time, typically measured in very few CPU clock cycles. Cache memory is a small but fast type of volatile computer memory used to hold frequently accessed data and instructions. It sits between the CPU and main memory, acting as a buffer to speed up memory access by storing copies of frequently accessed data. Cache memory is divided into levels, such as L1, L2, and sometimes L3, with each level offering progressively larger capacity and slightly slower access time than the previous level. Cache access times are typically measured in a few CPU cycles, making cache memory significantly faster than main memory. Main memory is a larger, cheaper, but slower type of volatile memory that stores data and instructions used by the CPU. It provides the working space for the operating system, applications, and data currently being processed by the CPU. Main memory is organized into memory modules, such as SDRAMs. Access times for main memory are typically tens to hundreds of nanoseconds, which is slower than cache but faster than secondary storage. Secondary storage refers to non-volatile storage devices, such as hard disk drives (HDDs) and solid-state drives (SSDs), used for long-term data storage. Recently, SSDs have been replacing HDDs. Secondary storage provides a much larger storage capacity than main memory but with much slower access times. Data stored in secondary storage persists even when the computer is powered off, making it suitable for long-term storage of files, applications, and system data. Access times for secondary storage are orders of magnitude slower than main memory, typically measured in milliseconds for HDDs and microseconds for SSDs.

6.11 Case Study: The Memory Hierarchy in an Intel Core i7

The memory hierarchy in an Intel Core i7 processor consists of several levels, each with different characteristics and access times (Fig. 6.12). An Intel Core i7 proces-

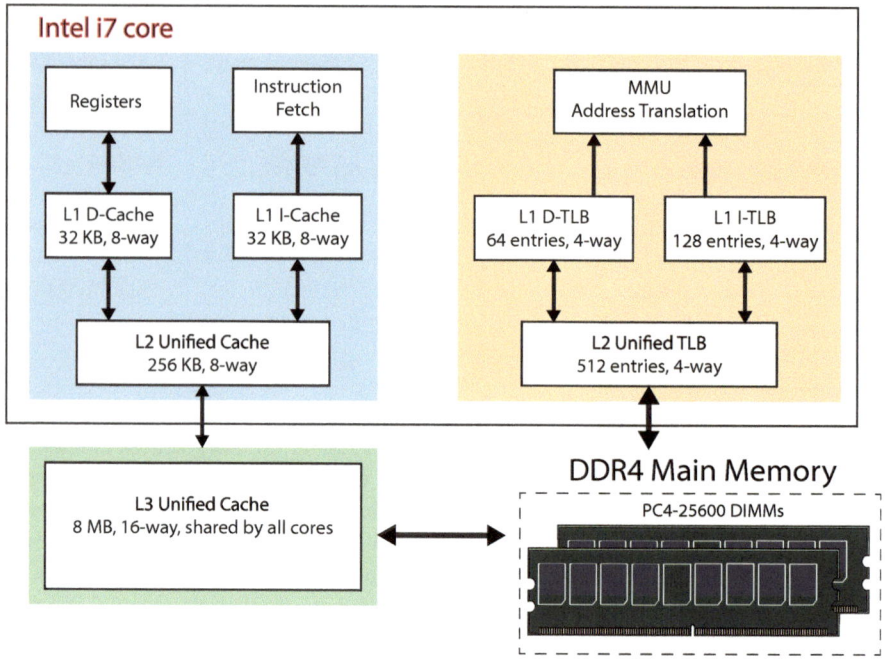

Fig. 6.12 The memory hierarchy in an Intel Core i7

sor typically includes multiple levels of cache, main memory (RAM), and TLBs (Translation Lookaside Buffers). Here's a detailed overview of the memory hierarchy typically found in an Intel Core i7 processor:

Registers

At the top of the memory hierarchy are registers, which are small, high-speed storage locations located directly within the CPU. Registers hold data that the CPU is actively processing, including operands, intermediate results, and memory addresses. Registers have extremely fast access times, typically measured in tenths of nanoseconds, making them the fastest form of memory in the system. Registers are organized into various types, including general-purpose registers, floating-point registers, and special-purpose registers for control and status information.

Cache Memory

Below registers are the cache memory levels, which provide a compromise between the speed of registers and the capacity of main memory. Intel Core i7 processors typically feature multiple levels of cache, including L1, L2, and L3 caches:

1. L1 cache (Level 1 cache) is the smallest and fastest cache, located directly on the CPU core. Intel Core i7 processors typically have separate instruction and data

caches, each with a size of 32 KB per core. The L1 cache operates at the speed of the CPU core, providing extremely fast access times.

2. L2 cache (Level 2 cache) is larger but slower than L1 cache. It is shared among multiple CPU cores and serves as a buffer between the CPU cores and the main memory. Intel Core i7 processors typically have a shared L2 cache with a size ranging from 256 KB to 1 MB per core.

3. L3 cache (Level 3 cache) is the largest and slowest cache in the hierarchy. It is shared among all CPU cores on the processor and acts as a centralized cache for frequently accessed data and instructions. Intel Core i7 processors feature a shared L3 cache, which can range in size from 4 MB to 16 MB or more, depending on the processor model and generation. The L3 cache operates slightly slower than the L1 and L2 caches but provides significantly larger capacity and serves as a shared cache for all CPU cores.

Translation Lookaside Buffers (TLBs)

TLBs are specialized caches used for storing virtual-to-physical address translations, improving the efficiency of memory access. The size and organization of TLBs in Intel Core i7 processors can vary depending on the specific model and generation. Typically, Intel Core i7 processors have multiple levels of TLBs, including small, fast L1 TLBs and larger, slower L2 TLBs. The exact size and organization of TLBs depend on factors such as the microarchitecture and specific features implemented in the processor.

Main Memory (RAM)

Main memory is the primary form of volatile memory used for storing data and program instructions that are actively being used by the CPU. In Intel Core i7 processors, the main memory is typically DDR4 memory, which offers higher capacities and bandwidth compared to earlier generations. Main memory has slower access times compared to cache memory but provides significantly larger storage capacity. Intel Core i7 processors support DDR4 and DDR5 memory modules, with DDR4 being more common in earlier generations and DDR5 becoming increasingly prevalent in newer models. DDR4 memory modules used with Intel Core i7 processors typically have speeds ranging from 2133 MHz to 3200 MHz, although higher-speed variants are available. DDR5 memory modules used with newer Intel Core i7 processors can have speeds ranging from 4800 MHz to 8400 MHz, offering higher bandwidth and improved performance compared to DDR4.

Storage Devices

Beyond main memory, the memory hierarchy may include various storage devices such as solid-state drives (SSDs) and hard disk drives (HDDs). Storage devices offer much larger storage capacities compared to main memory but have significantly slower access times. These devices are typically used for long-term storage of data and programs, with data being transferred to and from main memory as needed for active processing.

Index

© The Editor(s) (if applicable) and The Author(s), under exclusive license to Springer
Nature Switzerland AG 2024
P. Bulić, *Understanding Computer Organization*, Undergraduate Topics in Computer
Science, https://doi.org/10.1007/978-3-031-58075-8